International Migration in Europe

International Migration in Europe
Data, Models and Estimates

Edited by

James Raymer

School of Social Statistics, University of Southampton, UK

and

Frans Willekens

*Netherlands Interdisciplinary Demographic Institute,
The Hague, The Netherlands*

John Wiley & Sons, Ltd

Other Wiley Editorial Offices

John Wiley & Sons Inc., 111 River Street, Hoboken, NJ 07030, USA

Jossey-Bass, 989 Market Street, San Francisco, CA 94103-1741, USA

Wiley-VCH Verlag GmbH, Boschstr. 12, D-69469 Weinheim, Germany

John Wiley & Sons Australia Ltd, 42 McDougall Street, Milton, Queensland 4064, Australia

John Wiley & Sons (Asia) Pte Ltd, 2 Clementi Loop #02-01, Jin Xing Distripark, Singapore 129809

John Wiley & Sons Canada Ltd, 6045 Freemont Blvd, Mississauga, ONT, L5R 4J3

Wiley also publishes its books in a variety of electronic formats. Some content that appears in print
may not be available in electronic books.

Library of Congress Cataloging in Publication Data

International migration in Europe : data, models and estimates / edited by James Raymer and
    Division of Social Statistics, University of Southampton ; Frans Willekens.
        p. cm.
    Includes bibliographical references and index.
    ISBN 978-0-470-03233-6 (cloth)
    1. European Union countries—Emigration and immigration.   2. European Union
    countries—Emigration and immigration—Statistical methods.   3. European Union
    countries—Emigration and immigration—Mathematical models.   4. European Union
    countries—Emigration and immigration—Forecasting—Statistical methods.   5. Bayesian statistical
    decision theory. I. Raymer, James, 1972- II. Willekens, Frans. III. University of Southampton.
    Division of Social Statistics.
    JV7590.I56 2008
    304.8094—dc22

                                                                      2007045550

British Library Cataloguing in Publication Data

A catalogue record for this book is available from the British Library

ISBN 978-0-470-03233-6

Typeset in 10/12pt Times by Integra Software Services Pvt. Ltd, Pondicherry, India
Printed and bound in Great Britain by TJ International, Padstow, Cornwall
This book is printed on acid-free paper responsibly manufactured from sustainable forestry
in which at least two trees are planted for each one used for paper production.

This book is dedicated to our teacher Andrei Rogers of the University of Colorado at Boulder. Andrei pioneered multiregional demography and the indirect estimation of migration flows. We consider ourselves fortunate to be involved in the continuation of that endeavour.

Contents

8 Applying model migration schedules to represent age-specific migration flows

James Raymer and Andrei Rogers

9 Models for migration age schedules: a Bayesian perspective with an application to flows between Scotland and England

Peter Congdon

15 What happens when international migrants settle? Projections of ethnic groups in United Kingdom regions 329
Philip Rees

Contributors

JAKUB BIJAK   Central European Forum for Migration and Population Research, ul. Twarda 51/55, 00-818 Warsaw, Poland. Email: j.bijak@cefmr.pan.pl

MATTHEW J. BRIERLEY   School of Mathematics, University of Southampton, Southampton, SO17 1BJ, UK. Email: kismet_30@hotmail.com

ALESSIO CANGIANO   COMPAS (ESRC Centre on Migration, Policy and Society), University of Oxford, 58 Banbury Road, Oxford, OX2 6QS, UK. Email: alessio.cangiano@compas.ox.ac.uk

PETER CONGDON   Department of Geography, Queen Mary, University of London, Mile End Road, London, E1 4NS, UK. Email: p.congdon@qmul.ac.uk

JOOP DE BEER   Netherlands Interdisciplinary Demographic Institute, PO Box 11650, 2502 AR The Hague, The Netherlands. Email: beer@nidi.nl

JEAN-CHRISTOPHE DUMONT   Directorate for Employment, Labour and Social Affairs, Organisation for Economic Co-operation and Development, 2 rue André Pascal, 75775 Paris, Cedex 16, France. Email: jean-christophe.dumont@oecd.org

JONATHAN J. FORSTER   School of Mathematics, University of Southampton, Southampton, SO17 1BJ, UK. Email: j.j.forster@soton.ac.uk

ROEL JENNISSEN, Research and Documentation Centre (WODC), Dutch Ministry of Justice, PO Box 20301, 2500 EH The Hague, The Netherlands. Email: r.p.w.jennissen@minjus.nl

DOROTA KUPISZEWSKA   Central European Forum for Migration and Population Research, ul. Twarda 51/55, 00-818 Warsaw, Poland. Email: d.kupisz@cefmr.pan.pl

MAREK KUPISZEWSKI   Central European Forum for Migration and Population Research, ul. Twarda 51/55, 00-818 Warsaw, Poland. Email: m.kupisz@twarda.pan.pl

GEORGES LEMAÎTRE   Directorate for Employment, Labour and Social Affairs, Organisation for Economic Co-operation and Development, 2 rue André Pascal, 75775 Paris, Cedex 16, France. Email: georges.lemaitre@oecd.org

JOHN W. MCDONALD   Centre for Longitudinal Studies, Institute of Education, University of London, 20 Bedford Way, London, WC1H 0AL, UK. Email: John.McDonald@ioe.ac.uk

BEATA NOWOK   Population Research Centre, Faculty of Spatial Sciences, University of Groningen, PO Box 800, 9700 AV Groningen, The Netherlands. *Current address*: Netherlands Interdisciplinary Demographic Institute, PO Box 11650, 2502 AR The Hague, The Netherlands. Email: nowok@nidi.nl

JAMES RAYMER   Division of Social Statistics, School of Social Sciences, University of Southampton, Southampton, SO17 1BJ, UK. Email: raymer@soton.ac.uk

PHILIP REES   School of Geography, University of Leeds, Leeds, LS2 9JT, UK. Email: p.h.rees@leeds.ac.uk

ANDREI ROGERS   Population Program, Institute of Behavioral Science, University of Colorado, Boulder, CO 80309-0484, USA. Email: andrei.rogers@colorado.edu

PETER W.F. SMITH   Division of Social Statistics, School of Social Sciences, University of Southampton, Southampton, SO17 1BJ, UK. Email: p.w.smith@soton.ac.uk

LEO VAN WISSEN   Population Research Centre, Faculty of Spatial Sciences, University of Groningen, PO Box 800, 9700 AV Groningen, The Netherlands. Email: l.j.g.van.wissen@rug.nl

FRANS WILLEKENS   Netherlands Interdisciplinary Demographic Institute, PO Box 11650, 2502 AR The Hague, The Netherlands. Email: willekens@nidi.nl

Preface

This book is an outcome of the Research Training Network (RTN) *Demographic Sustainability and European Integration*, funded by the European Commission under the Fifth Framework Programme and coordinated by the Max Planck Institute of Demographic Research in Rostock, Germany (project number HPRN-CT-2001-00234). The primary objective of RTN is to promote training-through-research, especially of young researchers at pre-doctoral and postdoctoral levels, within the context of high quality transnational collaborative research projects. The ultimate aim of training and mobility of researchers is to improve the research potential and knowledge base of Europe and to create a European Research Area. The RTN programme supports the temporary appointment of young researchers coming from a country other than that of the research team and covers the costs of coordinating activities, such as the *Workshop on the Estimation of International Migration in Europe: Issues, Models and Assessment*, organized in September 2005 at the University of Southampton (UK) by the Southampton Statistical Sciences Research Institute (S3RI) and the Netherlands Interdisciplinary Demographic Institute (NIDI) in The Hague. Both young and senior scientists were invited to present their work on data issues, models and estimates associated with international migration in Europe. Many of the papers presented at the workshop are included in this book.

This book on data, models and estimation of international migration in Europe is published during an important stage of development of the European Statistical System. On 31 July 2007, *The Official Journal of the European Union* published the "Regulation (Ec) No 862/2007 of the European Parliament and of the Council of 11 July 2007 on Community statistics on migration and international protection and repealing Council Regulation (EEC) No 311/76 on the compilation of statistics on foreign workers". The Regulation establishes common rules for the collection and compilation of Community statistics on immigration and emigration, citizenship and country of birth, residence permits issued, entries into the Member States refused to third-country nationals, and third-country nationals found to be illegally present in Member States. An innovative and important element of the Regulation is that "As part of the statistics process, scientifically based and well documented statistical estimation methods may be used." and that "Member States shall report to the Commission (Eurostat) on the data sources used, the reasons for the selection of these sources and the effects of the selected data sources on the quality of the statistics, and on the estimation methods used, and shall keep the Commission (Eurostat) informed of changes thereto." We hope this research provides an important foundation for meeting the requirements of the new regulation.

James Raymer, Southampton
Frans Willekens, The Hague

Note

Additional information for some chapters in this volume may be found in appendices and/or supplementary material on the book's website at: http://www.wiley.com/go/raymer

1

Introduction and outline

James Raymer[†] and Frans Willekens[‡]

[†] *Division of Social Statistics, School of Social Sciences, University of Southampton*
[‡] *Netherlands Interdisciplinary Demographic Institute, The Hague*

1.1 Introduction

In response to a long history of inadequate, incomparable and missing migration data, a new regulation has been prepared by the European Parliament and the Council on Community Statistics that obliges Member States to produce and provide harmonised estimates of international migration and asylum flows (European Commission 2005). The aim of the new legislation is to increase the availability and comparability of migration statistics. Member States should use the best data sources available to supply Eurostat, the statistical office of the European Union, with detailed metadata. In addition, these data should be explained and assessed in how close they comply with the harmonised definitions. Harmonisation should reduce the impact of different migration definitions and sources on the published migration statistics. However, the guidelines and methodologies for providing harmonised estimates have yet to be developed or implemented.

A major research effort in support of the new regulation was the THESIM (*Towards Harmonised European Statistics on International Migration*) project, funded by the European Commission's Directorate-General for Research under the

International Migration in Europe: Data, Models and Estimates   Edited by J. Raymer and F. Willekens
© 2008 John Wiley & Sons, Ltd

Sixth Framework Programme (Poulain *et al.* 2006). One of the objectives of the project was to analyse the current functioning of migration statistics in the 25 European Union (EU) countries and to identify the problems that the countries might encounter in relation to the requirements of the new regulation. During the project, researchers met with experts from the national statistical institutes (NSIs) or national statistical offices (NSOs) and with the authorities responsible for the registration of migrants in EU countries. The meetings were a unique source of information, with explanations provided directly by a range of people involved in the statistics production process. The project pointed to the need for metadata that adequately document rules and procedures used during the processes of data collection and the production of statistics.

The work contained in this book builds on the THESIM study of the migration statistics production process and the resulting data situations. It can be considered an important starting point for the development of methodologies to convert raw data on migration into internationally comparable statistics. Issues, strategies and methods for obtaining consistent estimates of international migration flows are presented in the context of various data situations. Estimates of migration are needed so that national governments and international agencies have the means to improve their policies directed at supplying particular social services or at continuing, increasing or decreasing levels of international migration. Furthermore, our understanding of how or why populations change requires relatively detailed information about migration. Without these, the ability to predict, control or understand that change is limited.

Although all countries and regions of the world face similar challenges measuring international migration, this work focuses on European migration for practical, substantive and policy reasons. First, this work is produced within the context of the European Research Training Network (RTN) programme 'Demographic Sustainability and European Integration' funded by the European Commission (contract HPRN-CT-2001-00234). One of the objectives of the RTN programme was the development of a reliable database on population movements in Europe with internationally comparable data using current work at Eurostat and building on the European Monitoring System, which was initiated in 1992 by the Population Research Centre of the University of Groningen in the Netherlands and the International Institute for Applied Systems Analysis in Laxenburg, Austria.[1]

Second, international migration is the major factor contributing to population change in regions of the world and Europe in particular. In 2005, the United Nations estimated that worldwide there were about 191 million persons residing outside their country of birth, or nearly 3 % of the world's population (UNFPA 2006; IOM 2005). Excluding the former Soviet Union, Europe hosts about 19 % of all foreign-born persons in the world,[2] which is striking considering that the population of this area only represents 7–8 % of the world's population total. Since 1989, net immigration has contributed to at least 50 % of the European Union's annual

---

[1] For details the reader is referred to the RTN proposal and Willekens (1994).
[2] With the countries that were part of the former Soviet Union included, it is around 36 %.

population growth and, in 2002, it represented 85 % compared to 27 % in 1980 (Eurostat 2004). The underlying flows driving this demographic change, however, are largely unknown. The empirical evidence required to permit a truly integrated analysis of migration flows within Europe and between Europe and the rest of the world is not available.

Third, demographic change in Europe has never received the attention it is receiving today. Migration is at the core of the debate. Although there is a great demand for evidence-based policies, the evidence is limited because data are defective. The reason for this is primarily due to differences in data sources, availability, quality and measurement. The advantages to having a consistent and reliable set of migration flows are numerous.

Europe has always been affected by migration. Demographic growth, climatic change, trade, warfare, conquest and formation of nations, states and empires have all initiated population movement. However, since World War II, the levels and sources of migrants have changed considerably (Castles and Miller 2003). Foreign-born populations are growing rapidly as a result of very low or negative levels of natural increase among native-born populations and by increasing levels of immigration. The diversity of foreign-born populations is also expanding as European countries are becoming increasingly more connected with the world.

The reasons for migration are many. People move to other countries for employment, family reunion, protection or amenity reasons. Statistics on these flows, on the other hand, are relatively confusing or nonexistent. There are two reasons. First, no consensus exists on what exactly is a 'migration'. The theoretical concept of migration as a change of usual residence across administrative (national) boundaries may be generally accepted but implementation of this concept has yet to take place. Comparative analyses suffer from differing national views concerning who is a migrant. Second, the event of migration is rarely measured directly. More often it is inferred by a comparison of places of residence at two points in time that may be three months, six months, one year or several years apart. The challenge is compounded because individual countries use different methods of data collection and varying concepts of migration. International migration statistics may come from administrative data, decennial population censuses or surveys. Migrants may be defined according to previous or next country of residence, nationality or citizenship. Harmonisation of data collection processes and the data they generate is not even close to being realised. So, how does one overcome these obstacles to obtain an overall and consistent picture of the migration patterns occurring within Europe?

Solutions to deal with the above problems are set out in this book. First, we develop a classification of migration data types. Second, we develop models that are able to combine different data types to predict migration flows that are comparable across data types. Migrant stocks and migration flows measure migration differently and are examples of two data types. Additionally, differences in data sources, reliability and validity are accounted for in the models. The key aspect of the models is that they estimate migration flows using available information from one or more sources in combination with realistic assumptions about the relation between various data types and structures of the missing data. The outcomes are *synthetic data* that represent the best possible estimates of international migration during a given period. The approaches used to construct these synthetic data are rooted in statistical theory and were originally developed in studies of internal migration (Plane 1981, 1982; Rogers and Castro 1981; Rogers and von Rabenau 1971; Rogers *et al.* 2002, 2003; Sweeney 1999; Willekens 1977, 1980, 1982, 1983, 1999; Willekens *et al.* 1981). It was suggested that, to obtain estimates of international migration, one has to focus on data harmonisation and models, but such efforts have not been attempted until recently (Poulain 1994, 1995, 1999, 2001; Poulain *et al.* 2006; Willekens 1994; Raymer 2007).

This is the first volume that draws together modern statistical approaches to incomplete data and migration modelling. Issues dealing with data on migration are linked directly to models of migration in order to produce the best possible estimates of international migration in Europe. This link is essential both for the estimation of missing data and for the harmonisation of available data. International migration in Europe provides a good starting point for learning how to estimate international migration flows because it plays such a large role in its population change and the data are readily available for many of the countries in this region, albeit in various forms and in varying degrees of completeness.

The modelling approaches are rooted in statistics with an interdisciplinary audience in mind. For example, statistical methods are used to harmonise and estimate missing or incomplete data, demographic techniques are used to identify structures in the spatial and age patterns of migration, geographic methods are applied to model movement across space, and economic ideas about imbalances in the demand and supply of labour are used to produce more accurate estimates. The approach relies heavily on empirical regularities, which can be imposed in regression-type models. In the process of estimating migration flows, the population may be stratified by subpopulations that differ in migration behaviour, which can be incorporated in the modelling process. For the examples used in this book, the migrant subpopulations consist of various ethnicities, nativities and migrant types (i.e. labour, family-related, asylum seekers and so on).

This book is written for those carrying out research on topics that are directly or closely connected with migration patterns and processes. The level is written for upper undergraduate students and graduate-level students in the fields of demography, statistics, geography, regional science, sociology, economics and political science. We also hope that statisticians at national and international organisations will find the ideas and solutions presented in this volume useful.

1.2   Outline

1.2.1   Definitions

Migration is a complicated subject. In order to make sense of the various types of flows and definitions, we provide the following key definitions. First, a distinction between 'migration' and 'migrant' is important. Migration refers to the *event* of moving from one country to another. A migrant is a *person* who has changed his or her residential status, from one time to another. Population registers capture movements (i.e. migration events). Censuses and surveys capture migrants (i.e. status transitions).

There is also an important difference between migrant populations and migration flows. The foreign-born population contains persons who were born outside the country of current residence from the perspective of the settlement country. Expatriates are foreign-born persons residing in another country from the perspective of the origin country. Native-born persons are those born in the country of current residence. Foreigners are those who have not acquired citizenship in the country of residence but are residing there. Nationals may include both native-born and foreign-born persons.

Immigration refers to the flow of migration into a particular country (receiving country or destination country) and emigration refers to the flow of migration from a particular country (sending country or country of origin). Both immigration and emigration are associated with a particular time period. In this book, we find immigration flows that use anywhere from no specific time criteria (Germany) to permanent (Poland) as the time period for immigration or emigration flows. Mostly, we think of immigration or emigration flows being a one-year flow, in alignment with the United Nations recommended duration. These flows may contain native-born, foreign-born, foreigners and nationals. Finally, immigration and emigration data may be provided by the country of origin or the country of destination.

1.2.2 Frequentist and Bayesian approaches

Both frequentist and Bayesian models of migration are presented in this book. Bayesian methods remain relatively undeveloped for estimating migration patterns, as they are relatively complicated and intensive to implement. However, Bayesian methods are ideally suited to inadequate and missing data problems because they allow a direct combination of prior information or beliefs and empirical observation. This means that one can build a flexible model that incorporates the strengths of theory and empirical data, which is important for situations such as combining data from multiple sources, missing data and inconsistent data. The outcomes of Bayesian analysis include distributions of both parameters and predicted values, meaning that the user has much better measures of the uncertainties. This is very important when the data of interest are inadequate or missing.

1.2.3 Structure of the book

This book contains five parts aside from the 'Introduction and outline' and 'Conclusion'. Part I focuses on migration data and issues and includes four chapters – each addressing a particular migration data type. In Chapter 2, Jean-Christophe Dumont and Georges Lemaître focus on foreign-born and expatriate population stocks in OECD countries. They point out that these migration data are the most comparable and the most readily available, and therefore should be a starting point for comparative analyses of migration. Dorota Kupiszewska and Beata Nowok follow this in Chapter 3 with a review of international migration flow statistics produced by the 25 countries in the European Union (situation in 2006) and illustrate the many inconsistencies in the flow data caused by differences in concepts, measurements and the processing of migration data. Beata Nowok then extends this analysis in Chapter 4 with a detailed analysis of migration data in Central Europe over time. In Chapter 5, Alessio Cangiano focuses on Southern European countries and the evolution of their foreign populations. In particular, undocumented migrants are approximated using the information from regularisation programmes and other data sources.

The emphasis of Part II is on models for spatial and age patterns of migration. In Chapter 6, Frans Willekens provides an overview of migration models with complete and incomplete data. Matthew Brierley, Jonathan Forster, John McDonald and Peter Smith present a Bayesian approach in Chapter 7 to estimate international flows in Northern Europe in the context of data both missing and coming from different sources. In Chapter 8, James Raymer and Andrei Rogers review model migration schedules and provide a demonstration of how families of these schedules can be used to estimate age patterns of migration in the context of incomplete or missing data. Peter Congdon models age patterns of migration using Bayesian methods in Chapter 9, using migration between England and Scotland as the illustration.

In Part III, the above models and others are applied in two chapters to obtain current estimates of international migration flows in Europe and to estimate the impact of various asylum-seeker policies on other potential receiving countries. In Chapter 10, James Raymer applies the ideas set out in Chapters 6 and 8 in Part II to estimate age-specific international migration flows between countries in the European Union. In Chapter 11, Leo van Wissen and Roel Jennissen present a methodology for identifying potential redirection impacts of asylum seekers caused by restrictions imposed by particular countries of destination. A distinction is made between overall pattern (generation effect) and migration flows that result from redirection (substitution effect).

An overview of methods for forecasting international migration flows from traditional time series and Bayesian perspectives is provided in Part IV. First, in Chapter 12, Jakub Bijak provides a review of previous attempts to forecast migration before arguing for a Bayesian approach. For illustration, he forecasts migration flows from Poland to Germany. Joop de Beer then reviews extrapolation and time series models and forecasts different types of immigration and emigration flows in the Netherlands in Chapter 13.

Part V contains two chapters on population projections that emphasise the migration component. A multiregional model is presented in Chapter 14 by Marek Kupiszewski and Dorota Kupiszewska to estimate subnational and national populations in Central Europe. Then Philip Rees puts forward a projection model in Chapter 15 to estimate future regional ethnic populations in the United Kingdom.

The contributions of the book are many. First, a distinction is made between the different migration data types. This removes the need for *ad hoc* approaches to estimating flows. Second, the conventional methods (i.e. frequentist) for estimating incomplete migration data are complemented with a Bayesian perspective. The Bayesian perspective is better able to incorporate the additional information on migration and therefore to determine how better or more information results in better estimates of migration. Finally, this book demonstrates how these estimates can be incorporated in the projection of migration flows and the projection of populations.

References

Castles S and Miller MJ. 2003. *The age of migration: International population movements in the modern world*. New York: The Guilford Press.

European Commission. 2005. *Proposal for a Regulation of the Parliament and of the Council on Community statistics on migration and international protection*. Brussels. COM (2005) 375 final, 2005/0156 (COD). Available at: http://eurlex.europa.eu/LexUriServ/site/en/com/2005/com2005_0375en01.pdf

Eurostat. 2004. *Population statistics*, 2004 edn. Luxembourg: Office for Official Publications of the European Communities.

IOM. 2005. *World migration: Costs and benefits of international migration*. Geneva: International Organisation for Migration.

Plane DA. 1981. Estimation of place-to-place migration flows from net migration totals: a minimum information approach. *International Regional Science Review* 6(1):33–51.

Plane DA. 1982. An information theoretic approach to the estimation of migration flows. *Journal of Regional Science* 22(4):441–456.

Poulain M. 1994. *Internal mobility in Europe: the available statistical data*. Working Paper 17. Conference of European Statisticians, Commission of the European Communities (Eurostat), Mondorf-les-Bains, Luxembourg.

Poulain M. 1995. Towards a harmonisation of migration statistics within the European Community. In *Demographic consequences of international migration*. Voets S, Schoorl J and de Bruijn B, eds, pp. 11–25. The Hague: Netherlands Interdisciplinary Demographic Institute.

Poulain M. 1999. *International migration within Europe: towards more complete and reliable data?* Working Paper 37. Joint ECE–Eurostat Work Session on Demographic Projections, Perugia, Italy.

Poulain M. 2001. *Is the measurement of international migration flows improving in Europe?* Working Paper 12. Joint ECE–Eurostat Work Session on Migration Statistics, Geneva, Switzerland.

Poulain M, Perrin N and Singleton A, eds. 2006. *THESIM. Towards Harmonised European Statistics on International Migration*. Louvain-la-Neuve: UCL Presses Universitaires de Louvain.

Raymer J. 2007. The estimation of international migration flows: a general technique focused on the origin–destination association structure. *Environment and Planning A* 39:985–995.

Rogers A and Castro LJ. 1981. *Model migration schedules*. RR-81-30. Laxenburg: International Institute for Applied Systems Analysis.

Rogers A and von Rabenau B. 1971. Estimation of interregional migration streams from place-of-birth-by-residence data. *Demography* 8: 185–194.

Rogers A, Willekens FJ and Raymer J 2002. Capturing the age and spatial structures of migration. *Environment and Planning A* 34:341–359.

Rogers A, Willekens FJ and Raymer J. 2003. Imposing age and spatial structures on inadequate migration-flow datasets. *The Professional Geographer* 55(1):56–69.

Sweeney SH. 1999. Model-based incomplete data analyses with an application to occupational mobility and migration accounts. *Mathematical Population Studies* 7(3):279–305.

Willekens FJ. 1977. *The recovery of detailed migration patterns from aggregate data: An entropy maximizing approach.* Laxenburg: International Institute for Applied Systems Analysis.

Willekens FJ. 1980. Entropy, multiproportional adjustment and the analysis of contingency tables. *Systemi Urbani* 2:171–201.

Willekens FJ. 1982. Multidimensional population analysis with incomplete data. In *Multidimensional mathematical demography*, Land KC and Rogers A, eds, pp. 43–112. New York: Academic Press. Reprinted in *Readings in population research methodology*, Bogue DJ, Arriaga EE and Anderton DL, eds, pp. 22.49–22.82. Chicago: Social Development Center; and New York: UNFPA.

Willekens FJ. 1983. Log-linear modelling of spatial interaction. *Chapters of the Regional Science Association* 52:187–205.

Willekens FJ. 1994. Monitoring international migration in Europe: towards a statistical data base combining data from different sources. *European Journal of Population* 10(1):1–42.

Willekens FJ. 1999. Modeling approaches to the indirect estimation of migration flows: from entropy to EM. *Mathematical Population Studies* 7(3):239–278.

Willekens FJ, Pór A and Raquillet R. 1981. Entropy, multiproportional, and quadratic techniques for inferring patterns of migration from aggregate data. In *Advances in multiregional demography*. RR-81–6. Rogers A, ed., pp. 83–124. Laxenburg: International Institute for Applied Systems Analysis.

UNFPA. 2006. *State of the world population.* New York: United Nations Population Fund. Available at: http://www.unfpa.org/swp/2006/pdf/en_sowp06.pdf

Part I
Data Issues

2

Counting foreign-born and expatriates in OECD countries: a new perspective

Jean-Christophe Dumont and Georges Lemaître
Directorate for Employment, Labour and Social Affairs, OECD, Paris

2.1 Introduction

Since the end of the 1990s, international migration and international mobility of highly qualified workers have received increasing attention from policy-makers. This reflects, among other things, the recent rises in international movements that have occurred following the fall of the Iron Curtain and in conjunction with the globalisation of economic activity. In addition, demographic imbalances between developed and developing countries and large differences in wages have tended to encourage the movements of workers from economies where they are in surplus to those where they are most in need. Moreover, many OECD (Organisation for Economic Co-operation and Development) countries have been attempting to attract qualified human resources from abroad, which their increasingly knowledge-intensive economies need in order to sustain economic growth. Despite these increased movements and the heightened policy interest in this area, however, the quality and comparability of international data on migration have scarcely kept pace.

In particular, data that are generally available on migration do not provide a clear picture of the relative movements across countries. In some countries, the so-called

---

International Migration in Europe: Data, Models and Estimates   Edited by J. Raymer and F. Willekens
© 2008 John Wiley & Sons, Ltd

settlement countries (Australia, Canada, New Zealand and the United States), the focus tends to be on 'permanent' migrants, that is, persons who are admitted to the country and granted the right of permanent residence upon entry. Persons who are granted temporary permits may not even figure in the official migration statistics. In other countries, immigrants consist of persons who are enrolled in a population register, which is a file of persons residing in the country that is generally maintained at the municipal level. To be registered, a person entering from outside the country must intend to stay in the country for more than a specified minimum period and have a residence permit (if required) of at least the minimum duration. In some countries (e.g. Belgium, Japan), the minimum period is three months, in others (Sweden, Finland) one year. In practice, this means that international students, for example, will often be counted as immigrants in these countries. In the settlement countries, they would not figure in the official migration statistics. Although the solution would normally be to harmonise the statistics across countries, for a number of technical reasons, progress in this area is exceedingly slow.

As with international data on annual movements, those on the total immigrant population have suffered from differing national views concerning who is an 'immigrant'. In the settlement countries, the immigrant population consists of persons who are foreign-born, that is, who at some stage have immigrated into the country of residence. For these countries, the acquisition of nationality is relatively easy and it is rare to see statistics on persons of foreign nationality or citizenship. Although some foreign-born persons may have been born abroad with the citizenship of their current country of residence, these persons would not normally be considered as part of the immigrant population. This phenomenon can generally be ignored without risk of providing a distorted picture of the immigrant population.

In other countries, the immigrant population includes persons of foreign nationality. However, because persons born abroad can acquire the nationality of the country of residence and because persons born in a country do not necessarily acquire thereby the citizenship of the country of birth, statistics on the foreign population may not yield the same result as those on the foreign-born population. This would not be problematic if it were possible to produce data on both bases. But this was not the case for many countries until fairly recently, with the result that it was customary to see international statistics for two sets of generally non-overlapping countries: those applying the concept of a foreign country of birth to define the immigrant population, and those for whom foreign nationality was the determining criterion.

As immigrant populations have grown in many countries and naturalisations have become more common, estimates based on these different concepts have become less and less comparable across countries. While new arrivals of foreign citizens tend to increase the size of the foreign population, those already there may acquire the citizenship of the host country and become nationals. As a result, the magnitude of the population of foreign citizenship may tend to remain more or less stable or to grow slowly, while the number of foreign-born persons continues to increase.

In addition to the lack of comparability on immigrant populations, most OECD member countries have little information at their disposal on their emigrants (i.e.

expatriates); and those which have some information do not necessarily have a clear picture of the countries of destination or of the exact magnitudes of persons who have left the country. Finally, rare are the countries which have a precise picture of expatriates by the duration of stay abroad, the level of qualification and the occupation or branch of industry in which they are employed.

In developing countries, the question of the international mobility of highly qualified workers is generally manifested through a concern about the 'brain drain' and the loss of economic potential which could result from this. In OECD countries the retention of qualified persons and the return of expatriates constitute important challenges to which several countries have tried to respond by reinforcing tax incentives to promote return, seeking to enhance the environment for scientific and technical research or improving the status of certain professions. Several recent studies undertaken at the OECD have demonstrated that the question is more complex than is often depicted (OECD 2002; Dumont and Meyer 2003). These studies also highlight the deficiencies and the gaps in the statistical data available, making it difficult to grasp the complex international mobility patterns of highly skilled workers. To date, only two studies have attempted to estimate the percentage of persons living abroad by country of origin and by level of qualification (Carrington and Detragiache 1998; Adams 2003). Both studies apply the same methodology and include a number of biases, which limit their usefulness. The work by Carrington and Detragiache uses data from the 1990s, whereas more recent data are used in Adams (2003). As a result, current statistics tend to show a rather imperfect image of the actual extent of migration in general and of the movements of the highly skilled in particular, with respect to movements not only from developing to developed countries but also within the OECD area as well.

With the 2000 round of censuses, however, virtually all OECD countries have incorporated in their census a question on the country of birth of persons enumerated, as well as on their nationality. With this information, it is possible to provide, for the first time, a detailed, comparable and reliable picture of foreign-born populations within OECD countries, reflecting the cumulative effect of movements within and to the OECD zone over the past decades. Not only can foreign-born populations be compared on a common basis across countries, but also the extent of migration from a single source country to each OECD country as well as to OECD countries as a whole can be determined. With additional information on the educational attainment of migrants, flows of human capital can be depicted and, in particular, the conventional wisdom on movements of highly qualified persons confronted with actual data.

The remainder of this chapter is divided up as follows. Section 2.2 describes the new database that is the source of the information in this chapter. Section 2.3 presents the basic results with respect to the foreign and foreign-born populations, the geographic origin of foreign-born persons and their educational attainment. Section 2.4 discusses results concerning expatriation from individual OECD countries to other OECD countries, whereas Section 2.5 focuses on foreign-born from developing countries in OECD countries. Both present data on persons with tertiary education resident abroad. A summary and conclusions follow in the last section.

2.2   A new database on immigrant populations

The information presented in this chapter is based on a data collection addressed to 29 OECD national statistical offices (NSOs) and aimed at obtaining census data on the stock of the foreign-born population in OECD countries. The core objective of the project was to better measure and characterise foreign-born populations and to obtain, by aggregating across OECD receiving countries, data on expatriate populations in OECD countries. Further information on this data collection is provided in the Appendix.[1]

Most population censuses in member countries were conducted around the year 2000. Owing to their comprehensive coverage, censuses are particularly useful for identifying and studying small population groups. In several countries, however, there is no population census and it has been necessary to turn to data from population registers or from large-sample surveys. Census data were actually used for 23 of the 29 participating countries and other sources for the remainder – see an earlier version of this chapter (Dumont and Lemaître 2005) for more detailed information. The database currently includes data on the foreign-born in OECD countries by detailed place of birth, nationality and educational attainment (three levels).

The database covers 227 countries of origin and 29 receiving countries within the OECD zone. For 0.46 % of the total population of all OECD countries, the place of birth was not recorded, and for 0.24 % the specific country of birth (either a region was specified or no answer was given). The level of education was reported for more than 98 % of the population of age 15 years or older. Finally, complete information (i.e. detailed education and detailed place of birth) is available for 97.8 % of the OECD population aged 15+. The percentage of expatriates living abroad has been calculated for more than 100 countries.

Data adjustments have been necessary for only two situations. First, data for Japan and Korea were not available by country of birth. For these two countries, it has been assumed that the country of nationality is the country of birth. This seems a reasonable assumption for the foreign-born, given the very low rate and number of naturalisations in these two countries. However, it will tend to overestimate the number of foreign-born relative to other countries, because persons born in Japan or Korea to foreigners will tend also to be recorded as foreign and thus be classified as foreign-born.

The second situation concerns Germany, where the available source was the Microcensus, a large-scale household sample survey (the last German census was conducted in 1987). This source identifies whether or not a person was born abroad, but not the country of birth. Equating country of birth and country of nationality for Germany would have attributed 'Germany' as the country of birth to naturalised foreign-born persons, whose numbers are not negligible, as well as to the numerous 'ethnic' German immigrants who obtained German nationality upon entry into Germany. Another data source (the German Socio-Economic Panel) was used to

---

[1] The appendices and/or supplementary material for this chapter are available on the book's website at: http://www.wiley.com/go/raymer

adjust the data for Germany where this was possible – see Dumont and Lemaître (2005) for more details.

The new database on immigrants and emigrants in OECD countries is the first internationally comparable dataset with detailed information on the foreign-born population for almost all member countries of the OECD and on the prevalence of highly skilled emigration for about 100 countries. This provides a broad view of the significance of highly skilled emigration, for both OECD and less developed countries.

2.3   Immigrant populations in OECD countries

2.3.1   The foreign and foreign-born populations

The percentages of the foreign and the foreign-born populations in OECD countries are compared in Table 2.1. Further information on all the data in this chapter is provided in the Appendix. It is in three of the four settlement countries (Australia, Canada and New Zealand, the United States being the exception), as well as in Luxembourg and Switzerland, that the percentage of the foreign-born is highest, with numbers close to or exceeding 20 % in all of these. In addition, in some European countries (e.g. Austria, Germany, the Netherlands and Sweden), the percentage of foreign-born is almost as high as, and in some cases higher than, that recorded in the United States (approximately 12 %). In a number of other European countries (Belgium, France, Greece and Ireland), the percentage of the foreign-born population also exceeds 10 % of the total population. These figures are appreciably higher than those generally presented for the immigrant population, measured on the basis of foreign nationality and which never exceed 10 %, except for Luxembourg and Switzerland. It is clear that many European countries have admitted immigrants in considerable numbers over the past decades, certainly more than is evident from statistics of the resident foreign population.

The first waves of these immigrants to Europe arrived during the 1950s and 1960s, with the strong economic expansion accompanying and following post-war reconstruction and widespread shortages of low-skilled labour. With the first oil crisis in the 1970s, the European countries which had been importing labour closed the door to further labour migration. Many resident migrants stayed on and brought in their families; others sought and brought in spouses from their former countries. The situation more or less stabilised until the late 1980s and early 1990s, when the fall of the Iron Curtain brought freedom to travel to millions of persons who had formerly been unable to leave their countries. Other movements included persons fleeing dictatorship and ethnic conflict and seeking asylum in European countries. The accumulated effect of these various sources of migration has resulted in significant immigrant populations in European countries, even if there has been little active labour recruitment in most of these countries for the last 30 years.

Not all OECD countries were receiving immigrants over this period, however. The percentage of immigrants in countries that historically have been, and in some

Table 2.1 Percentage of foreign-born and of foreigners in the total population in OECD countries and acquisition of citizenship by the foreign-born.

Country of residence	Total population	Percentage foreign-born	Percentage foreign	Percentage of foreign-born with citizenship of country of residence
Australia	18 769 242	23.0	7.4	68.4
Austria	8 032 926	12.5	8.8	40.9
Belgium	10 296 350	10.7	8.2	40.8
Canada	29 639 055	19.3	5.3	72.6
Czech Republic	10 230 060	4.5	1.2	79.8
Denmark	5 368 354	6.8	5.0	40.3
Finland	5 181 115	2.5	1.7	41.6
France	58 520 688	10.0	5.6	53.1
Germany	82 229 250	12.5	8.9	. . . [b]
Greece	10 934 097	10.3	7.0	41.5
Hungary	10 198 315	2.9	0.9	71.1
Ireland	3 858 495	10.4	5.9	45.2
Italy	56 995 744	3.9	2.3	47.5
Japan[a]	126 920 100	1.0	1.0	. . . [b]
Korea[a]	46 136 101	0.3	0.3	. . . [b]
Luxembourg	439 539	32.6	36.9	13.0
Mexico	97 483 412	0.5	. . . [b]	. . . [b]
Netherlands	15 987 075	10.1	4.2	65.0
New Zealand	3 737 229	19.5	. . . [b]	. . . [b]
Norway	4 552 252	7.3	4.3	47.6
Poland	38 230 080	2.1	0.1	96.1
Portugal	10 356 117	6.3	2.2	66.3
Slovak Republic	5 379 455	2.5	0.5	84.2
Spain	40 847 371	5.3	3.8	30.9
Sweden	8 975 670	12.0	5.3	62.5
Turkey	67 785 786	1.9	. . . [b]	79.2
United Kingdom	58 789 205	8.3	4.5	. . . [b]
United States	281 421 941	12.3	6.6	46.4
Simple average		**9.4**	**6.1**	**54.9**
Weighted average		**7.6**	**4.8**	**51.9**

[a] In the absence of place-of-birth data for Japan and Korea, it has been assumed that all foreigners are foreign-born and nationals native-born.
[b] . . . Not available.

cases still are, emigration countries, such as Mexico, Turkey, Italy, Spain and Portugal, is much lower, generally less than 5 %. The same holds for the countries in Central and Eastern Europe, where economic development has been lower relative to the rest of Europe and which have never experienced the same kind of labour shortages that occurred elsewhere.

Some caution, however, needs to be exercised in interpreting the data in Table 2.1 for a number of countries. In France and in Portugal, for example, the foreign-born population includes a significant proportion of persons born abroad as citizens and repatriated from former colonies. About 1.6 million people born with French nationality outside of France (mainly in Algeria) were counted in the population census of 1999. A similar situation occurs for the United States, because of persons born overseas of American parents (for example, to military personnel stationed abroad).

For certain countries in Table 2.1, in particular the United States, Australia or Canada, statistics on non-citizens have seldom been compiled or published. This is in contrast to many European countries, which tend not to distinguish persons by country of origin once they have acquired the citizenship of the host country. As illustrated in Table 2.1, 6.6 % of the population in the United States does not have United States citizenship. The figure for Australia is 7.4 %, and that for Canada 5.3 %. These are levels for the foreign population that are comparable to those recorded in some European countries such as France, Sweden, Denmark and the Netherlands.

In general, measures of the foreign population tend to be much smaller than those of the foreign-born population. The differences between the statistics on foreigners and on the foreign-born are partly attributable to the varying requirements across countries for obtaining the citizenship of the country of residence, and to the fact that, in many countries, persons born in the country to parents of foreign nationality do not automatically acquire the citizenship of the host country (see Table 2.1). In Australia and in Canada, but also in Sweden and the Netherlands, a large share of the foreign-born acquires the citizenship of the host country (some 60–75 %). In Portugal a high proportion of the foreign-born has Portuguese citizenship, but in this case the result is largely attributable to persons repatriated from former colonies in the mid-1970s. In the United States, citizenship is also relatively easy to acquire but undocumented immigrants account for a significant proportion of the immigrant stock.

The acquisition of citizenship is more difficult and less common in Switzerland. Belgium and Luxembourg also have a low uptake of citizenship by immigrants, many of whom are from other European Union countries and have less incentive to take out the nationality of the host country since European Union law accords them more or less the same rights as nationals of these countries. Acquisition of citizenship is high, on the other hand, in the Czech and Slovak republics, Hungary and Poland, because immigrants to these countries are largely ethnic nationals who have emigrated from neighbouring countries over the past 15 years and for whom acquisition of citizenship was either automatic or facilitated.

In summary, statistics on the foreign population exclude foreign-born persons who have acquired the citizenship of the host country and may suggest that migration is much less prevalent than is in fact the case. Statistics on the foreign-born population thus provide a more accurate picture of the extent of migration in countries as well as of developments over time (unlike nationality, the place of birth is an unvarying characteristic).

One might also expect that, because naturalisation tends to be associated with a longer period of residence in the host country and with certain legal advantages, among them access to civil service employment, statistics based on the foreign population would tend to give a less favourable picture of the labour market outcomes of immigrants. This is because the exclusion of naturalised persons removes from the population those persons who can be expected to have longer durations of residence in the host country and more favourable outcomes. However, in practice this does not seem to be generally the case. Although unemployment rates of foreigners are indeed higher than those of the foreign-born in most countries, the difference is generally not large (OECD 2004).

Still, the distinction between foreign-born and native-born remains a useful and indeed necessary one. Education, training and labour market policies will clearly not be the same for disadvantaged persons of immigrant background who have been born, raised and educated in the host country as for immigrants who arrive with language deficiencies or educational credentials and labour market experience obtained elsewhere, which may not always be recognised. Statistics that are able to distinguish between these two groups are clearly needed if appropriate diagnoses of labour market problems are to be made and the appropriate policies developed for addressing them.

2.3.2   The geographic origin of immigrants

All geographic areas of the planet are represented as source regions for immigrants in OECD countries (see Table 2.2), which have acted as magnets for populations from less prosperous countries. Recent decades have seen a broadening of the geographic origin of immigrants, some of it driven by the removal of restrictions in some countries to migration from non-European countries, some by ethnic conflict, and some by increased opportunities in higher education in OECD countries for persons from developing countries. Irregular migration has also increased with expanded possibilities for international travel and easier communication about labour market opportunities in OECD countries.

However, persons from other OECD countries still account for some 46 % of the stock of immigrants in the OECD area. Most of this represents movements which took place from the 1950s through the 1970s. The exception concerns movements from Mexico to the United States which continue to be significant. More recent movements have tended to come from outside the OECD area. On the whole, Asians and Latin Americans (excluding Caribbean countries) each account for more than 15 million persons in OECD countries. Spain, a recent immigration country, alone has received more than 740 thousand people from Latin America,

Table 2.2 Foreign-born persons by region of origin, OECD countries.

	Africa	of which North African countries	Asia	of which China and Chinese Taipei	Latin America	North America	Caribbean	Oceania	EU25	Other Europe	Unspecified
Australia	191 501	2 573	1 115 655	232 320	74 893	81 018	32 000	423 428	1 889 893	264 819	6
Austria	19 934	3 560	57 236	8 254	6 054	9 029	3 976	1 931	364 624	527 007	16 717
Belgium	247 515	139 799	68 494	9 410	20 387	18 071		1 468	621 471	117 787	12
Canada	323 580	52 485	2 040 590	657 930	336 570	287 465	285 295	53 215	2 014 255	375 710	335
Czech Republic	2 374	588	21 365	1 251	870	2 687	595	341	344 256	75 989	
Denmark	31 875	6 520	110 454	4 590	9 208	11 123	785	2 249	118 004	77 355	
Finland	9 713	1 783	18 375	2 120	1 817	4 086	261	750	51 681	44 764	1
France	2 862 569	2 296 979	444 774	36 831	79 987	58 398	24 836	6 211	1 978 923	412 539	
Germany[a]	175 665	51 230	567 021		47 578	81 308	1 128		2 552 578	5 244 548	1 587 387
Greece	58 275	1 416	75 854	671	5 486	35 683	367	21 111	191 038	733 183	882
Hungary	2 687	517	10 730	4 002	773	3 199	688	298	65 057	209 815	5
Ireland	26 650	1 238	27 768	7 449	2 793	25 624	482	8 406	291 340	16 408	339
Japan	5 742	421	969 799	253 096	232 248	45 871		8 801	25 299	6 098	1
Korea[b]			116 732	56 272	1 562	14 408	274	719	3 246		15 707
Luxembourg	5 692	1 134	4 382	1 202		1 399		133	116 309	11 855	1 046
Mexico	1 214	262	10 765	2 001	71 644	349 366	9 922	811	44 396	4 096	403
Netherlands	280 007	163 658	367 987	34 754	221 626	29 826	93 326	13 226	340 220	269 158	1
New Zealand	39 351	273	175 302	62 736	3 651	21 126	17 100	156 078	271 008	14 724	207
Norway	31 278	5 665	100 274	5 869	15 133	17 017	1 268	1 489	116 637	49 868	805

Table 2.2 (*continued*).

	Africa	of which North African countries	Asia	of which China and Chinese Taipei	Latin America	North America	Caribbean	Oceania	EU25	Other Europe	Unspecified
Poland	2 962	741	9 479	667	920	10 566	202	671	248 868	483 223	18 391
Portugal	349 859	1 596	16 859	2 397	74 949	14 627	914	1 256	159 008	34 000	
Slovakia	404	50	1 400	142	154	945	77	64	99 931	16 097	
Spain	423 082	343 819	86 669	28 848	744 221	25 141	95 979	4 443	597 948	194 676	42
Sweden	78 039	9 962	244 246	12 106	59 965	17 627	2 840	3 376	456 262	215 241	
Switzerland	68 801	21 153	101 599	8 318	48 327	29 319	8 834	4 787	854 305	352 962	101 822
Turkey	12 686	1 627	83 657	1 802	1 010	15 006	216	3 265	447 739	695 795	1
United Kingdom	838 459	26 088	1 579 133	154 111	95 357	238 043	232 940	170 278	1 493 235	175 577	42 541
United States	988 253	58 530	8 402 240	1 550 070	13 476 759	965 485	4 469 340	288 391	4 594 095	1 442 654	7 574
Total	7 078 167	3 193 667	16 828 839	3 139 219	15 633 942	2 413 463	5 283 645	1 177 196	20 351 626	12 065 948	1 794 230

[a] Data for EU25 are limited to 16 countries (BEL, DNK, FIN, FRA, GRC, IRL, ITA, LUX, NLD, AUT, PRT, SWE, POL, ESP, HUN and GBR) in data provided by Germany.
[b] Data for EU25 are limited to three countries (DEU, FRA and GBR) in statistics provided by Korea.

and the United States has received approximately 13.5 million. In the former case, migration has clearly been driven by the common language and the facilitated integration which it conveys; in the latter, by geographic proximity, historical ties and the pull of a dynamic labour market.

It is continental Europe (including Turkey and Central and Eastern Europe) which accounts for the largest number of immigrants in OECD countries (over 32 million). There are, for example, nearly two million persons from the enlarged European Union (EU25) in each of Canada, Australia, France and Germany, and over 4.5 million in the United States. Migrants from the United Kingdom account for over 20% of this, and from Germany, Italy and Poland close to 15% each. Note the absence of France among the large European countries which has only about a 3% share of persons born in EU25 countries and residing in other OECD countries.

Among immigrant populations in OECD countries, persons born in North Africa (Algeria, Tunisia and Morocco) are at least as numerous as persons born in China. Geographic proximity to Europe and/or the existence of former colonial links undoubtedly have a lot to do with this. Migrant populations originating from North Africa are concentrated in France, Spain and the Netherlands.

2.3.3   The educational attainment of immigrant populations

The decision to emigrate is a function, among others, of the trade-off between the expected returns from migration and the cost of moving. The decision, however, is not an entirely free one, because of host country restrictions on entry and work. The nature of these restrictions and national policies regarding migration as well as labour market needs can have a significant influence on the educational composition of migrants. The picture across OECD countries is a diverse one on this front, with some countries showing a high percentage of persons with high educational credentials among the immigrant population and some a high percentage with low qualifications (see Table 2.3).

Countries which practice a selective immigration policy based on human capital criteria (Australia, Canada and New Zealand) are among the countries with the highest percentages of highly qualified immigrants). The data presented here incorporate an implicit assumption, namely that persons born abroad were educated abroad. This is not necessarily the case.

The percentage of highly educated immigrants is also high in the United Kingdom, Ireland, Korea and Norway, where 30–41% of foreign-born have a higher degree. In addition, in a number of countries, foreign-born persons with a doctoral degree account for a high proportion of all persons holding such degrees in the host country. In the United States, even if a significant part of the foreign-born are not highly qualified, more than 440 thousand hold a PhD. This accounts for approximately 25% of the total stock of PhDs in the country. The proportion of foreign-born doctorates in Sweden is comparable, and in Australia and Canada it stands even higher, at 45% and 54%, respectively.

Table 2.3 Distribution (percentages) of the population by educational level and place of birth in OECD countries (persons aged 15+).

Country	Native-born residents				Foreign-born residents				Native-born emigrants
	Less than upper secondary (ISCED 0/1/2) (1)	Upper secondary and post-secondary non-tertiary (ISCED 3/4) (2)	Tertiary (ISCED 5/6) (3)	of which PhD (ISCED 6) (4)	Less than upper secondary (ISCED 0/1/2) (6)	Upper secondary and post-secondary non-tertiary (ISCED 3/4) (7)	Tertiary (ISCED 5/6) (8)	of which PhD (ISCED 6) (9)	Tertiary: (ISCED 5/6) (5)
Australia	45.8	15.7	38.6	1.55	38.3	18.8	42.9	3.53	44.4
Austria	33.4	55.7	10.9		49.4	39.3	11.3		30.4
Belgium	46.8	30.3	22.9	0.44	54.2	24.2	21.6	1.11	31.9
Canada	31.6	36.9	31.5	0.32	30.1	31.9	38.0	1.29	40.3
Czech Republic	22.8	67.0	10.2	0.37	38.4	48.8	12.8	0.71	24.9
Denmark	41.0	40.2	18.8	0.20	48.6	31.9	19.5	0.20	37.4
Finland[a]	40.3	36.3	23.4	0.54	52.7	28.4	18.9	0.97	26.2
France	45.8	37.4	16.9		54.8	27.2	18.1		33.7
Germany	23.7	56.8	19.5		43.7	40.8	15.5		29.3
Greece	54.4	32.2	13.4	0.89	44.8	39.9	15.3	0.91	17.7
Hungary	45.1	44.2	10.7		41.1	39.1	19.8		29.5
Ireland	47.8	29.5	22.7	0.26	29.6	29.3	41.0	1.17	27.5
Italy	63.6	28.3	8.1		54.3	33.5	12.2		13.1
Japan	25.3	47.3	27.4		25.9	44.2	29.9		49.9
Korea	36.1	37.2	26.7	1.56	23.8	44.0	32.2		44.2
Luxembourg	28.7	58.6	12.8		36.7	41.6	21.7		25.0
Mexico	72.3	16.8	10.9	0.60	36.5	25.7	37.8	5.96	5.6
Netherlands	40.7	39.8	19.5		53.0	29.4	17.6		36.5
New Zealand	30.1	42.7	27.2		18.7	50.4	31.0		44.8
Norway	21.2	55.6	23.2	0.32	18.3	50.6	31.1	1.45	33.9
Poland	31.2	58.4	10.4	0.34	47.9	40.3	11.9	0.86	26.5
Portugal	80.0	12.2	7.7	0.13	54.7	25.9	19.3	0.52	6.8
Slovak Republic	28.0	62.0	10.0		29.3	56.1	14.6		14.0
Spain	63.9	16.7	19.4	0.51	55.4	22.8	21.8	0.99	19.3
Sweden	25.0	52.2	22.8	0.70	29.6	46.2	24.2	1.53	40.1
Switzerland	25.6	56.3	18.1		41.6	34.7	23.7		24.8
Turkey	79.4	15.2	5.4		49.3	34.1	16.6	1.13	6.8
United Kingdom[b]	51.2	28.7	20.1		40.6	24.5	34.8		41.1
United States	21.9	51.2	26.9	0.70	39.8	34.3	25.9	1.40	49.2

[a] For Finland, 'less than upper secondary' includes 'unspecified' educational attainment.
[b] Educational levels for the United Kingdom are for people aged 16–74 years.

The situation in Belgium, Finland, France, Italy, the Netherlands, Portugal and Spain differs significantly from this. In these countries, at least 50 % of the foreign-born have less than upper secondary education. In Belgium, France and the Netherlands, this reflects the nature of labour recruitment in the past, as well as the educational level of family members joining the original migrants. Migration to Italy, Portugal and Spain is of more recent origin and has been essentially low-skilled, with migrants taking on jobs which natives are unwilling to fill.

Emigration countries such as Mexico, Ireland, Portugal and Turkey tend to have immigrant populations that are more qualified than the domestic population, whereas in Germany, the Netherlands and the United States, it is the reverse, with, however, small differences with respect to persons with tertiary qualifications. Luxembourg and Switzerland, on the other hand, are characterised by an immigrant population that has a higher percentage of both low and highly educated persons.

2.4   Expatriates of OECD member countries residing in other member countries

Within OECD countries, much attention has been directed in recent years at the emigration of highly qualified persons, attracted to countries where job opportunities are more prevalent. Broad empirical evidence regarding the extent of this phenomenon has been notably absent from the public debate. Although the database described here does not allow one to remedy this as yet with respect to recent departures, it does provide a broad overall picture of emigration over the past decades.

2.4.1   The extent of expatriation in OECD countries

Identifying and counting expatriates is not without difficulties and different methods may produce different estimates. There are three main types of estimates, each of them with its advantages and shortcomings: (i) statistics of people registered in embassies and consulates overseas, (ii) emigration surveys in origin countries and (iii) compilation of statistics from receiving countries.

Administrative data from embassies and consulates provide an interesting source for estimating the stock of nationals abroad. Indeed, in most cases expatriates need to register to receive social benefits or pension payments, to pay taxes, to vote overseas, to renew identity papers, or simply to report their presence in the country. Unfortunately, because registration is not always compulsory or enforced, the data coverage is not perfect and may vary a lot from one country to another. For instance, the estimate of French citizens living in other OECD countries by the Ministry of Foreign Affairs (1.4 million in 1999) is more than double the number of official registrations at consulates. Furthermore, because people do not necessarily deregister and because some people may register even for short stays abroad (especially in countries where there is some risk), overestimation is also a problem.

Several countries have included specific questions on residents temporarily overseas in censuses or have implemented specific surveys to identify their nationals abroad. It is possible to ask an interviewed household member how many usual members of the household are currently abroad. This type of estimate, however, covers only short stays abroad (including those for reasons of tourism) and excludes many long-term emigrants, because the situations in which the entire household has settled overseas are not covered.

For the data sources considered here, the expatriate population is identified by compiling the data on the foreign-born in all OECD countries for a specific place of birth. The estimate does not include expatriates in non-OECD countries. It is thus not directly comparable to the other sources mentioned previously. One of the problems with this approach is that it is not always possible to identify foreign-born persons who were citizens of their current country of residence at birth (e.g. children born overseas of national parents). These persons will be recorded as foreign-born, when they would more accurately be characterised as foreign-born nationals. This situation can be particularly problematic for countries which have had important communities abroad. Another problem arises from the fact that some people do not report their place of birth in censuses. Persons not specifying a place of birth represent 10 % of the total population in the Slovak Republic, about 5.7 % in Australia, and 4 % in New Zealand and Switzerland. Furthermore, some censuses do not identify systematically all countries of origin (e.g. Korea only records 17 foreign nationalities in its census). Consequently, the estimates presented in this chapter on expatriates should be considered a lower bound.

The complete data on emigrants from OECD countries to other OECD countries is presented in Table 2.4. It shows the stock of persons born in one OECD country and residing in another. Since these data summarise the cumulative effects of migration over recent decades, they may not reflect current trends. In particular, certain former high expatriation countries (Ireland, Portugal and Spain), in which income levels were relatively low in the 1960s and 1970s, have become immigration countries in the 1990s.

In the 29 OECD countries currently under review, 36.3 million persons, i.e. 46 % of the total foreign-born population, come from another OECD country. In certain host countries, such as Luxembourg, the Slovak Republic, Ireland, Mexico, the Czech Republic and to a lesser extent Switzerland and Belgium, the share of the foreign-born from other OECD countries is very high (between 65 % and 85 %). At the other extreme, it is close to 24 % in Hungary, Poland and Korea and only 11 % in Japan.

The largest number of expatriates is from Mexico, with nearly 9.5 million people, of whom the vast majority are resident in the United States. The number of persons born in Germany and in the United Kingdom residing in other OECD member countries is also large, more than three million people for each of them. The number of persons born in Turkey, Italy and Poland and residing in other OECD countries amounts to over two million persons each.

Table 2.4 Persons born in an OECD country and residing in another member country (total population).

Country of origin		AUS	AUT	BEL	CAN	CZE	DNK	FIN	FRA	DEU	GRC	HUN	IRL	ITA	JPN	KOR
AUS	Australia	—	1686	1136	20155	230	1663	656	4216	133341	20449	258	6107	17872	6148	719
AUT	Austria	19313	—	3166	22585	7358	1464	312	12171	22702	2252	3716	533	16982	293	
BEL	Belgium	4900	1523	—	20990	755	1249	206	124709		4671	520	1141	42977	324	
CAN	Canada	27289	1658	4145	—	490	2752	1181	18913		12477	632	4081	23487	7067	2468
CZE	Czech Republic	6973	54627	77	16500	—	292	39	3438		3725	2494	1189	7126	113	
CSFR	Czechoslovakia[a]	—		3152	13415	136	2320	298	6262	36877						
DNK	Denmark	9089	1090	2973	18400	332	—	708	5482	17594	830	100	697	2593	311	
FIN	Finland	8258	1300	2761	14395		3575	—	3525	11067	849	343	687	2230	512	
FRA	France	18827	5903	151976	80965	3633	4038	1089	—	74131	6723	1738	6815	131293	3768	1142
DEU	Germany	108220	140099	83386	191140	9647	26559	3582	215167	—	101425	10173	8770	195309	3407	920
GRC	Greece	116431	3060	15089	76900	1806	1066	468	11872	261329	—	1228	345	14853	165	
HUN	Hungary	22752	30953	5486	50830	6200	1604	873	10543	38309	1586	—	456	6073	266	
ISL	Iceland	463	135	164	500	20	5855	120	333		32	5	55	173	31	
IRL	Ireland	50235	546	2999	26430	67	1091	200	5316	7946	498	48	—	2206	618	
ITA	Italy	218718	26099	132466	319230	1035	3364	958	409190	429313	5929	935	3705	—	1127	
JPN	Japan	25471	1957	3850	27245	193	1364	640	14261		560	324	716	5002	—	13398
KOR	Korea[b]	38900	1446	4049	82890	76	8056	132	15852		204	144	166		618	—
LUX	Luxembourg	141	514	10459	560	15	245	32	9895	4540	99	17	85	4732	8	

Country of residence

Table 2.4 (*continued*).

Country of origin		AUS	AUT	BEL	CAN	CZE	DNK	FIN	FRA	DEU	GRC	HUN	IRL	ITA	JPN	KOR
MEX	Mexico	1154	721	1150	44190		524	153	6360	68459	363	45	314	4338	1222	
NLD	Netherlands	83324	5248	97165	119310	549	4833	731	27618		3083	513	3512	10046	604	
NZL	New Zealand	355765	245	301	9920	35	538	86	1071		506	35	2256	602	2401	
NOR	Norway	4324	742	1295	6505	107	16386	954	2838		459	288	441	1071	280	
POL	Poland	58110	41671	19894	182155	24707	10723	1173	106650	1170711	15468	2685	2167	33968	468	
PRT	Portugal	15441	950	21371	155980	39	686	141	579465	94258	292	28	590	4158	368	
SVK	Slovak Republic	2984	15981	30	10740	285372	135	17	2149		411	37439	332	2803	107	
ESP	Spain	12662	2072	36840	10785	170	2851	779	342071	86160	972	139	4632	20433	1183	
SWE	Sweden	6818	3214	3991	7725	210	18706	28040	8658	10783	5428	394	1315	5091	798	
CHE	Switzerland	10753	11713	4274	21595	385	1910	615	75598	28945	3567	616	882	194370	677	
TUR	Turkey	29821	125026	70793	17810	222	30175	2150	179392	1610735	76561	696	545	8149	915	
GBR	United Kingdom	1036245	6786	26176	624305	1436	13615	2731	84493	85058	13303	1186	248515	50888	10411	1184
USA	United States	53694	7371	13925	278570	2197	8367	2903	39464	81308	23091	2567	21541	50781	38804	11940
OECD foreign-born		2347075	494336	724539	2472720	347422	176006	51967	2326972	4273566	305813	69306	322590	859606	82396	31771
Percentage of total foreign-born from OECD countries		*57.6*	*50.1*	*65.9*	*53.3*	*77.5*	*48.8*	*39.5*	*39.7*	*51.8*	*27.3*	*23.7*	*80.7*	*38.5*	*10.8*	*23.5*

Table 2.4 (continued).

| Country of origin | | Country of residence | | | | | | | | | | | | | | | |
		LUX	MEX	NLD	NZL	NOR	POL	PRT	SVK	ESP	SWE	CHE	TUR	GBR	USA	Total
AUS	Australia	96	281	9529	56142	1101	608	1192	52	3913	2525	3420	2938	107871	75314	346181
AUT	Austria	624	500	6746	1200	1040	4312	391	808	4100	5967	54616	14335	19503	70560	407564
BEL	Belgium	14770	735	46003	513	907	2797	2879	179	28200	1356	10738	8751	21668	41705	393098
CAN	Canada	305	5768	8427	7770	2290	1555	7326	115	3810	2471	7519	1427	72518	945060	1172696
CZE	Czech Republic	253	225	121	663	567	6200	130	75585	1891	522	11021	1026	12220	24865	231629
CSFR	Czechoslovakia[a]			4984		317					7330				45245	120200
DNK	Denmark	1522	245	3242	1446	23326	704	387	17	5749	40921	4122	3372	18695	34064	196293
FIN	Finland	701	126	2379	372	7027	192	312	11	5378	189341	3842	1672	11322	22865	294673
FRA	France	18864	5751	19338	2283	3069	34647	95282	1393	156681	6155	98352	16048	96281	204238	1231559
DEU	Germany	12847	5595	123110	8382	12880	101633	24283	735	135638	40217	181984	273535	266136	1241450	3513382
GRC	Greece	865	298	7375	942	636	2793	125	26	1132	10853	6295	59217	35169	178155	807628
HUN	Hungary	293	239	5333	987	1507	1344	217	17293	1460	13794	12403	520	13159	94095	338282
ISL	Iceland	309	16	385	84	3941	41	34	1	306	3811	151	43	1552	9805	28056
IRL	Ireland	641	192	4425	6726	499	71	533	2	4342	1349	1542	538	537108	164435	819962
ITA	Italy	12254	3904	17207	1440	1506	4292	1958	117	26578	6584	234634	2843	107244	536370	2496746
JPN	Japan	289	2936	5879	8622	932	230	280	16	3154	2502	4388	2003	37535	497945	661403
KOR	Korea[b]	513	2100	5305	17934	6347	125	74	1	2158	9574	1613	513	12310	156085	365966
LUX	Luxembourg	—	15	827	30	93	125	3313		1029	139	1436	46	1222	2690	42307

Table 2.4 (continued).

Country of origin

	Country of residence														
	LUX	MEX	NLD	NZL	NOR	POL	PRT	SVK	ESP	SWE	CHE	TUR	GBR	USA	Total
MEX Mexico	61	—	1454	243	471	116	214	9	20949	1328	2863	154	5049	9336530	9429914
NLD Netherlands	3284	773	—	22239	4389	964	3250	32	23153	5150	16771	21823	40438	105920	669897
NZL New Zealand	33	77	3582	—	345	50	48	3	331	763	1148	290	58286	26350	465034
NOR Norway	152	134	2499	465	—	315	283	9	5922	45087	1818	3554	13798	36340	145914
POL Poland	1006	971	17351	1938	6702	—	358	3473	16423	41608	10679	3415	60711	477450	2311629
PRT Portugal	41690	288	10218	141	760	60	—	4	56359	2533	100975	225	36555	212115	1294000
SVK Slovak Republic	93	23	67	138	306	1514	30	—	1217	374	3736	315	5273	15945	387438
ESP Spain	2120	21114	18279	339	1782	1111	13966	30	—	5470	61679	1209	54482	114190	815400
SWE Sweden	984	425	3642	960	32939	703	741	23	9424	—	6878	5335	22525	54435	239201
CHE Switzerland	787	1478	5792	2763	1507	506	12897	51	53484	2557	—	10369	16010	49445	512759
TUR Turkey	290	246	181865	396	8410	452	106	30	986	34083	58546	—	54079	90595	2582784
GBR United Kingdom	3167	2688	45691	218394	14332	2630	10068	87	107794	16428	25378	18939	—	823279	3492040
USA United States	1094	343597	21356	13344	14725	9010	7301	829	21320	15143	21775	13579	158434	—	1276936
OECD foreign-born	119907	400740	582411	376896	154653	179012	187978	100931	702881	515935	950322	468034	1897153	15687540	37090571
Percentage of total foreign-born from OECD countries	84.7	81.4	36.1	54.0	46.5	23.7	28.9	84.8	32.4	47.9	64.7	37.2	39.4	47.5	47.0

[a] CSFR stands for 'Czechoslovakia not included elsewhere'.

[b] Data for Korea are partial as several OECD countries do not systematically distinguish the Democratic Republic of Korea and the People's Republic of Korea (e.g. 529 408 people in Japan and 743 260 in the United States).

Expressed as a percentage of the total population of the given country, almost 24 % of people born in Ireland are currently living in another OECD member country (see Figure 2.1). Other significant expatriate communities include persons born in New Zealand (16 %), Portugal (13.7 %), Luxembourg (12.8 %) and Mexico (9.9 %). By contrast, few persons born in the United States, Japan, Korea and France tend to expatriate (less than 2.5 % in all cases and barely 0.5 % for the United States and Japan). All other things being equal, expatriation tends to be a small-country phenomenon, with the percentage expatriates living abroad being negatively associated with (the log of the) population (correlation $= -0.53$).

2.4.2 The educational attainment of expatriates

Interest in expatriate populations tends to focus on the educational levels, rather than on the sizes of the communities. Speculation on this point regularly feeds the media in certain countries, generally without credible statistical evidence. Some national studies exist (e.g. Ferrand 2001; Hugo *et al.* 2003; Saint-Paul 2004), but the issue has not been treated extensively.

The distributions of educational attainment for OECD expatriates living in other OECD countries are set out in Table 2.5. Here, the relative shares of highly qualified migrant populations (i.e. persons with tertiary education) can be compared. The United States and Japan exhibit the highest proportions of expatriates with tertiary education (almost 50 %). The selectivity of expatriation with respect to qualifications, measured by the difference between the proportion of expatriates and that of the native-born with tertiary-level attainment, highlights several European countries, notably France, Austria and Switzerland (at least 20 percentage point difference). Hungary and Denmark also have a relatively significant proportion of their emigrant populations who are graduates of higher education institutions compared to the native-born. On the other hand, expatriation originating from Portugal, Turkey, Mexico or the Slovak Republic is essentially low-skilled.

With the notable exceptions of some Central and Eastern European countries as well as Mexico, Ireland, Korea and Finland, highly educated expatriates in OECD countries from the rest of the world systematically exceed highly educated expatriates from OECD countries (see Figure 2.2). On this measure (and provided that expatriates with high skill levels to non-OECD countries can be assumed to be relatively uncommon), most OECD countries would seem to benefit from the international mobility of the highly skilled. Note, however, that for some countries such as Ireland, which appears as a net loser in these data, an analysis of recent net flows of migrants would produce a rather different picture.

Within the OECD area, only the United States, Australia, Canada, Switzerland, Spain, Sweden, Luxembourg and Norway (in this order) are net beneficiaries of highly skilled migration from other OECD countries. The United Kingdom has 700 thousand more highly skilled expatriates in OECD countries than it has from other OECD countries. Comparable figures exceed 500 thousand for Germany, 400 thousand for Mexico and 300 thousand for Poland. France and Belgium have

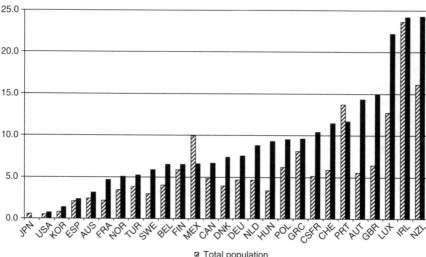

	Total population	Highly educated		Total population	Highly educated
JPN	0.5		DEU	4.6	7.5
USA	0.5	0.8	NLD	4.6	8.8
KOR	0.8	1.4	HUN	3.4	9.3
ESP	2.1	2.3	POL	6.2	9.5
AUS	2.4	3.1	GRC	8.1	9.6
FRA	2.1	4.6	CSFR	5.1	10.4
NOR	3.4	5.0	CHE	5.9	11.5
TUR	3.9	5.2	PRT	13.7	11.7
SWE	3.0	5.9	AUT	5.6	14.4
BEL	4.0	6.5	GBR	6.4	14.9
FIN	5.8	6.5	LUX	12.8	22.2
MEX	9.9	6.5	IRL	23.7	24.2
CAN	4.8	6.7	NZL	16.1	24.2
DNK	3.9	7.4			

Figure 2.1   Emigrants as a percentage of all native-born, total population and highly educated aged 15+. *Notes:* CSFR stands for 'Former Czechoslovakia'. Data for Korea are partial, as several OECD countries do not systematically distinguish South and North Korea

Table 2.5 Number and distribution (percentages) of OECD expatriates by level of education.[a]

Country of origin	Tertiary	Upper secondary and post-secondary non-tertiary	Less than upper secondary	Total(number of persons)
Australia	45.9	33.1	21.0	267314
Austria	30.0	47.0	23.0	366024
Belgium	34.6	33.1	32.2	321544
Canada	40.6	40.0	19.4	1044978
CSFR	30.1	42.5	27.4	109984
Czech Rep.	25.2	50.5	24.3	215879
Denmark	37.4	38.7	23.9	173009
Finland	26.3	42.4	31.3	265245
France	36.4	32.8	30.8	1013581
Germany	30.4	42.1	27.5	2933757
Greece	16.6	26.7	56.8	735430
Hungary	29.6	42.4	28.0	314922
Iceland	36.1	39.7	24.2	23070
Ireland	27.5	21.2	51.2	792316
Italy	13.0	26.8	60.3	2430339
Japan	49.7	38.9	11.4	575992
Korea	44.2	38.2	17.6	312538
Luxembourg	27.9	32.3	39.8	27164
Mexico	5.6	24.4	70.0	8431381
Netherlands	36.1	35.0	28.9	616909
N. Zealand	44.6	22.5	32.9	410663
Norway	33.9	39.0	27.1	122079
Poland	26.6	42.0	31.4	1276482
Portugal	6.7	24.0	69.2	1268726
Slovak Rep.	14.0	45.5	40.5	374570
Spain	18.7	27.8	53.5	763013
Sweden	40.1	38.3	21.6	206604
Switzerland	36.5	36.9	26.5	262456
Turkey	6.4	21.7	71.9	2195645
UK	41.2	32.8	26.0	3229676
USA	49.9	28.3	21.8	809540

[a] Population aged 15 and over. Percentage calculations do not take account of unspecified cases.

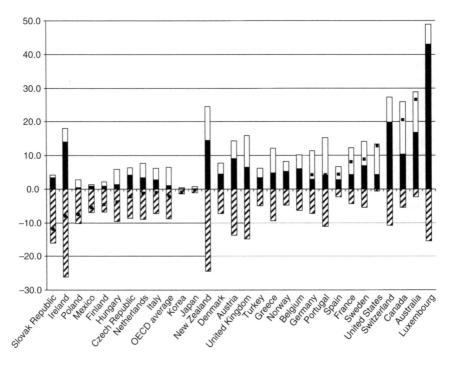

■ Expatriates from other OECD countries ▨ OECD Expatriates to other OECD countries
□ Expatriates from the rest of the world ■ "Net" persons with tertiary attainment as a result of migration

	Expatriates from other OECD countries	OECD expatriates to other OECD countries	Expatriates from the rest of the world	'Net' persons with tertiary attainment as a result of migration
Slovak Republic	3.3	−16.0	0.9	−11.9
Ireland	14.0	−26.1	4.0	−8.1
Poland	0.4	−10.2	2.3	−7.6
Mexico	0.8	−6.9	0.5	−5.6
Finland	0.9	−6.8	1.3	−4.6
Hungary	1.4	−9.7	4.5	−3.8
Czech Republic	4.1	−8.7	2.2	−2.3
Netherlands	3.3	−8.9	4.4	−1.2
Italy	2.8	−7.3	3.3	−1.2
OECD average	1.1	−8.8	5.4	−2.3
Korea	0.2	−1.4	0.2	−1.0
Japan	0.2	−1.1	0.5	−0.4
New Zealand	14.6	−24.4	10.0	0.2

Figure 2.2 Migrants with tertiary attainment as a percentage of all residents with tertiary attainment, circa 2000.

Denmark	4.4	−7.3	3.2	0.3
Austria	9.1	−13.8	5.2	0.5
United Kingdom	6.5	−14.9	9.4	1.0
Turkey	3.4	−4.9	2.7	1.2
Greece	4.8	−9.4	7.3	2.7
Norway	5.2	−4.9	3.0	3.2
Belgium	5.9	−6.4	4.2	3.7
Germany	2.8	−7.3	8.6	4.1
Portugal	4.1	−11.2	11.2	4.1
Spain	2.7	−2.3	3.8	4.2
France	4.2	−4.4	8.2	8.0
Sweden	6.9	−5.4	7.3	8.8
United States	4.2	−0.7	9.2	12.7
Switzerland	20.0	−10.8	7.3	16.4
Canada	10.3	−5.4	15.5	20.4
Australia	16.8	−2.4	12.1	26.5
Luxembourg	43.1	−15.4	5.8	33.5

Notes: Data are largely from the 2000 round of population censuses in OECD countries. Tertiary attainment is classified according to the International Standard Classification of Education (ISCED). The education in question may not have been received in the origin country. 'Net' appears in quotes in the fifth column heading because emigration to non-OECD countries is not included. This would have required collecting census data from these countries.
Source: See *Trends in International Migration* 2004 (http://www.oecd.org.els/migration) for more details.
Migrants with tertiary attainment as a percentage of all residents with tertiary attainment, circa 2000.

Figure 2.2   *(continued).*

about the same number of expatriates residing in other OECD countries as they have expatriates from OECD residing in their countries. This of course gives only a partial picture of movements of the highly educated, because it does not include movements of the highly skilled between non-OECD and OECD countries. When movements from all countries to the OECD are included, the picture changes significantly.

For each OECD country, Figure 2.2 contrasts the population of highly skilled foreign-born residing in that country with the corresponding expatriates living in other OECD countries. The picture is thus not entirely complete, because it excludes outflows outside the OECD area.

The difference between the number of highly skilled expatriates in OECD countries and highly skilled foreign-born from all countries is largely positive in the United States, Canada and Australia, but also in France and Germany, even though these countries have a significant number of highly skilled expatriates in other OECD countries. Highly skilled immigrant populations expressed as a percentage of the total highly skilled workforce are particularly significant (over 20 %) in Australia, Luxembourg, Switzerland, Canada and New Zealand. The percentage of the highly skilled who are expatriates is below 10 % for most OECD countries (see Figure 2.1) and particularly low in Japan, the United States, Spain and Australia. Conversely, more than 10 % of the highly skilled born in Switzerland, Portugal, Austria, or the United Kingdom are living in other OECD countries. This percentage is over 20 % for two countries: Ireland (24 %) and New Zealand (24 %). Table 2.3 clearly confirms the selective character of migration (in favour of the highly skilled) in OECD countries. This phenomenon is the result of pull factors attributable to selective migration policies in receiving countries, but also to other factors such as the fact that highly qualified persons are more tuned in to the international labour market (i.e. because of their social capital, language skills, access to information, and so on) and have more resources to finance a move.

2.5   Highly skilled expatriates from non-member countries in OECD countries

2.5.1   Introduction

Until the construction of the dataset described in this chapter, there was limited data on the extent of international mobility of the highly educated, in particular from developing countries. One study by Carrington and Detragiache (1998), which has recently been updated by Adams (2003), relies on United States census data on the foreign-born and OECD immigrant stock data from the *Trends in International Migration* database[3] to construct a database for expatriation by level of education and by country of origin. The authors use the United States 1990 Census data to determine the educational profile of the immigrant population by country of birth and apply it to estimate the level of education of immigrant populations (in many cases, foreigners) in other OECD countries. The Barro and Lee (1993) database on educational attainment levels is the source for the population by level of education and place of birth. This then becomes the denominator of reference to estimate the percentage of persons at each education attainment level living abroad.

The estimates based on this methodology are subject to a number of limitations. One significant problem concerns the assumptions made because of data availability limitations. In particular, the foreign-born population in EU countries is assumed to be the foreign population and foreigners of a particular nationality are considered to have the same educational profile as the foreign-born of the United States. As a result the estimates tend to be problematical for small source countries and countries whose citizens tend to migrate to countries other than the United States. In addition, Cohen and Soto (2001) have shown that the Barro and Lee (1993) database on educational attainment is of uneven quality.

The database on immigrant populations and expatriates in OECD countries, which is the basis of this paper, has direct measures of the educational attainment of immigrant populations for all OECD receiving countries and can avoid making the assumptions of previous studies. The percentage of persons living abroad can be produced by level of education qualification and country of origin. This percentage for each country and educational level is calculated by dividing the number of

---

[3] See Chapter 3 for a brief description of the database.

expatriates of a given country of origin and level of education by the total native-born population of the same country and level of education (note: native-born = expatriates + resident native-born). Three levels of qualification are considered: less than upper secondary, upper secondary and tertiary, according to the International Standard Classification of Education. Highly skilled persons correspond to those with a tertiary level of education.

Two sets of estimates of the *resident native-born* using two reference databases for the structure of education of the population aged 15+ in origin countries have been produced. The first makes use of an updated version of Barro and Lee (1993) for the year 2000 which covers 113 countries (Barro and Lee 2000). The second database covers 95 countries (Cohen and Soto 2001). The authors of the latter have used the OECD education database plus some other sources for non-member countries to construct a new database on human capital stock in 2000. Data for the total population come from the World Development Indicators of the World Bank. A spearman rank correlation test confirms that the two calculations produce a similar classification ($\rho = 0.94$), but there exist significant differences for some countries (e.g. Argentina, Chile, Zimbabwe, Singapore and Uruguay).

Because of differences in the population stocks between the World Bank figures and those obtained directly from OECD censuses (partly attributable to differences in reference years) and differences in the specification of levels of education, some differences appear when comparing the percentages of persons resident abroad calculated for OECD countries from these two datasets with those discussed and presented earlier for OECD countries alone, based on census data. The OECD database giving these results is available at www.oecd.org/migration/census database.

More recently, a second database similar to the one described here, due to Docquier and Marfouk (2005), has become available on the Internet, based largely on census sources as well. It differs from this one in a number of respects: (1) it covers, with the exception of eight OECD countries, the population aged 25+ rather than 15+; (2) it includes data for two years, 1990 and 2000; (3) it classifies attainment levels into low/medium/high on the basis of years of schooling, whereas this database uses the International Standard Classification of Education (OECD 1999); and (4) it includes the percentage of the highly educated native-born living abroad for some 75 additional countries for which there are no available educational attainment data, by assuming that attainment levels for each such country are identical to those of the neighbouring country with the closest human development index.

2.5.2   Results

Among non-member countries the biggest expatriate community in OECD countries is that originating in the former USSR with 4.2 million people, followed by the former Yugoslavia (2.2 million), India (1.9 million), the Philippines (1.8 million),

China (1.7 million), Vietnam (1.5 million), Morocco (1.4 million) and Puerto Rico (1.3 million). Among persons with tertiary education, the former USSR still ranks first (1.3 million) with India having the second largest expatriate community (1.0 million). (See also Table A6 in Dumont and Lemaître (2005).)

Table 2.6 Highly skilled expatriates from selected non-OECD countries:[a] percentage of all highly persons with tertiary education.

Cohen and Soto (2001)	Highly skilled aged 15+	Barro and Lee (2000)	Highly skilled aged 15+
(a) 15 non-OECD countries with the lowest percentage of highly skilled 15+ expatriates in OECD countries			
Brazil	1.7	Brazil	1.2
Myanmar	1.7	Thailand	1.4
Indonesia	1.9	Indonesia	1.5
Thailand	1.9	Paraguay	1.8
Bangladesh	2.0	Argentina	1.8
Paraguay	2.0	China	2.4
Nepal	2.1	Myanmar	2.4
India	3.1	Peru	2.7
Bolivia	3.1	Nepal	2.9
China	3.2	Bangladesh	3.0
Jordan	3.2	Bolivia	3.1
Venezuela	3.3	India	3.4
Costa Rica	4.0	Egypt	3.4
Syria	4.3	Venezuela	3.5
Egypt	4.4	Swaziland	3.5
(b) 15 non-OECD countries with the highest percentage of highly skilled 15+ expatriates in OECD countries			
Guyana	83.0	Guyana	76.9
Jamaica	81.9	Jamaica	72.6
Haiti	78.5	Guinea-Bissau	70.3
Trinidad and Tobago	76.0	Haiti	68.0
Fiji	61.9	Trinidad and Tobago	66.1
Angola	53.7	Mozambique	52.3
Cyprus	53.3	Mauritius	50.1
Mauritius	53.2	Barbados	47.1
Mozambique	47.1	Fiji	42.9
Ghana	45.1	Gambia	42.3
United Republic of Tanzania	41.7	Congo	33.7
Uganda	36.4	Sierra Leone	32.4
Kenya	35.9	Ghana	31.2
Burundi	34.3	Kenya	27.8
Sierra Leone	33.3	Cyprus	26.0

[a] Two different sources for the educational attainment of non-OECD countries have been used. They are identified at the top of each column. See text and bibliography for the detailed references.

To estimate the percentage of persons living abroad by level of qualification for non-member countries, information on the level of education of the relevant population in the country of origin is required. The two sets of estimates compiled are presented in Table 2.6 for the 15 countries with the lowest percentages of persons abroad for the highly qualified aged 15 and over as well as for the 15 countries with the highest percentages. Most OECD countries, which are not included in Table 2.6, would tend to fall among countries having lower rates.

Among countries with a low percentage of highly qualified expatriates (i.e. less than 5 %), we find most of the large countries included in the database (i.e. Brazil, Indonesia, Bangladesh, India and China). At the other end of the spectrum, smaller countries, a number of which are islands such as Jamaica, Haiti, Trinidad and Tobago, Mauritius or Fiji, have more than 40 % of their highly skilled populations abroad and sometimes as much as 80 %. As for OECD countries, the importance of the size of the origin country is confirmed by simple correlation analysis.

The world map (see Figure 2.3) presents the expatriate percentages for the highly skilled for all countries, with African countries standing out as those having particularly high expatriate percentages. Anglophone African countries as well as Portuguese-speaking countries (e.g. Mozambique and Angola, but also Cape Verde) record the highest percentages. Expatriation of the highly skilled is also quite significant in Central America but more moderate in Asia, with the relative exceptions of Hong Kong and Singapore. The former USSR faces intensive migration from former Soviet republics towards Russia, which unfortunately it is not possible to illustrate here. However, emigration of the highly skilled from

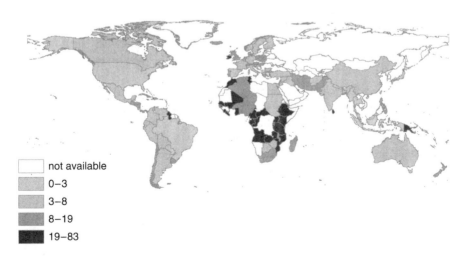

Figure 2.3   Percentage of emigrants to OECD countries among highly educated born in the country. Calculations based on Cohen and Soto's (2001) data on highly educated persons in countries of origin.

countries of the former USSR, considered as a whole, towards OECD countries remains moderate relative to the total stock of qualified persons in these countries.

2.6   Summary and conclusions

The data strategy used for the results presented here consists of producing estimates of expatriate populations in OECD countries for a particular country of origin by compiling and aggregating data on the foreign-born from that country for all OECD countries. Although it does not give a complete picture of expatriation (non-OECD countries were not included in the host country data collection), it does provide a comprehensive picture of immigrant populations in developed countries, one that would not generally be possible to obtain by examining individual national data sources alone. In addition it has made it possible to provide comparable data for the first time on the size and educational attainment of immigrant populations in OECD countries.

According to this new database on immigrant populations and expatriates in OECD countries, the following conclusions can be drawn:

- The percentage of the foreign-born in European OECD countries is generally higher than the percentage of foreigners. This is not entirely surprising, because of the economic incentives associated with the acquisition of the host-country nationality, among them access to civil service employment without restriction. Migration to a number of European countries (e.g. Sweden, Germany, Austria, the Netherlands, Greece or France) is significantly higher than is generally reported and approaches levels that are as high as, if not higher in relative terms than, observed, for example, in the United States. In short, relatively high income levels in Europe have attracted migrants from developing countries as much as they have to elsewhere in the OECD area.

- The stock figures shown here reflect migration waves over a long period. Although recent migration to OECD countries tends to come largely from non-OECD countries, migration between OECD countries continues to have a significant impact. This migration is quite selective towards highly skilled migrants, underlining the effects of the current competition between member countries to attract 'the best and the brightest' from other countries, both inside and outside of the OECD area.

- In most OECD countries, the number of immigrant populations with tertiary education exceeds the number of highly qualified expatriates in other OECD countries. On this measure, most OECD countries would appear to benefit from the international mobility of the highly skilled. This conclusion, however, must be considered as tentative, because the database described here does not cover expatriates in non-member countries.

- Among non-member countries the impact of the international mobility of the highly skilled is diverse. The largest developing countries (China and India) do not seem to be significantly affected and indeed may benefit from indirect effects associated with this mobility (return migration, technology transfers and remittances). At the other end of the spectrum, some of the smallest countries, especially in the Caribbean and in Africa, have significant percentages of their highly educated persons living abroad. Further analysis is needed to better understand the determinants, the dynamics and the impact of the international mobility of the highly skilled on these countries.

The approach described illustrates the merits of relying on the data of other countries to produce data for a particular country that would not otherwise be available. However, it does rest on the assumption that the data so compiled can be reasonably aggregated, that is, that the data definitions of the resident foreign-born populations are comparable. In the case of population censuses, this is very nearly the case, because any differences with respect to the definition of the resident population will generally bear on temporary residents, whose numbers can be expected to be small compared to the total stock of foreign-born persons. This would not be the case for flow data.

An extension of the database described here is planned for release, which will include more detailed demographic information (age and sex distributions) as well as extensive information on the occupations, the sectors of activity and the fields of study of both immigrant populations and expatriates.

Acknowledgements

This chapter is adapted from a previous working paper published in OECD Economic Studies, No. 40, 2005/1 as 'Counting immigrants and expatriates in OECD countries: a new perspective'. The authors would like to acknowledge the contribution of national participants in the data collection effort and of John Martin, Martine Durand and Jean-Pierre Garson, who have provided comments and advice on a preliminary version of this paper.

References

Adams RH. 2003. *International migration, remittances, and the brain drain: a study of 24 labor-exporting countries.* World Bank Policy Research Working Paper 3069. World Bank, Washington, DC.

Barro RJ and Lee JW. 1993. International comparisons of educational attainment. *Journal of Monetary Economics* 32:363–94. Available at: http://www.nuff.ox.ac.uk/Economics/ Growth/barlee.htm

Barro RJ and Lee JW. 2000. *International data on educational attainment: updates and impli- cations.* NBER Working Paper 7911. National Bureau of Economic Research, Cambridge, MA. Available at: http://www2.cid.harvard.edu/ciddata/

Carrington W and Detragiache E. 1998. *How big is the brain drain?* IMF Working Paper WP/98/102. International Monetary Fund, Washington, DC.

Cohen D and Soto M. 2001. *Growth and human capital: good data, good results.* OECD Development Centre Working Paper 179. Organisation for Economic Co-operation and Development. Available at: http://www.oecd.org/dataoecd/33/13/2669521.xls

Docquier F and Marfouk A. 2005. International migration by educational attainment, 1990–2000. In *International migration, remittances and the brain drain.* Özden Ç and Schiff M., eds. New York: World Bank and Palgrave Macmillan.

Dumont JC and Lemaître G. 2005. *Counting immigrants and expatriates in OECD countries: a new perspective.* OECD Social, Employment and Migration Working Paper 25. Organisation for Economic Co-operation and Development, Paris. Available at: http://www.oecd.org/dataoecd/27/5/33868740.pdf

Dumont JC and Meyer JB. 2003. The international mobility of health professionals: an evaluation and analysis based on the case of South Africa. In *Trends in international migration.* Paris: Organisation for Economic Co-operation and Development.

Ferrand A. 2001. *Rapport d'information au Sénat chargé d'étudier l'ensemble des questions liées à l'expatriation des compétences, des capitaux et des entreprises.* Rapport 386.

Hugo G, Rudd D and Harris K. 2003. *Emigration of Australians: recent developments and policy issues.* National Centre for Social Application GIS, University of Adelaide.

OECD. 1999. *Classifying educational programmes – Manual for ISCED – 97 Implementation in OECD Countries.* Paris: Organisation for Economic Co-operation and Development.

OECD. 2002.*The international mobility of the highly skilled.* Paris: Organisation for Economic Co-operation and Development.

OECD. 2003. *Trends in international migration.* Paris: Organisation for Economic Co-operation and Development.

OECD. 2004. *Migration for employment: Bilateral agreements at a crossroad.* Paris: Organisation for Economic Co-operation and Development.

Saint-Paul G. 2004. *The brain drain: some evidence from European expatriates in the United States.* IZA DP 310. Institute for the Study of Labor, Bonn, Germany.

3

Comparability of statistics on international migration flows in the European Union

Dorota Kupiszewska[†] and Beata Nowok[‡]

[†] *Central European Forum for Migration and Population Research, Warsaw*
[‡] *Population Research Centre, University of Groningen*

3.1 Introduction

With a growing interest in international migration, there is an increasing need for statistical data on international migration flows. Such data are prepared by national statistical institutes (NSIs) or other relevant institutions in most European Union (EU) countries. However, currently available statistics are not internationally comparable. The differences stem mostly from disparities in the legislation, the efficiency of the registration systems and the statistical methodologies used when processing the data. The lack of data comparability, together with the lack of good metadata, is a serious problem for data users, be it journalists, policy-makers, or academic researchers.

In recent decades there have been many efforts to harmonise migration statistics (see review by Herm 2006), including the preparation and publication of the *Recommendations on Statistics of International Migration* by the United Nations (United Nations 1998). However, the results of these efforts are far from satisfactory. In

International Migration in Europe: Data, Models and Estimates   Edited by J. Raymer and F. Willekens
© 2008 John Wiley & Sons, Ltd

the EU, a new *Regulation on Community Statistics on Migration and International Protection* has recently been prepared (European Commission 2005). It is expected to be voted on by the European Parliament and to come into force in the first half of 2007. The final version of the *Regulation* will be a compromise between an ambitious plan to collect a wide range of well harmonised data and realistic practical capabilities of the Member States who will be legally bound to supply the data (Thorogood 2006).

This chapter reviews the availability and comparability of international migration statistics in the EU based on recent information (up to 2005). In Section 3.2 the incomparability problems are demonstrated through a direct comparison of data on flows between pairs of countries as reported by countries of origin with those reported by the destination countries. Double entry migration matrices are discussed as well as graphs illustrating how the differences in the flows reported by selected sending and receiving countries change over time. Section 3.3 attempts to explain the observations presented earlier by comparing the sources and definitions used in various EU countries. Specific problems relating to the comparability of data published in international sources are discussed in Section 3.4. Finally, Section 3.5 presents some recommendations for data providers and data users, encouraging the latter to take every effort (no doubt it will be needed) to better understand the data.

3.2   Empirical observations

In this section, we look at available data as seen by the end user. We investigate immigration and emigration data by organising them in a way that allows us to compare data reported by sending and receiving countries and to evaluate international comparability of data provided by individual countries. In particular, we analyse two types of information: (i) the double entry matrix containing the flows between 25 EU countries (without Bulgaria and Romania) in one year and (ii) graphs presenting a time series of flows between selected pairs of countries. We do not attempt to explain the observations, but rather pretend to be in the position of a researcher who is looking for data and finds some numbers but not much description of the data. The explanations come in Section 3.3, together with the information on data sources and definitions.

3.2.1   Double entry matrices

In order to illustrate the problems with data on international migration flows, we have constructed a double entry matrix for the year 2003, covering the 25 EU countries (Table 3.1). Most of the data in the matrix have been provided by the NSIs in response to the 2004 round of data collection conducted jointly by five international organisations and coordinated by Eurostat, the statistical office of the European Communities.

The idea of presenting international migration flow data in the form of double entry matrices is more than 30 years old. Two main promoters of such

Table 3.1 Migration flows between 25 EU countries[a] according to the receiving (I) and sending countries (E) in 2003.[b,c]

Sending country		BE	CZ	DK	DE	EE	EL	ES	FR	IE	IT	CY	LV	LT	LU	HU	MT	NL	AT	PL	PT	SI	SK	FI	SE	UK
BE	I	—	80	587	4 291	...	...	3 037	...	...	...	44	11	10	804	...	...	5 348	267	64	105	19	2	184	399	...
	E	—	...	...	...	...	...	...	...	...	...	...	...	...	...	...	...	...	...	...	...	...	...	...	...	...
CZ	I	...	—	232	9 258	...	...	388	...	...	...	79	2	11	3	...	...	333	1 201	46	9	7	650	47	123	...
	E	78	—	47	950	2	66	70	283	31	197	32	12	13	6	35	5	149	315	1 040	27	9	18 262	57	43	455
DK	I	...	65	—	2 693	...	...	764	...	...	...	14	22	81	14	...	...	474	203	17	58	5	0	371	4 603	...
	E	511	180	—	2 540	133	229	1 720	1 333	264	782	24	348	596	131	120	14	609	231	548	174	29	79	403	4 582	4 317
DE	I	...	1 228	3 221	—	...	...	13 746	...	...	...	195	79	257	436	...	...	7 921	12 239	2 261	645	242	106	807	2 872	...
	E	4 623	8 909	2 712	—	597	18 106	16 236	19 060	2 415	33 802	306	1 474	2 011	1 510	15 429	120	8 616	15 976	82 910	8 880	2 346	9 546	2 380	3 786	15 550
EE	I	...	4	169	947	—	...	60	...	...	...	0	69	53	2	...	...	53	37	0	2	0	0	1 292	311	...
	E	...	...	...	...	—	...	...	...	...	...	...	...	...	...	...	...	...	...	...	...	...	...	...	...	...
EL	I	...	57	278	12 959	...	—	273	...	...	...	4 971	4	8	5	...	...	882	465	58	15	6	2	63	585	...
	E	...	...	...	...	...	—	...	...	...	...	...	...	...	...	...	...	...	...	...	...	...	...	...	...	...
ES	I	...	103	1 665	14 647	...	...	—	...	...	...	38	2	85	28	...	...	2 794	615	85	743	6	2	608	1 234	...
	E	647	34	130	2 109	4	38	—	2 474	487	801	1	1	31	89	41	6	600	93	144	627	12	16	102	164	2 335
FR	I	...	462	1 488	18 133	...	...	8 847	—	...	...	99	12	40	987	...	...	2 919	741	191	458	23	9	312	931	...
	E	...	...	...	...	...	...	...	—	...	...	...	...	...	...	...	...	...	...	...	...	...	...	...	...	...

Table 3.1 (continued).

Sending country		Receiving country																							
		BE	CZ	DK	DE	EE	EL	ES	FR	IE	IT	CY	LV	LT	LU	HU	MT	NL	AT	PL	PT	SI	SK	FI	SE	UK
IE	I	..	45	306	2046	..	..	1 649	..	—	..	64	6	39	3	..	..	615	138	12	37	0	1	146	230	0
	E	..	..	..	..	..	..	..	..	—	..	..	..	..	..	..	..	..	..	..	..	..	..	..	..	..
IT	I	..	274	895	23 702	..	..	5 796	..	—	—	11	17	47	68	..	..	1 661	1 460	229	312	49	14	209	473	..
	E	..	..	..	..	..	..	..	..	—	—	..	..	..	..	..	15	..	..	..	..	..	..	..	..	261
CY	I	..	35	33	260	..	..	19	..	—	..	—	0	2	..	..	..	32	21	3	4	1	0	22	44	..
	E	22	16	14	80	0	604	14	59	0	16	—	0	0	0	19	15	19	19	15	0	0	0	0	21	261
LV	I	0	17	381	1 966	..	..	207	..	..	..	12	—	177	1	1	..	74	68	4	11	0	0	63	182	..
	E	..	3	40	170	44	2	2	28	11	32	1	—	80	..	0	0	19	10	15	37	0	3	33	45	40
LT	I	..	27	701	3 457	..	..	1 401	..	..	..	15	146	—	1	..	..	173	120	60	11	0	1	44	232	..
	E	51	53	158	1 204	34	15	465	143	276	184	2	192	—	1	7	0	95	41	123	55	0	3	112	191	980
LU	I	..	3	196	1 728	..	..	89	..	..	..	0	0	1	—	—	..	166	57	3	10	2	0	34	78	..
	E	1 119	7	119	747	3	22	73	1 254	44	208	..	2	1	—	17	5	97	22	11	521	3	7	33	74	171
HU	I	..	58	170	14 965	..	..	271	..	..	..	125	2	5	8	—	—	498	2 595	20	17	6	25	76	244	..
	E	..	..	..	..	..	..	..	..	..	..	..	..	..	..	—	—	..	..	..	..	..	..	..	..	..
MT	I	..	3	18	98	..	..	5	..	..	..	12	0	1	..	..	—	44	4	0	0	1	1	1	33	..
	E	..	..	..	..	..	..	..	..	..	..	..	..	..	..	..	—	..	..	..	..	..	..	..	..	..
NL	I	..	245	820	13 015	..	..	3 567	..	..	..	65	8	40	25	..	..	—	655	72	264	8	11	239	707	..
	E	9 284	172	430	9 822	18	482	3 365	3 373	459	1 274	37	21	41	150	242	33	—	470	622	666	40	65	292	648	7 022

	I/E																								
AT	I	…	339	262	13 456	…	…	554	…	68	17	4	14	7	…	…	510	—	140	33	55	48	92	333	…
	E	177	837	100	4 422	17	340	300	426	852	13	25	70	44	1 752	6	295	—	1 904	180	295	1 329	196	396	668
PL	I	…	1 653	995	104 924	…	…	3 498	…	20	123	15	113	10	…	…	2 106	2 974	—	36	2	36	89	1 134	…
	E	138	46	68	15 013	0	56	139	251	311	0	4	7	14	6	0	275	355	—	5	0	10	11	117	282
PT	I	31	170	…	7 699	…	…	5 505	…	0	0	2	13	512	…	…	1 619	330	13	1 850	3	2	56	143	…
	E	0	0	…	955	0	0	0	849	0	0	0	0	770	0	0	0	0	0	—	0	0	0	0	2 187
SI	I	16	31	…	2 053	…	…	71	…	2	0	2	1	…	…	…	60	372	1	1	—	1	6	22	…
	E	40	12	4	463	0	8	12	47	127	4	0	0	17	5	0	14	188	7	3	—	4	1	24	32
SK	I	…	24 385	84	10 684	…	…	324	…	0	63	5	5	5	…	…	191	2 330	19	1	5	—	12	56	…
	E	7	448	0	199	0	2	8	17	38	0	0	0	5	18	0	8	134	10	0	0	—	1	6	52
FI	I	…	55	421	2 204	…	…	802	…	110	11	38	32	2	…	…	362	251	6	15	0	0	—	3 395	…
	E	245	34	397	761	311	56	792	284	210	19	21	20	57	96	2	217	76	23	26	2	4	—	3 428	1 070
SE	I	…	83	2 705	3 397	…	…	1 537	…	205	46	40	58	11	…	…	638	474	91	31	18	7	3 438	—	—
	E	411	77	2 585	1 580	99	510	1 356	946	441	54	58	38	66	127	19	499	238	216	92	10	23	3 386	—	3 676
UK	I	…	488	3 707	13 197	…	…	34 177	…	…	2 870	35	122	37	…	…	5 872	261	947	16	33	914	3 022	—	—
	E	…	…	…	…	…	…	…	…	…	…	…	…	…	…	…	1 180	…	…	…	…	…	…	…	…

[a] List of country codes: BE – Belgium, CZ – Czech Republic, DK – Denmark, DE – Germany, EE – Estonia, EL – Greece, ES – Spain, FR – France, IE – Ireland, IT – Italy, CY – Cyprus, LV – Latvia, LT – Lithuania, LU – Luxembourg, HU – Hungary, MT – Malta, NL – Netherlands, AT – Austria, PL – Poland, PT – Portugal, SI – Slovenia, SK – Slovakia, FI – Finland, SE – Sweden, UK – United Kingdom.

[b] … – Data not available.

[c] — – Not applicable.

Source: Prepared by the authors based on data from Eurostat and NSIs.

matrices have been John Kelly of the Statistics Department of the United Nations Economic Commission for Europe (ECE) and Michel Poulain – see review papers by Kelly (1987) and Poulain (1999). To the best of our knowledge the first such matrix was constructed for ECE countries for the year 1972 and presented at the Meeting on Migration Statistics organised by the Conference of European Statisticians in 1975. It was later reproduced in a special article on international migration statistics published in the 1977 issue of the United Nations *Demographic Yearbook* (United Nations, 1978:3–16). Matrices presented regularly at several subsequent meetings reflected the first attempts to harmonise the definitions.

The idea of double entry migration matrices is to present the data on immigration, reported by the receiving countries, and those on emigration, reported by the sending countries, in one table. Accordingly, in Table 3.1, the cells representing migration from country A to country B contain two entries: the upper one includes immigration (I) from country A reported by country B; and the lower one includes emigration (E) to country B reported by country A. In order to better understand the data in the double entry matrix, we have calculated I/E ratios and $I-E$ differences (not shown), where I and E are the flows reported by the receiving and by the sending country, respectively. The 2003 I/E ratios are presented in Table 3.2 (the rows and columns for which the calculation was not possible due to missing data have been omitted). The cells where a nonzero flow was reported by the receiving country and null flow by the sending country ($E=0$ and $I \neq 0$) are marked with the infinity sign (∞).

A number of problems can be noticed when analysing the double entry matrix, the I/E ratios and $I-E$ differences. There is either no data or only one figure in a number of cells of the double entry matrix (see Table 3.1). The figures reported by the receiving country are often several times (or even more) higher than those reported by the sending country (see Table 3.2). In the extreme case of the flow from Latvia to Spain in 2003, the Spanish data show 100 times higher migration than Latvian data. Particularly large I/E ratios have also been observed for flows from Slovakia to the Czech Republic and Germany ($I/E = 54$ in both cases), from Slovakia to Spain ($I/E = 41$), and also from Poland to the Czech Republic ($I/E = 36$) and Spain ($I/E = 25$). The absolute differences between the flows reported by the receiving and sending countries (the values of $I-E$), which might be more important for policy considerations or population projections, especially in the case of large flows, are also significant and can sometimes be huge. This concerns, in particular, flows to and from Germany. In 2003, the highest difference observed was migration from Poland to Germany, with a difference of almost 90 thousand ($I = 104\,924$, $E = 15\,013$). Other striking examples are the flows from Slovakia to the Czech Republic ($I = 24\,385$, $E = 448$), from Spain to Germany ($I = 14\,647$, $E = 2109$) and from Slovakia to Germany ($I = 10\,684$, $E = 199$).

Despite the general belief that immigration data are better than those concerning emigration, the numbers reported by receiving countries are often smaller than those reported by sending countries. Flows for which $I < E$ constituted 40 % of

Table 3.2 Ratios of flows reported by the receiving and sending countries[a] (I/E) in 2003.[b-d]

Sending country	Receiving country															
	CZ	DK	DE	ES	CY	LV	LT	LU	NL	AT	PL	PT	SI	SK	FI	SE
CZ	—	4.94	9.75	5.54	2.47	0.17	0.85	0.50	2.23	3.81	0.04	0.33	0.78	0.04	0.82	2.86
DK	0.36	—	1.06	0.44	0.58	0.06	0.14	0.11	0.78	0.88	0.03	0.33	0.17	0.00	0.92	1.00
DE	0.14	1.19	—	0.85	0.64	0.05	0.13	0.29	0.92	0.77	0.03	0.07	0.10	0.01	0.34	0.76
ES	3.03	12.81	6.94	—	38.00	2.00	2.74	0.31	4.66	6.61	0.59	1.19	0.50	0.13	5.96	7.52
CY	2.19	2.36	3.25	1.36	—	0/0	∞	...	1.68	1.11	0.20	∞	∞	0/0	∞	2.10
LV	5.67	9.53	11.56	103.50	12.00	—	2.21	...	3.89	6.80	0.27	0.30	0/0	0.00	1.91	4.04
LT	0.51	4.44	2.87	3.01	7.50	0.76	—	1.00	1.82	2.93	0.49	0.20	0/0	0.33	0.39	1.21
LU	0.43	1.65	2.31	1.22	...	0.00	1.00	—	1.71	2.59	0.27	0.02	0.67	0.00	1.03	1.05
NL	1.42	1.91	1.33	1.06	1.76	0.38	0.98	0.17	—	1.39	0.12	0.40	0.20	0.17	0.82	1.09
AT	0.41	2.62	3.04	1.85	1.31	0.16	0.20	0.16	1.73	—	0.07	0.18	0.19	0.04	0.47	0.84
PL	35.93	14.63	6.99	25.17	∞	3.75	16.14	0.71	7.66	8.38	—	7.20	∞	3.60	8.09	9.69
PT	∞	∞	8.06	∞	0/0	∞	∞	0.66	∞	∞	∞	—	∞	∞	∞	∞
SI	1.33	7.75	4.43	5.92	0.00	∞	∞	...	4.29	1.98	0.14	0.33	—	0.25	6.00	0.92
SK	54.43	∞	53.69	40.50	∞	∞	∞	1.00	23.88	17.39	1.90	∞	∞	—	12.00	9.33
FI	1.62	1.06	2.90	1.01	0.58	1.81	1.60	0.04	1.67	3.30	0.26	0.58	0.00	0.00	—	0.99
SE	1.08	1.05	2.15	1.13	0.85	0.69	1.53	0.17	1.28	1.99	0.42	0.34	1.80	0.30	1.02	—

[a] See footnote a in Table 3.1 for the list of country codes.
[b] ∞ Cells where a nonzero flow was reported by the receiving country and a null flow by the sending country.
[c] ... Data not available.
[d] — Not applicable.

Source: Authors' computations based on data in Table 3.1.

all nonzero flows in 2003. The situation is similar to the one observed in the 1995 matrix for 33 European countries, for which Poulain (2001:18) reported that: 'The immigration figure is higher than the emigration one in one case out of two, while in the other case, the emigration figure is higher.' In the 2003 double entry matrix, some immigration numbers are considerably lower, in both relative and absolute sense, than the corresponding emigration data. For example, in the case of the flow from Germany to Slovakia in 2003, Slovakia (the receiving country) reported a value 90 times smaller than Germany (106 according to Slovakia and 9456 according to Germany). The flow from Germany to Poland, according to Poland, was only 2261, almost 81 thousand less than the value reported by Germany.

To assess which countries are performing worst in measuring migration, we have checked, for each sending country, which countries were least effective (in terms of I/E) in recording the immigration from this country. Slovakia and Poland, followed by Portugal, Luxembourg, Latvia and Slovenia, were the countries most often recording much less incoming migrants than reported by the sending country. To measure which countries have recorded the highest proportion of immigration flow, we have checked, for each sending country, which receiving countries had the highest I/E ratio. The 'winner' was Germany, followed by Denmark and Spain.

To assess the emigration statistics, we have checked, for each receiving country, which sending countries had the highest I/E ratios and which the lowest ones. The countries recording the lowest proportion of emigration are Slovakia, Portugal and Poland, followed by Spain, Italy and Latvia. The highest proportion of emigration flows is recorded in Germany and Denmark, followed by Austria. Portugal reported zero emigration to individual countries more often than any other country. Another feature specific to Portugal is that it reported nonzero immigration from Portugal to Portugal (1850 in 2003).

3.2.2   Evolution of migration flows over time

Other interesting observations can be made by examining the evolution of the flows between specific pairs of countries over time, as reported by both sending and receiving countries (proposed by Poulain 2001). Such graphs illustrate the data incomparability problems faced by the end users of migration statistics who attempt to understand or forecast international migration trends. Figures 3.1–3.9 present 16 out of 600 graphs that could be theoretically prepared for a system of 25 countries (i.e. 25×24). They have been prepared using data from the Eurostat database and chosen from amongst those with a reasonable number of data points and a significant level of flows to illustrate various observed situations. The following points are brought out by the graphs shown:

- Sometimes, data reported by the receiving country and the sending country have opposite trends, e.g. migration from Lithuania to Poland (Figure 3.1).

- A group of countries have exceptionally good agreement between their data. This applies to flows between Nordic countries, e.g. migration from Denmark to Sweden and from Sweden to Finland (Figure 3.2).

- For most pairs of countries, flows reported by one country are higher in both directions, e.g. migration between Germany and Finland (Figure 3.3).

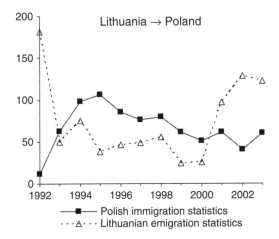

Figure 3.1   Migration flows from Lithuania to Poland according to data from the receiving and sending country.

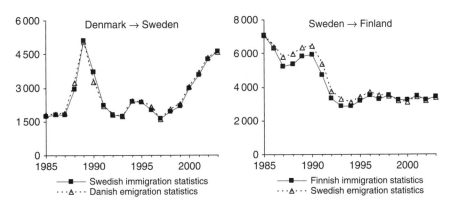

Figure 3.2   Migration flows from Denmark to Sweden and from Sweden to Finland according to data from the receiving and sending country.

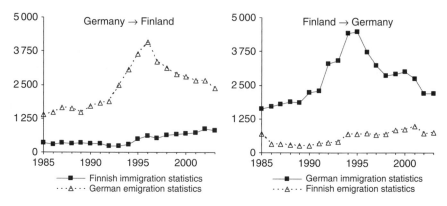

Figure 3.3 Migration flows between Germany and Finland according to data from the receiving and sending country.

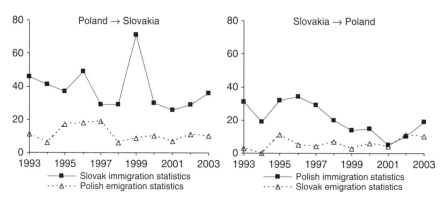

Figure 3.4 Migration flows between Slovakia and Poland according to data from the receiving and sending country.

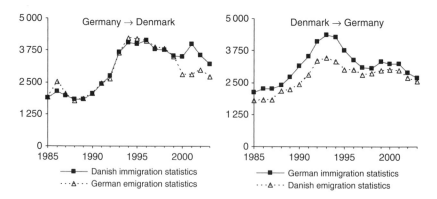

Figure 3.5 Migration flows between Germany and Denmark according to data from the receiving and sending country.

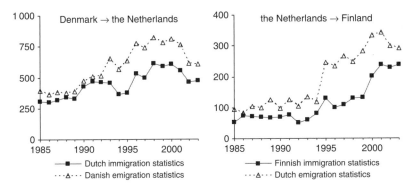

Figure 3.6 Migration flows from Denmark to the Netherlands and from the Netherlands to Finland according to data from the receiving and sending country.

- For some pairs, data reported by the receiving country are higher than data from the sending country in both directions, e.g. migration between Poland and Slovakia (Figure 3.4).

- Data reported by one country may be comparable for flows to and from some countries but not for others, e.g. German data are comparable to Danish (Figure 3.5) but higher than Finnish (Figure 3.3).

- Flows reported by one country may be consistently lower than those reported by another country or consistently higher, e.g. Dutch figures are lower than Danish but higher than Finnish (Figure 3.6).

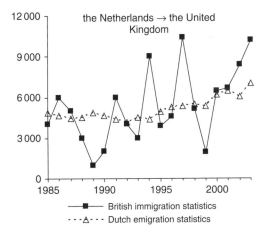

Figure 3.7 Migration flows from the Netherlands to UK according to data from the receiving and sending country.

Note that, due to large errors, the UK's Office for National Statistics would not normally publish flow statistics at this level of disaggregation.

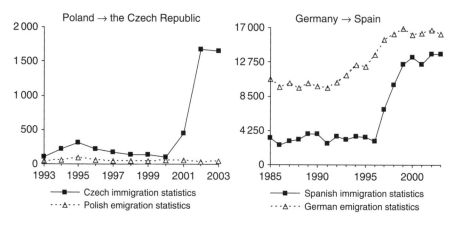

Figure 3.8 Migration flows from Poland to the Czech Republic and from Germany to Spain according to data from the receiving and sending country.

- Flows reported by one country oscillate much more than corresponding flows reported by other countries. This is observed for migration data from the United Kingdom (Figure 3.7).

- A sudden jump or a steep increase not reported by the partner country occurs in the levels reported by some countries, e.g. in Czech data on immigration from Poland and in Spanish data on immigration from Germany (Figure 3.8).

- Sometimes, a very low level of flows in both directions is reported by one country but not the other, e.g. by Slovakia for migration to the Netherlands and by Poland for migration from Germany (Figure 3.9).

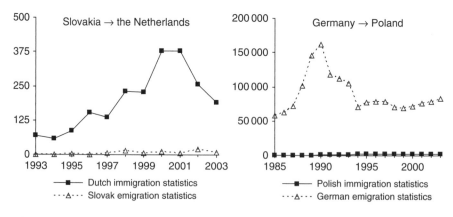

Figure 3.9 Migration flows from Slovakia to the Netherlands and from Germany to Poland according to data from the receiving and sending country.

3.3 Data sources and definitions

To explain the origin of the problems presented in Section 3.2, one would have to understand how the national statistical institutes (or other responsible bodies) produced their statistics. The relevant issues are, among others, the sources of data and the definitions of the terms *migration* and *migrant*, in particular, the time criterion that might appear in the definition. When talking about the definitions, we are referring to the rules applied (explicitly or implicitly) in the measurement of international migration.

Statistics on international migration flows are affected by the procedures (including definitions) adopted by particular countries. There are three basic stages to consider:

Stage 1. Collection of raw data in the primary data source, e.g. in a population register, in statistical forms or in survey forms.

Stage 2. Production of statistics.

Stage 3. Dissemination of statistics.

Differences between the countries occur at all the stages.

Stage 1 is strongly dependent on the legislation and on the attitudes of migrants towards the legal rules. It determines the availability of data (through the recorded variables) and their coverage (e.g. the data collection might cover nationals or foreigners[1] only). Stage 1 is also important for the definitions, as it preconditions who might be potentially included in the migrant count. For example, generally only legal migrants are covered by the official statistics, the only exception being Spain, which includes irregular migrants in its population register (*Padrón*) (Breem and Thierry 2006; see also Chapter 5). Theoretically, some illegal migrants may also be included in the passenger surveys carried out in Cyprus and the United Kingdom (Nowok *et al.* 2006). Stage 1 may be crucial for the reliability and validity of the data; for example, the reliability of the data in population registers depends to a large extent on the willingness of people to register and deregister. Generally, the under-registration concerning emigration is higher than in the case of immigration.

The methodologies applied in Stage 2 determine how the raw data are used to produce the statistics. Appropriate selection rules might facilitate the compliance of the statistics with internationally agreed definitions. For example, if the UN definitions[2] were adopted and the register includes information on all migration events, then only those referring to stays lasting at least one year should be counted

---

[1] Foreigners are the persons who are not citizens of the considered country. Nationals are persons who are citizens of the considered country.

[2] According to the UN recommendations a long-term migrant is defined as a person 'who moves to a country other than that of his or her usual residence for a period of at least a year (12 months), so that the country of destination effectively becomes his or her new country of usual residence' (United Nations 1989: Box 1).

in the long-term migration statistics. On the other hand, statistics are not always produced even though the underlying raw data are available.

In Stage 3, statistics are disseminated through various channels. There are differences in terms of availability and correctness of the statistics published in various sources. A particularly important aspect at this stage is the appropriate documentation. Unfortunately, the metadata are very often inadequate, for example, information on data coverage and definitions are not provided.

The problems with the availability and comparability of statistics arising at Stage 1 are certainly more difficult to overcome than those related with Stages 2 and 3. Generally, major changes in statistics require changes in the legislation concerning the primary data sources. This is probably one of the main reasons why efforts to harmonise international migration statistics have not been very successful.

There are two sources of data on international migration: primary and secondary. Primary data are raw microdata that can be used for compiling international migration statistics. Primary data sources include, for example, population registers tracking changes in residence (centralised or run by local authorities) and registers of foreigners run by the ministries of interior. Usually, they are not accessible to the end user. Secondary data are the processed data (i.e. macrodata) produced from the raw data. Secondary data sources include the databases, publications and unpublished documents prepared by the institution responsible for producing the statistics (usually the national statistical office), as well as other sources in which the statistics are further disseminated (e.g. publications prepared by international organisations).

3.3.1   Primary data sources

Table 3.3 presents the sources used in the 25 EU countries to produce statistics on immigration and emigration, separately for nationals and foreigners. In almost all the countries the statistics on immigration and emigration, if both are available, are produced using the same source. The exceptions are statistics on nationals in Malta (see below) and statistics on foreigners in Portugal. Also statistics on nationals and foreigners, if both available, come usually from the same source, with the exception of the Czech Republic, Hungary and Slovenia, where data on nationals come from population registers and data on foreigners come from registers of foreigners, run separately (see Section 4.2 in Chapter 4 for more details).

In the majority of countries (i.e. 16 countries), statistics on flows of *nationals* are compiled using data from population registers, either central or local. There are the following exceptions. Cyprus, Ireland, Portugal and the UK base their statistics on data from sample surveys. These are passenger surveys conducted at the borders in the case of Cyprus and the UK, and household surveys in Portugal and Ireland. In the UK, data from its International Passenger Survey (IPS) are supplemented by information from the Irish Statistical Office concerning flows from and to Ireland. Poland and Slovakia implemented special statistical forms filled in at the time of registration or deregistration in local authorities (see Section 4.2 in Chapter 4). In Malta, immigrants fill in special forms at customs when crossing the border. Maltese

Table 3.3 Primary sources of data on international migration flows in 25 EU countries.[a,b]

		Population register (central or local)		Register of foreigners or residence permit register	Sample survey		Statistical form or another type of source	
		NAT	FOR	FOR	NAT	FOR	NAT	FOR
Belgium	*I*	x	x					
	E	x	x					
Czech	*I*	x		x				
Republic	*E*	x		x				
Denmark	*I*	x	x					
	E	x	x					
Germany	*I*	x	x					
	E	x	x					
Estonia	*I*	[x]	[x]					
	E	[x]	[x]					
Greece	*I*			[x]				
	E							
Spain	*I*	x	x					
	E	x	x					
France	*I*			x		[x]		
	E							
Ireland	*I*				x	x		
	E				x	x		
Italy	*I*	x	x					
	E	x	x					
Cyprus	*I*				x	x		
	E				x	x		
Latvia	*I*	x	x					
	E	x	x					
Lithuania	*I*	x	x					
	E	x	x					

Table 3.3 (*continued*).

		Population register (central or local)		Register of foreigners or residence permit register	Sample survey		Statistical form or another type of source	
		NAT	FOR	FOR	NAT	FOR	NAT	FOR
Luxembourg	I	x	x					
	E	x	x					
Hungary	I	x		x				
	E	x		x				
Malta	I						x	x
	E						x	
Netherlands	I	x	x					
	E	x	x					
Austria	I	x	x					
	E	x	x					
Poland*c*	I						x	x
	E						x	x
Portugal	I			x	[x]			
	E				x	x		
Slovenia	I	x		x				
	E	x		x				
Slovakia	I			x			x	x
	E			x			x	x
Finland	I	x	x					
	E	x	x					
Sweden	I	x	x					
	E	x	x					
UK	I				x	x	x	x
	E				x	x	x	x

a I – immigration; *E* – emigration; NAT – nationals; FOR – foreigners.
b [ ] Information referring to data currently not disseminated, but potentially available in future.
c Since 2006, Poland has prepared statistics on flows based on the data taken from the population register.
Source: Prepared and revised by the authors based on Nowok and Kupiszewska (2005).

data on emigration of nationals used to be obtained from foreign embassies, but currently the only data are those concerning emigration to the UK, received from the British High Commission. Greece and France have data neither on immigration nor emigration of nationals. Portuguese statistics on immigration of nationals are used for internal purposes only and are not published.

In the case of statistics on flows of *foreigners*, population registers are also the most frequent source (i.e. they are used in 13 countries). The Czech Republic, Hungary, Slovenia, France and Portugal (immigration only for the latter two) use data from the registers of foreigners or the residence permit registers. Slovakia has used this source to provide data on flows by citizenship since 2003. Sample surveys are used, as for nationals, in Cyprus, Ireland and the UK, and for emigration statistics in Portugal. In the UK, figures based on the International Passenger Survey are adjusted using information from the Irish Statistical Office (as for nationals) as well information on asylum seekers provided by the Home Office. Polish and Slovak[3] flow data, as well as Maltese data on immigration, come from statistical forms, as used for nationals. France and Malta produce no statistics on emigration of foreigners, and Greece has no statistics on flows of foreigners at all.

Knowing the primary sources of data, we may begin to understand some of the peculiarities found in Tables 3.1 and 3.2 and in Figures 3.1–3.9. The consequences of using sample surveys are clearly seen in Figure 3.7 for flows to and from the UK: the immigration and emigration figures reported by the UK show strong fluctuations,[4] compared with the much smoother curves reported, for example, by the Netherlands. We may also explain the specific observations for Portuguese data. Portuguese emigration statistics are based on data from household surveys, with the sample not large enough to catch relatively small flows to some countries; hence zero values are reported for these flows. For immigration, the Portuguese data cover foreigners only and refer to the number of residence permits. The figures reported for flows from Portugal to Portugal represent foreign children born in Portugal who received residence permits.

Missing sources on international migration flows, identified in Table 3.3, explain only some empty cells in the matrix: the lack of Estonian and Greek data and French emigration data. Other empty cells are related to the lack of information on the previous/next country of residence in the primary data source or to the lack of data in the secondary data sources.

3.3.2 Definitions

Differences in definitions are crucial for understanding the differences in the data reported by receiving and sending countries. The definitions specifying who is

---

[3] In Slovakia statistical forms are used to prepare statistics on flows by country of previous/next residence and were used till 2002 for the figures on flows by citizenship.

[4] Owing to the small sample size, the UK's Office for National Statistics would not normally publish migration estimates from the International Passenger Survey at this level of disaggregation. Only aggregated figures are considered reliable.

included in international migration statistics are usually not stated explicitly or are expressed in a very vague way. Generally, the definitions might be identified by analysing (i) the rules governing the collection of data in the primary data sources, for example the administrative rules for reporting changes of place of residence in population registers, and (ii) the selection rules applied to the raw data when preparing the statistics. These rules differ not only between countries, but also between statistics on nationals and foreigners and between immigration and emigration. Therefore, even if the same rules were applied in two countries, and both countries had reliable statistics, the difference between criteria applied in the immigration and emigration statistics would result in different flows reported by both countries.

The differences in the definitions of migration are related to the different concepts of place of residence and to different duration of stay criteria. Generally, the *de jure* approach is applied everywhere, which means that the legal place of residence is important, not the actual one. The law regarding the registration of place of residence might differ between nationals and foreigners and also between various groups of foreigners (EU and non-EU citizens).

The largest differences regard the time criterion which specifies the minimum duration of stay in the destination country required for the change of residence to be counted as international migration. Table 3.4 summarises the time criteria applied in the 25 EU countries. This table was particularly difficult to compile and might be questioned, although its contents have been consulted with most of the countries concerned. The table indicates that there are very few countries that comply with the UN recommendations and use the one-year duration of stay criterion. The following generalisations can be distinguished by analysing the information in Table 3.4:

1. Duration of stay is not taken into account. An example is Germany, where everybody taking up a residence as an owner–occupier, tenant or subtenant is counted. Another example is Irish migration data, which are based on the survey question about the place of residence one year earlier and there is no question about the intended duration of stay or the time already spent in the country.

2. A minimum period of stay criterion applies that might be three months, six months or one year. Specific time limits are in use in the Netherlands: four out of six months for immigration and eight out of 12 months for emigration. Generally, in all the EU countries this time criterion refers to the intended duration of stay as opposed to the actual duration – the latter was taken into account in the immigration statistics in the Czech Republic from 2001 onwards (Hola 2006). The time limits presented in Table 3.4 have various meanings. For example, they might refer to the period of stay related with the obligation to register (or deregister) as specified by law governing population registers, or to the duration of validity of residence permits. They might also refer to the selection rules applied when statistics are produced.

3. The concept of 'permanent migration' or migration for permanent stay is in operation (see Section 4.3 in Chapter 4 for the details of this concept). Temporary changes of residence are not counted, only those declared as permanent ones being included. This option applies to the former socialist countries, Poland and Slovakia,[5] and statistics on flows of nationals (as well as foreigners till 2000) in the Czech Republic.

4. In several countries, permit expiry is used as a criterion in the statistics on emigration of foreigners, in which case the duration of stay in the destination country is not taken into account. Usually, expired permits are counted in addition to the number of persons who deregistered from the population register. This measure is used in order to prevent the under-registration of emigration. The main problem of using this option is that it does not give any information on the country of destination. In the Czech Republic, this problem is solved by assuming that the destination is the same as the country of citizenship. In Lithuania, the destination is assumed to be the same as the country of previous residence.

Apart from the differences in the duration of stay criterion, another time-related problem is that the date (year) to which a migration event is assigned might not be the one when the move took place. It concerns all four options listed above: e.g. there might be a delay between the arrival and the registration, between the arrival and the issuance of the permit that is counted in the statistics or between the departure and the date of permit expiry.

A specific situation exists in Denmark, Finland and Sweden due to the Inter-Nordic Migration Agreement, which covers these three countries plus Norway and Iceland. Migration movement between these countries is first registered in the country of destination and then the information is transferred to the country of origin, so that statistics concerning emigration to each of these countries follow the rules governing immigration statistics in the partner countries.

Having in mind the differences in the definitions one may attempt to explain further features in the double entry matrix and flow time series. First, the excellent agreement in the flows between the Nordic countries is no longer a mystery. Second, the figures reported by the receiving country (immigration data) are higher than those reported by the sending country (emigration data) when both countries have comparable definitions, as in the case of Poland and Slovakia (Figure 3.4). When the definitions are different, then the country with a wider definition reports higher numbers than the partner country for the flows in both directions. For example, the relation between the figures reported by Germany, the Netherlands and the Nordic countries corresponds to the differences in the definitions. The German one is the widest, and therefore their flow figures are usually the highest. The time criterion

---

[5] In the case of 2003 data for Slovakia, the permanent migration concept refers to data on flows of nationals and data on flows of foreigners by previous/next country of residence, while data on flows of foreigners by citizenship cover also temporary migration of foreigners with permits for more than three months.

Table 3.4   Duration of stay criteria in the international migration definitions in 25 EU countries.[a–e]

		None		Three months		Six months		Other below one year		One year		Permanent		Permit expiry
		NAT	FOR	NAT	FOR	NAT	FOR	NAT	FOR	NAT	FOR	NAT	FOR	FOR
Belgium	*I*			x	x									
	E			x	x									p
Czech Republic	*I*				x^{EEA}					x^{nEEA}		x		
	E											x	x	p
Denmark	*I*	x			x^{nEEA}		x^{EEA}							
	E^f					x	x							
Germany	*I*	x	x											
	E	x	x											
Estonia	*I*	[x]				x								
	E	[x]	[x]											
Greece	*I*										[p]			
	E													
Spain	*I*	x	x											
	E	x	x											
France	*I*										p^g			
	E													
Ireland	*I*	x	x											
	E	x	x											
Italy	*I*	x	x^{EEA}				x^{nEEA}							
	E									x	x			
Cyprus	*I*									x	x			
	E									x	x			
Latvia	*I*	x	x							x				
	E					x	x							p
Lithuania	*I*					x	x			x				
	E					x	x							p
Luxembourg	*I*	x	x											
	E	x	x											
Hungary	*I*			x	x^{EEA}					x^{nEEA}				
	E			x									x	p

Table 3.4   *(continued).*

		None		Three months		Six months		Other below one year		One year		Permanent		Permit expiry
		NAT	FOR	NAT	FOR	NAT	FOR	NAT	FOR	NAT	FOR	NAT	FOR	FOR
Malta	*I*											x	x	
	E											x		
Netherlands	*I*							x^h	x^h					
	E							x^i	x^i					
Austria	*I*			x	x					[x]	[x]			
	E			x	x					[x]	[x]			
Poland	*I*											x	x	
	E											x	x	
Portugal	*I*										p			
	E									x	x			
Slovenia	*I*			x^j	x							x		
	E			x^j									x	p
Slovakia	*I*				p							x	x	
	E											x	x	p
Finland	*I*	x									x			
	E^f									x	x			
Sweden	*I*									x	x			
	E^f									x	x			
UK	*I*									x	x			
	E									x	x			

^{a} I – immigration; *E* – emigration; NAT – nationals; FOR – foreigners.
^{b} [ ] Information referring to data that might be available in future.
^{c} p Migration data based on issued or expired residence permits.
^{d} EEA refers to EEA citizens.
^{e} nEEA refers to non-EEA citizens.
^{f} Registration of emigration to the Nordic countries follows the rules applied for registration of immigration in the receiving country.
^{g} Only foreigners with right of long-term settlement are included.
^{h} Four out of six months.
^{i} Eight out of 12 months.
^{j} Refers to nationals holding permanent residence status.
Source: Prepared by the authors based on Nowok and Kupiszewska (2005).

used in the Netherlands is longer than in Germany and Denmark, but lower than in Finland and Sweden, which is reflected in registered flow levels presented in Figure 3.6. Slovakia and Poland have the lowest levels of reported flows because they include migration for permanent stay only.

In several cases our expectations based on the definitions do not agree with the observed *I/E* ratios, for example for Luxembourg and Slovenia. In Luxembourg, it might be due to the fact that information on the country of previous residence was available only for a fraction (25 % in 2003) of flows. In Slovenia, data disaggregated by previous/next country of residence refer to nationals only. A large percentage of flows with unknown origins or destinations was also recorded in 2003 in Spain: 60 % for emigration and 30 % for immigration. This resulted in the undercounting in the Spanish data on emigration by country of next residence, despite the broad definition currently used in this country (see high I/E ratios for the outflows from Spain, i.e. the ES row in Table 3.2). In the case of immigration to Spain, the overcounting due to the broad definition was stronger than the undercounting due to the lack of data on country of previous residence, hence high I/E ratios for the flows to Spain (see the ES column in Table 3.2).

As concerns the sudden jumps observed in the Spanish and Czech data (Figure 3.8), they might be explained by the changes in the definitions. In the Czech Republic until 2000 the statistics covered permanent migration only, as derived from the statistical forms, similarly to Poland and Slovakia. Since 2001, data from the register of foreigners were used as well. Immigration statistics covered persons who stayed over one year (the exact criteria varied over time) and emigration statistics included data on permits that expired, along with self-reported departures for permanent stay abroad[6] (see Section 4.5 in Chapter 4 for further analysis). In Spain emigration statistics until 2001 covered assisted emigration of nationals only. Since 2002 all emigration events registered in the population register are counted and the total emigration figure increased from 134 in 2001 to 36 605 in 2002. For Spanish immigration statistics, Figure 3.8 suggests that some changes in the administrative or statistical procedures must have taken place between 1996 and 2000, because the increases in immigration flows reported by Spain are much higher than those reported by the countries of origin; however, we do not know the nature of these changes. More information on sources and definitions used for international migration statistics in the 25 European Union countries may be found in Nowok and Kupiszewska (2005) and in Poulain *et al.* (2006).

3.4   Secondary data sources and data availability

Statistical data available to the end users usually do not come directly from the primary data sources described earlier. Usually, the data are processed by the national statistical office or another body responsible for statistics production. The statistics are then disseminated in yearbooks and other publications either in a traditional printed form or, increasingly, through the Internet. Also, statistics are collected from individual countries and disseminated by international organisations. The final effect is that a number of secondary sources, differing in the scope of presented data, are available to those looking for international migration flow statistics concerning the EU. They include the following sources:

---

[6] The increase in the level of registered flows in CZ is also due to the amendments in the Act on Residence of Foreigners in July 2001.

- official websites of national statistical institutes;

- Eurostat online database and printed publications;

- Council of Europe (CoE) publications 'Recent demographic developments in Europe';

- SOPEMI[7] reports published by the Organisation for Economic Co-operation and Development (OECD);

- annual reports on asylum and migration prepared by the European Commission's Directorate-General for Justice, Freedom and Security (DG JLS).

The data presented in various sources might differ and, because of insufficient documentation, it is difficult to understand the origin of these differences. We have compared figures on total immigration flows and total emigration flows in the period 1999–2002 for the above listed sources and found that the figures are fully consistent across the sources only for seven out of the 25 EU countries: the Czech Republic, Denmark, Finland, Poland, the Slovak Republic, Slovenia and Sweden. For the other countries some discrepancies have been identified, sometimes significant. Table 3.5 shows numbers for selected countries, presenting various types of problems. In the table, the differences between the data coming from the same type of source (e.g. the CoE report) but published in different years are not shown: in the case of differences, the data from a more recent edition are quoted. Similarly, if data collected for the DG JLS report 2002 differed from those published in the Annual Report 2001, then the former are presented.

The examples presented in Table 3.5 indicate that the differences might have various origins. For Lithuania, the differences result from the post-census revisions, with the revised emigration figures much higher than non-revised ones (21 816 vs 2616 in 2000). For the United Kingdom, either figures from the International Passenger Survey only are presented, or more complete estimates referred to by the Office for National Statistics as Total International Migration (TIM). The TIM estimates are produced based on the data from the IPS, adjusted for asylum seekers, visitor and migrant switchers (persons who, respectively, extend or shorten their intended length of stay) and flows from Ireland (Office for National Statistics 2004). For the Netherlands some sources present emigration figures excluding administrative corrections. In Italy there are two sources of data (both based on the population register) differing in methodology. In the case of Hungary, there is a mixture of provisional and final data, as well as data concerning either total flow or flow of foreigners only. In Malta, immigration data concern either total flow or migrants of Maltese origin only. In all the examined sources, the footnotes that should give some explanation are often missing or incorrect.

The official country statistics published by the *national statistical institutes* on their *websites* are usually the most recent and the most reliable data that are

---

[7] SOPEMI – the French acronym used to refer to OECD's Continuous Reporting System on Migration (Système d'observation permanente des migrations).

Table 3.5   Migration flows in selected countries according to various sources.[a]

	1999	2000	2001	2002
Immigration to Hungary				
Eurostat Population Yearbook 2004	...	20 184	21 233	...
Eurostat Migration Yearbook 2002	*18 456*	...	...	...
CoE	*18 216*	20 184	22 079	17 558
DG JLS	20 151	20 184	20 308	[b]15 675
SOPEMI 2004	*20 200*	*20 200*	*20 300*	*15 700*
NSI – total (final data)	21 422	21 726	22 079	...
NSI – total (provisional data)	...	...	21 233	17 558
NSI – non-nationals (final data)	*20 151*	*20 184*	*20 308*	...
NSI – non-nationals (provisional data)	*18 456*	*14 484*	*19 462*	*15 675*
Emigration from Italy				
Eurostat Population Yearbook 2004	...	56 601	...	...
Eurostat Migration Yearbook 2002	76 500	...	...	...
CoE	...	...	...	41 756
DG JLS	56 707	76 483	66 821	49 383
SOPEMI 2004	...	...	...	...
Istat balance survey	76 483	66 821	67 125	49 383
Istat individual survey	64 873	56 601	56 077	41 756
Emigration from Lithuania				
Eurostat Population Yearbook 2004	...	21 816	7 253	...
Eurostat Migration Yearbook 2002	1 369	...	...	...
CoE	1 369	2 616	7 253	7 086
DG JLS	...	2 616	...	...
SOPEMI 2004	...	...	...	...
NSI (yearbook)	23 418	21 816	7 253	7 086
Immigration to Malta				
Eurostat Population Yearbook 2004	...	965	1 002	915
Eurostat Migration Yearbook 2002	339	...	...	...
CoE	339	450	472	535
DG JLS	708	965	1 002	...
SOPEMI 2004	...	...	...	...
NSI website – Maltese origin	339	450	472	382
NSI website – non-Maltese nationals	*369*	*515*	*530*	*533*
Emigration from the Netherlands				
Eurostat Population Yearbook 2004	...	[c]78 977	[c]82 566	[c]96 918
Eurostat Migration Yearbook 2002	59 023	...	...	...
CoE (tables)	59 023	...	63 318	66 728

DG JLS	78 779	78 977	82 566	96 918
SOPEMI 2004	[c]78 800	[c]79 000	[c]82 600	[c]96 900
NSI – emigration	59 023	61 201	63 318	66 728
NSI – emigration including net administrative corrections	78 779	78 977	82 566	96 918
Immigration to the United Kingdom				
Eurostat Population Yearbook 2004	. . .	483 400	479 600	512 800
Eurostat Migration Yearbook 2002	[d]354 077	. . .	. . .	. . .
CoE	[d]354 077	[d]364 370	. . .	. . .
DG JLS	[e]453 800	[e]483 400	[e]479 600	[e]512 800
SOPEMI 2004	[f]453 800	[f]483 400	[f]479 600	[f]512 800
NSI website[g]	453 800	483 400	479 600	512 800

[a] Italic – non-nationals only; . . . data not available.
[b] Provisional data.
[c] Including corrections.
[d] International Passenger Survey.
[e] Total International Migration (data from IPS, adjusted for asylum seekers and flows from Ireland).
[f] Data from IPS (flows from Ireland not included), adjusted for asylum seekers and visitor switchers.
[g] Total International Migration (data from IPS, adjusted for asylum seekers, visitor and migrant switchers and flows from Ireland).

publicly available. The scope of the data, in particular on international migration, and the form of their presentation vary significantly among the EU countries. Very few countries provide comprehensive databases containing detailed data that can be consulted, printed and downloaded free of charge (e.g. Denmark and the Netherlands) and some do not present data on international migration (e.g. Hungary). In the majority of cases only a short time series of total immigration and emigration are disseminated on the official websites.

Migration data from different European countries are disseminated by international organisations. In order to fulfil this task and to reduce the burden on national statistical offices, five international organisations decided to conduct jointly a comprehensive data collection. The collection, initiated in 1993 by Eurostat and the United Nations Economic Commission for Europe (UNECE), was later joined by the United Nations Statistical Division (UNSD), the Council of Europe (CoE) and the International Labour Office (ILO) (Herm 2006). Currently, the Joint Eurostat–UNECE–UNSD–CoE–ILO Questionnaire on International Migration Statistics is sent annually to 55 countries, which provide their responses on a voluntary basis. As concerns Europe, Eurostat processes and disseminates data received from 37 countries. In the data collections conducted in 2000–2004 (requesting flow data for 1999–2003 respectively) the questionnaire included the following tables concerning long-term (as defined in the UN recommendations) flow data: (i) immigration and emigration by sex and previous/next country of residence and (ii) immigration and emigration by sex, citizenship and five-year age group. There was also a table concerning aggregated figures on major categories of inflows and outflows, including visitors, diplomatic and military personnel, short-term migrants, border workers and so on, but most countries either did not complete this table or provided only a few figures. In 2005, the tables concerning long-term migration by previous/next country of residence have been extended to include the age dimension.

The statistics on international migration flows sent by the NSIs in response to the 2003–2005 Joint Migration Questionnaires are presented in Table 3.6. No flow data were delivered by Estonia and Greece. Hungary and Italy provided their flow statistics with a one-year delay (e.g. flow data for 2002 were sent in 2004). Provisional data have been sent by Hungary (for 2002 and 2003), Ireland and, for immigration, by Portugal. As regards Belgian and Hungarian data, and French data on immigration, although data on flows are available, the statistics disaggregated by country of previous or next residence are not produced either because there is no information in the database or because of data quality problems. However, the majority of countries sent their data on time, albeit with many not corresponding to the UN definition.

The scope of data disseminated by various international organisations varies. *Eurostat* is potentially the most comprehensive source of data on international migration in the EU Member States. For a number of years it has maintained the NewCronos database, with data available online and on CD-ROMs. Since 1 October 2004 the Eurostat data may be accessed free of charge using a new interface, at the address epp.eurostat.cec.eu.int. At the time of writing this chapter, the part of the database dedicated to international migration was under review and not all data could be consulted. When fully updated, the website will provide access to all or most of the data on international migration flows that are collected through the Joint Eurostat–UNECE–UNSD–CoE–ILO Questionnaire on International Migration. A relatively large amount of data on international migration may be found in Eurostat statistical yearbooks (Eurostat 2002, 2004, 2006), especially on the accompanying CDs.

The *Recent demographic developments in Europe* reports published every year by the *Council of Europe* cover all Member States and some non-member states. In the 2005 edition, 46 Member States and Belarus were included (Council of Europe 2006). As for international migration statistics, the hard copy of the report contains figures on net migration rates only. Figures on immigration and emigration flows are provided in the country-specific tables on the CD-ROM. One of the problems with the reports published before 2006 is that it is not always clear which disaggregation has been provided, by citizenship or by country of previous/next residence, because the headers and footnotes are not consistent and sometimes misleading. For most countries, figures by citizenship have been presented, but in several cases they are accompanied by footnotes stating that 'Nationality shown because country of origin and destination not available for international migration', even though in fact both types of data are available. In the 2004 edition, the data in the International Migration table for the Netherlands have been mixed up: the 2002 column includes data by country of previous/next residence, while the 2003 column includes data by citizenship.

Table 3.6 Migration data provided by national statistical institutes in response to the Joint Eurostat–UNECE–UNSD–CoE–ILO Questionnaire on International Migration sent out in 2003–2005.[a]

Country	Immigration by sex and country of previous residence	Immigration by sex, citizenship and age	Emigration by sex and country of next residence	Emigration by sex, citizenship and age
Belgium	na	±y	na	±y
Czech Republic	+	+	+	+
Denmark	+	+	+	+
Germany	+	+	+	+
Estonia	na	na	na	na
Greece	na	na	na	na
Spain	large Unk	+	large Unk	+
France	na	non-EU; −age; −sex	na	na
Ireland	p; ±origin	p; ±ctz; broad age	p; ±dest	p; −ctz; broad age
Italy	±y; delay	±y; delay	±y; delay	±y; delay
Cyprus	+	+	+	+
Latvia	+	+	+	+
Lithuania	+	+	+	+
Luxemburg	large Unk	+	large Unk	+
Hungary	na	delay	na	delay
Malta	±y; nat; ±origin	±ctz	±y; nat; to UK	na
Netherlands	+	+	+	+
Austria	±y	±y	±y	±y
Poland	+	na	+	na
Portugal	p, for	p, for	−sex	na
Slovenia	nat	+	nat	±age
Slovakia	+	+	+	+
Finland	+	+	+	+
Sweden	+	+	+	+
United Kingdom	aggreg	aggreg; ±age	aggreg	aggreg; ±age

[a]*Notes*: +,data provided to Eurostat for all or most years; −age, no disaggregation by age; ±age, disaggregation by age provided for a few citizenship (or origin/destination) categories; aggreg, data available for citizenship (or origin/destination) aggregates only; broad age, data disaggregated by broad age groups; −ctz, no disaggregation by citizenship; ±ctz, data provided for a few citizenship categories; delay, data provided with one year delay; ±dest, data provided for a few destination categories; for, data provided for foreigners only; large Unk, large proportion of unknown origin/destination; na, data not available; nat, data provided for nationals only (for immigration to Malta: Persons of Maltese origin); non-EU, data provided for non-EU citizens only; ±origin, data provided for a few origin categories; p, provisional data; −sex, no disaggregation by sex; to UK, data provided for emigration to the UK only; ±y, data provided for some years only.

Each year the *OECD* publishes a report prepared using its Continuous Reporting System on Migration (*SOPEMI*): 'Trends in International Migration' until 2005 and 'International Migration Outlook' in 2006. The SOPEMI reports cover the Member States of the OECD as well as selected non-member countries, i.e. the Baltic States, Bulgaria and Romania in the 2004 edition (OECD 2005). The migration flows tables in the Statistical Annex refer to flows of foreigners, usually disaggregated by citizenship, and are provided for selected countries only. Most of the data published in the Statistical Annex are supplied by the SOPEMI correspondents appointed by the OECD Secretariat in each country, so they do not necessarily represent the official national statistics. The overview part of the SOPEMI report 2006 contains internationally harmonised data on a subset of long-term immigration flows of foreigners. The harmonisation was not done based on the UN definition and the one-year duration of stay criterion, but based on the legal status of immigrants. The statistics cover persons holding permanent residence permits or permits of limited duration but indefinitely renewable (including those who arrived earlier on a short-term permit, but were granted long-term status during the reference year), as well as some persons arriving within the free movement regime (OECD 2006).

The European Commission's *Directorate-General for Justice, Freedom and Security* publishes on its website Statistical Annual Reports on Migration and Asylum, covering 25 Member States, Bulgaria, Romania, Iceland and Norway. The first report concerned the situation in 2001 and was published in early 2004. At the time of writing this chapter, the reports for 2002 and 2003 have been prepared. As for flow data, the 2003 report contains tables and graphs with total annual immigration and emigration in 2000–2003, as well as flows by citizenship in 2003, distinguishing nationals and foreigners subdivided into EU15 citizens, EU10 citizens, ten main groups of immigrants from third countries, and others (DG JLS 2006).

Finally, it is worth mentioning the recent data collection initiative by UNSD. Until 2001, UNSD collected data on international migration flows through the annual International Migration and Travel Statistics Questionnaire sent out within the preparation of the UN Demographic Yearbook. Selected data have been published on a very irregular basis, the last time in the *Demographic Yearbook 1989* (United Nations 1989). Recently the UNSD conducted a review of their data collection (UNSD 2004a) and proposed a new questionnaire that would comply with the latest UN recommendations (UNSD 2004b). Compared with the previous UNSD questionnaire and the Joint Migration Questionnaire, the new questionnaire introduced a number of new tables: inflows of foreigners disaggregated by reason for admission and duration of stay, inflows of citizens by purpose and duration of stay abroad, outflows of foreigners by current status in the country, outflows of citizens by purpose of travel abroad and sex. It is worth noting an asymmetric treatment of citizens and non-citizens in the inflows and outflows tables; this asymmetry would make it difficult to compare data provided by receiving and sending countries. The trial questionnaire was sent out to the European countries together with the Joint Eurostat–UNECE–UNSD–CoE–ILO Questionnaire on International Migration Statistics 2005. It turned out that most countries were not able to provide the requested data.

3.5   Conclusions

Achieving comparability of international migration statistics is a difficult task. The legislation and administrative procedures concerning registration, which is the main source of information on migration flows in most of the EU countries, will continue to differ. This is unfortunate, particularly since the data already contained in the registers should allow the compiling of statistics on flows that would be more complete and more comparable.

It should be noted that the lack of comparability of statistics on international migration flows is strictly linked with the lack of comparability of statistics on population stocks; therefore both problems should be addressed simultaneously. Population stocks are related with migration flows through the population balance equation. If flows figures are not internationally comparable then the annual stock figures are usually not comparable either. This applies in particular to those countries which estimate their population using the component method (this method is used in the majority of the EU countries).

The Implementing Measures for the forthcoming *Regulation of the European Parliament and of the Council on Community statistics on migration and international protection* should provide detailed guidelines on methodologies of producing statistics based on different sources available in the countries. Mathematical and statistical models are also needed to provide consistent, internationally comparable estimates of flows (see Chapters 6 and 7). Much more attention should also be paid to the proper description of the statistics – first by the NSIs when providing the data, and later when the statistics are disseminated.

In the meantime, there are important conclusions for all users of the data on international migration. First, try to find out what the real content of the data is. Second, do not rely on one source. If interested in the flows between a pair of countries, look for data from both the receiving and sending countries and check why the numbers differ. Third, when comparing the flows in a group of countries, do not draw conclusions without taking the definitions into account. These remarks seem obvious but we have to keep repeating them until the situation improves.

Acknowledgements

The material presented in this chapter has been collected among others during two projects in which the Central European Forum for Migration Research (CEFMR) took part in 2003–2005: *Annual International Migration Statistics* and *Towards Harmonised European Statistics on International Migration* (THESIM). The former was a subproject of the *Annual Demographic Statistics, Annual Migration Statistics and Annual Regional Statistics* project conducted by NIDI in cooperation with CEFMR on behalf of Eurostat (NIDI 2006). One of the tasks performed by CEFMR has been checking and processing the data supplied by European national statistical institutes in response to the Joint Eurostat–UNECE–UNSD–CoE–ILO Questionnaires on Migration Statistics, which proved a good opportunity to get a comparative picture of the information supplied by different countries.

The THESIM project, coordinated by Professor Michel Poulain from the Université Catholique de Louvain, was funded by the European Commission's Directorate-General for Research within the Sixth Framework Programme (Poulain *et al.* 2006).

One of the objectives of the project was to support the implementation of the forth-coming *Regulation of the European Parliament and of the Council on Community statistics on migration and international protection*, in particular to analyse the current functioning of migration statistics in the 25 EU countries and to identify the problems that the countries might encounter in relation to the requirements of the new regulation. During the project, in which seven scientific teams[8] partici-pated, meetings with experts from the NSIs and with the authorities responsible for the administrative procedures relevant to population registers and registration of migrants have been organised in all 25 EU countries. The meetings were a unique source of information, with explanations provided directly by a range of people involved in the statistics production process. In addition to these two projects, the *Quality Review of MIGRAT in NewCronos* project, conducted in 2003 by NIDI and CEFMR on behalf of Eurostat, allowed us to gain an important insight into the problem of the quality of data on international migration flows and stocks.

References

Breem Y and Thierry X. 2006. Country report for Spain. In *THESIM. Towards Harmonised European Statistics on International Migration*. Poulain M, Perrin N and Singleton A, eds, pp. 447–455. Louvain-la-Neuve: UCL Presses Universitaires de Louvain.

Council of Europe. 2006. *Recent demographic developments in Europe, 2005*. Strasbourg: Council of Europe Publishing.

DG JLS. 2006. *Annual report on asylum and migration 2003*. European Commission's Directorate-General for Justice, Freedom and Security. Available at http://ec.europa.eu/justice_home/doc_centre/asylum/statistics /doc_annual_report_2003_en.htm

European Commission. 2005. *Proposal for a Regulation of the Parliament and of the Council on Community statistics on migration and international protection*. Brussels. COM (2005) 375 final, 2005/0156 (COD). Available at http://eurlex.europa.eu/LexUriServ/site/en/com/2005/com2005_ 0375en01.pdf

Eurostat. 2002. *European Social Statistics*. Migration, 2002 edn. Luxembourg: Office for Official Publications of the European Communities.

Eurostat. 2004. *Population statistics*, 2004 edn. Luxembourg: Office for Official Publications of the European Communities.

Eurostat. 2006. *Population statistics*, 2006 edn. Luxembourg: Office for Official Publications of the European Communities.

Herm A. 2006. Recommendations on international migration statistics and development of data collection at an international level. In *THESIM. Towards Harmonised European Statistics on International Migration*. Poulain M, Perrin N and Singleton A, eds, pp.77–106. Louvain-la-Neuve: UCL Presses Universitaires de Louvain.

Hola B. 2006. Lessons to be learnt from the use of administrative sources in a recent immigration country. Presented at the *UNECE Workshop on International Migration Statistics for Countries of CIS and South-Eastern Europe*, Edinburgh, 23–24 November. Available at http://www.unece.org/stats/documents/ece/ces/ge.10/2006/mtg1/9.e.ppt

---

[8] The following teams participated in the THESIM project: GEDAP UCL (Belgium), NIDI (the Nether-lands), INED (France), ICMPD (Austria), CEFMR (Poland), ICStat (Italy) and A. Herm from Estonia.

Kelly JJ. 1987. Improving the comparability of international migration statistics: contributions by the Conference of European Statisticians from 1971 to date. *International Migration Review* 21:1017–1037.

NIDI 2006. *Annual demographic statistics, annual migration statistics and regional population statistics*. Report on behalf of Eurostat.

Nowok B and Kupiszewska D. 2005. *Official European statistics on international migration flows: availability, sources and coverage*. CEFMR Working Paper 5/2005. Available at http://www.cefmr.pan.pl/docs/cefmr_wp_2005-05.pdf

Nowok B, Kupiszewska D and Poulain M. 2006. Statistics on international migration flows. In *THESIM. Towards Harmonised European Statistics on International Migration.* Poulain M, Perrin N and Singleton A, eds, pp. 203–231. Louvain-la-Neuve: UCL Presses Universitaires de Louvain.

OECD. 2005. *Trends in international migration*. Annual Report. 2004 edn. Paris: OECD.

OECD. 2006. *International migration outlook*. Paris: OECD.

Office for National Statistics. 2004. *International migration*, Series MN, no. 29. London: Office for National Statistics.

Poulain M. 1999. *International migration within Europe: towards more complete and reliable data?* Working Paper 37. Joint ECE–Eurostat Work Session on Demographic Projections, Perugia, Italy.

Poulain M. 2001. *Is the measurement of international migration flows improving in Europe?* Working Paper 12. Joint ECE–Eurostat Work Session on Migration Statistics, Geneva, Switzerland.

Poulain M, Perrin N and Singleton A, eds. 2006. *THESIM. Towards Harmonised European Statistics on International Migration*. Louvain-la-Neuve: UCL Presses Universitaires de Louvain.

Thorogood D. 2006. Improving the quality and availability of migration statistics in Europe. Presented at the *UN Expert Group Meeting on Measuring International Migration: Concepts and Methods*, United Nations, New York. ESA/STAT/AC.119/5. Available at http://data.un.org/unsd/demographic/meetings/egm/migrationegm06/DOC%205%20 Eurostat.pdf

United Nations. 1978. *Statistics of international migration, Demographic yearbook 1977*. UN Publication, Sales No. E/F.78.XIII.1. New York: United Nations.

United Nations. 1989. *United Nations Demographic Yearbook 1989*. UN Publication, Sales No. E/F.90.XIII.1. New York: United Nations.

United Nations. 1998. *Recommendations on statistics of international migration*, Revision 1. Statistical Papers, Series M, No. 58. UN Publication, Sales No. E.98.XVII.14. New York: United Nations.

UNSD. 2004a. *United Nations 'Demographic yearbook' review: National reporting of international migration data – Implications for international recommendations*. New York: United Nations Statistics Division. ESA/STAT/2004/4. Available at http://unstats.un.org/unsd/demographic/products/dyb/dybtr.htm

UNSD. 2004b. *Trial questionnaire on international travel and migration statistics*. New York: United Nations Statistics Division. Available at http://unstats.un.org/unsd/demographic/products/dyb/dybquest.htm

4

Evolution of international migration statistics in selected Central European countries

Beata Nowok

Population Research Centre, University of Groningen

4.1 Introduction

Two historical events have profoundly influenced international migration in Central European post-Soviet countries. In the late 1980s and early 1990s, they were transformed from communist societies into democratic market economy societies; and, in 2004, they became Member States of the European Union (EU). During communist times, migratory movements were shaped primarily by restrictive exit policies. With the end of communism, international movements were then principally affected by the immigration policies of destination countries. The relaxed regulation on international movements resulted in increases in both the volume and diversity of migration flows. Furthermore, the entry into the EU and access to new labour markets led to increased opportunities and mobility. Yet, despite the growing importance of international migration, accurate migration statistics are very often missing or inadequate (see also Chapter 3). One of the foremost problems for post-Soviet countries in Central Europe relates to the continued use of outdated concepts, inherited from the previous regime, to produce their international migration statistics. In this chapter, the Czech Republic, Hungary, Poland, Slovakia and

International Migration in Europe: Data, Models and Estimates   Edited by J. Raymer and F. Willekens
© 2008 John Wiley & Sons, Ltd

Slovenia are used to illustrate how statistical agencies in post-Soviet countries in Central Europe have responded to capture changes in their international migration patterns.

The main objective of this chapter is to assess the extent to which the Czech Republic, Hungary, Poland, Slovakia and Slovenia can satisfy the requirements on international migration flow statistics specified in the forthcoming EU Regulation (European Commission 2005). This chapter is supplementary to Chapter 3 and is included to better understand the full complexity of the problems faced by some countries in their attempt to provide adequate and comprehensive statistics on migration flows. The issues covered focus mostly on the concept of residence and the registration system used to capture changes in international movements. A strong emphasis is placed on the analysis of time trends in the migration statistics. In Section 4.2, an overview of the available data sources is presented. In Section 4.3, a description of the current rules used to provide international migration statistics is carried out. Immigration and emigration patterns over time are presented in Section 4.4 to show how they are influenced by changes in definition and coverage. Finally, the quantitative impacts resulting from changes in migration definitions are set out in Section 4.5. Note that, unless otherwise indicated, the current information in Sections 4.2 and 4.3 relates to the year 2005 for the Czech Republic, Poland, Slovakia and Slovenia and to the year 2004 for Hungary. As these statistics are subject to on-going improvements and development, some changes in the system are likely to have already been introduced in some of these countries.

4.2   Data sources and availability of statistics on international migration flows

The Czech Republic, Hungary, Poland, Slovakia and Slovenia all have centralised population registers. Such registers usually provide the best comprehensive statistics on international migration. Unfortunately, the population registers in the above countries are currently inadequate and unable to provide all the following core tabulations of international migration flows: citizenship by age and sex, country of birth by age and sex, and country of previous/next residence by age and sex. The main problems include limited coverage of foreign populations, lack of detailed information in the available data, and the citizens of these countries generally not complying with registration rules.

All of the five countries maintain centralised registers of foreigners that require authorisation to enter or stay in the country. These registers constitute a valuable source of foreign immigration data. There are, however, two important items that must be checked before they can be considered reliable. First, the registers should be able to keep track of permits issued to a given individual in order to count migrants and not just the number of documents processed. Second, they should be able to identify the actual presence of a document holder in the country, which is usually achieved by the registration obligation. In the five countries considered in this chapter, most foreigners are registered in both a population register and a register of foreigners. The exchange of information between these two databases, however, differs depending on where they have to register their place of stay after arriving. The registration can take place either in the municipality, which runs the population register, or in a police office, which runs the register of foreigners – for more details see Kupiszewska and Nowok (2006) and Kupiszewska *et al.* (2006). Finally, some countries derive their migration statistics directly from forms that are filled out when persons (both nationals and foreigners) register or deregister their permanent place of residence. Countries use this system to augment administrative sources, in particular, to add information regarding the characteristics of migrants and information about their migrations. Note that two types of data collection systems, one based on forms and the other based on two different registers (i.e. a population register for nationals and a register of foreigners for non-nationals), are characteristic of these countries (and not of other EU Member States).

Poland (until 2005) and Slovakia use forms to obtain information about international migration. Slovakia also makes use of its register of foreigners to obtain statistics on citizenship. In 2006, Poland began to use its population register to derive migration data for both nationals and foreigners. In Hungary, the Czech Republic and Slovenia, population registers and registers of foreigners are used to produce international migration statistics for nationals and foreigners, respectively. In Hungary, however, the population register does not cover the whole target population of foreigners; only permanent migrants are included. The term 'permanent' is used here (and throughout this chapter) in the literal sense; that is, once a move was made, migrants were not expected to migrate again. In the Czech Republic and Slovenia, the population registers provide complete coverage. When the population registers were extended to include all major categories of foreigners, the missing data have been transferred from the register of foreigners. However, in both countries, the register of foreigners provides better quality data. Note that, up until 2004, the Czech Republic obtained statistics on international migration of nationals from statistical forms, which were inherited from the time when Czechoslovakia existed. These forms are still in use in Slovakia.

The data sources described above are used to produce and disseminate annual official statistics on immigration and emigration for the five countries considered in this chapter. Data on the characteristics of migrants vary considerably across these countries, as shown in Table 4.1. However, despite efforts to combine data from different sources, none of these countries are able to produce all of the tabulations required by the forthcoming EU Regulation. The missing characteristics are either not collected or of poor quality.

Statistics on *immigration and emigration flows by citizenship* are available for all the countries, except Poland. In Poland, information about citizenship is collected, but due to a large number of missing values, it has been decided not to provide such data to the public. In Slovenia, where figures for emigration of foreigners are the results of estimates, statistics by citizenship are not disaggregated by age. There are also many gaps in available *immigration and emigration flows by country of birth*. Currently, they are available only in Hungary (for both nationals

Table 4.1 Availability of statistics on international migration flows in selected post-Soviet Central European countries.[a]

Country	Total	By citizenship	By country of birth	By country of previous/next residence
Czech Republic	+	+	−	+
Hungary	+	+	+	−
Poland	+	−	−	+
Slovakia	+	+	±[b]	+
Slovenia	+	±[c]	−	±[d]

[a] Information referring to disaggregation by age and sex; + available; − not available; ± some data are missing.
[b] Data on nationals are not available.
[c] Data on foreigners are not disaggregated by age.
[d] Data on foreigners are not available.

and non-nationals). In Slovakia, these data are available for foreigners only. In the Czech Republic, Slovakia and Slovenia, these data may be available in the near future. However, in Poland, there are no concrete plans to provide these data.

Statistics on *immigration by country of previous residence* and *emigration by country of next residence* are not compiled at all for foreigners in Hungary and Slovenia, as the data are either not complete or not collected. In the Czech Republic, Poland and Slovakia, these tabulations are available. However, in the Czech Republic, emigration data are based on expiry of residence permits, and the assumption is made that all foreigners migrate to their countries of citizenship. Because of inadequate data collection systems, a consistent picture of the migration patterns for these five countries is hindered.

4.3 Definition of international migration in official flow statistics

None of the five countries currently adhere to the United Nations' one-year definition of international migration (United Nations 1998). The reason for this is that these countries are still in the process of transitioning from being former communist countries. The concept of permanent residence inherited from the communist era, and the concept of permanent international migration that follows, does not fit in with the contemporary migration patterns. Unlike in the Soviet period, citizens in the Czech Republic, Hungary, Poland, Slovakia and Slovenia are now free to move within their countries and between countries, as well as to return.

The reliability of statistics is also affected by the coexistence of permanent and temporary residences in the registration systems. A permanent resident of the country has one primary residence that does not change when a temporary move is made. For international movements, this means that persons choose whether their departure is reported as temporary or permanent. If there is no upper limit for a temporary stay, then there are no incentives for giving up a permanent residence in the country. Finally, the most problematic issue is actually the general non-participation in the population registration process, a consequence of strict population control policies in the past. This mainly refers to the migration of nationals, as migration of foreigners is more restricted.

Despite the difficulties above, some improvements have been made. Most of these are aimed at complying with the United Nations' recommendation of providing international migration data. The Czech Republic and Slovenia are the countries that have made the most progress. Often, compliance may be made by some relatively minor adjustments to existing data collection systems or by using alternative data sources. Except in the Czech Republic, the coexistence of permanent and temporary places of residence in the registration systems is still in force. Hungary simply changed the title of the residence definition, rather than the definition itself. Now, a 'permanent place of residence' is called a 'place of residence' and a 'temporary place of residence' is called a 'place of stay'. Differences in residence concepts between countries refer to the minimum duration of stay that qualifies as a temporary residence. In Hungary and Slovakia, a stay longer than three months constitutes a temporary residence, irrespective of the total duration of stay. In Poland and Slovenia, the required duration of stay is more than two months and more than 60 days, respectively. However, in Slovenia, a definition of a temporary stay abroad differs from the definition of a temporary residence in the country. It is defined as a stay for more than three months.

In Hungary, Poland, Slovakia and Slovenia, persons with right of permanent residence are generally obliged to report both permanent and temporary movements. However, there are some exceptions. In Hungary, arrival after a temporary stay abroad has to be reported only if the person does not have a place of residence in the country or wants to have a new one. In Hungary and Slovakia, temporary residents do not have to report when leaving the country. It is also the case in Poland, unless the actual duration of stay differs from the intended one.

In the Czech Republic, a person may have only one place of residence. For nationals, it is considered their permanent place of residence. Official registration of temporary residence, alongside a permanent place of residence, was abandoned in the first half of the 1990s. If nationals want to leave the country to settle abroad, but do not want to give up their permanent residence in the country, they do not have to report their departure. If nationals go abroad for a temporary stay they do not have to report their departure either. Nationals coming for a temporary or permanent stay in the Czech Republic have to register only if they do not have a permanent residence in the country (also true for Hungary). Thus, in the case of international movements, the removal of temporary place of residence has not led to an improvement in the statistics. Temporary residents of the Czech Republic do not have to report their departure abroad (also true for Hungary and Slovakia).

The definition of international migration is different for foreigners and nationals in the Czech Republic, Hungary and Slovenia, where population registers and registers of foreigners are used to capture these populations, respectively. Furthermore, the definitions of immigration and emigration are not consistent for foreigners (in Slovenia, this is also the case for nationals).

In the *Czech Republic*, data on immigration and emigration of its citizens refer to self-declared changes of permanent place of residence. Figures on the immigration of foreigners are currently produced by the Czech Statistical Office based on residence permits granted to foreigners in a reference year. There are three types of residence permits that are included: long-term residence permits granted after a minimum stay in the country, temporary residence permits, and permanent residence permits granted as a first residence permit (without the condition of a minimum duration of stay). These permits are issued to different groups of migrants and they are connected with different durations of stay in the country. Long-term residence permits are granted to non-EU citizens after a one-year stay in the Czech Republic based on long-term residence visas. Thus, the actual duration of stay of non-EU citizens with long-term residence permits is by definition longer than one year. Temporary residence permits are granted to EU citizens if they intend to stay in the Czech Republic for more than three months. Permanent residence permits that are not conditional on stay in the country are generally granted to refugees. Refugees registered for permanent stay are automatically included and the date of granting the status is taken as the date of immigration. Therefore, the duration of their stay is not taken directly into account. Foreigners are treated as emigrants on the date of cancelling their permanent place of residence in the country or on the expiry date of their residence permit (or a card corresponding to the permanent residence permit), provided they had not been renewed.

In *Hungary*, official statistics on the international migration of its citizens cover temporary and permanent changes in residence. Data on immigration include citizens who were born abroad or who have lived permanently abroad and citizens who came back after a temporary stay abroad. Data on emigration refer to nationals who left Hungary with the intention of living temporarily or permanently abroad. Statistics on the immigration of foreigners cover those who received various residence permits. Refugees are not included. A temporary residence permit is issued only to non-EEA citizens staying in Hungary with a valid residence visa. Thus, a non-EEA citizen who intends to stay in Hungary for more than 90 days must enter the country with a valid residence visa, which is granted for up to one year. Then, he or she may apply for a temporary residence permit. EEA nationals are exempted from the visa requirement, but if the stay exceeds 90 days, they must apply for an EGT residence permit that certifies their entitlement to stay. A permanent residence permit may be granted after three years of continuous stay in the country. It gives the right to stay in Hungary for indefinite duration. Currently it is called a settlement permit and it replaced the previous immigration permit. Data on the emigration of foreigners include those who reported their departure for permanent stay abroad, those who did not renew their residence permits, and those whose place of residence is unknown. Refugees leaving the country are not included.

Slovenian citizens are considered immigrants if they register for a permanent residence or if they gave notice of their arrival for a permanent or temporary (more than 60 days) stay after returning from a temporary (more than three months) stay abroad that had been reported before their departure. Slovenian citizens are considered emigrants if they notify the authorities of their intention to live permanently or temporarily (for more than three months) abroad. Foreigners are included in the immigration statistics when they register their residence in the country. All foreigners with temporary residence permits, even for less than one year, who registered are included. The temporary residence permits are required for a stay exceeding 90 days. Emigration is estimated on the basis of changes in the stock of foreigners in the register of foreigners, vital statistics and immigration data (taking into account the acquisition of citizenship). Foreigners are removed from the foreign population if they declare their departure for a permanent stay abroad or if their residence permit expires. Foreigners with permanent residence who leave the country temporarily are not included, even if they go for more than one year.

In *Poland*, the definition is consistent for nationals and non-nationals and for inflows and outflows. The annual statistics represent the permanent migration concept of migration. Temporary migrations are also recorded, though they are not included in the published statistics. This is due to the fact that the quality of temporary migration is highly questionable. This is particularly true for the emigration data. Despite the legal obligation for every person going abroad for a period of two months or more to declare their departure, most do not bother.

In *Slovakia*, there are two types of migration statistics, derived from different sources of data. The annual statistics released by the statistical offices as figures for immigration and emigration flows, and used to estimate population size between censuses, are prepared based on statistical forms and refer to moves from or to abroad of both nationals and non-nationals who change their place of permanent residence. Thus, the migration concept is exactly the same as in Poland. Supplementary migration data for foreigners are obtained from the register of foreigners. They represent a different definition with figures that are not comparable with the data obtained from the statistical forms. These immigration data refer to all residence permits issued to foreigners, which are required for stays exceeding three months. Note that a foreigner may have been granted more than one residence permit during a given year. Emigration data represent residence permits that expired in a particular year.

In summary, there are two types of international migration statistics provided by the five Central European countries covered in this chapter. The first represents flows of permanent migration; a concept carried over from communist times and consistent for both nationals and foreigners. The second represents flows of different types of migration, depending on the particular group of interest (i.e. foreigners or nationals). Neither situation is perfect. The concept of permanent migration, which is understood literally, is one of the most problematic among the concepts applied in the EU countries (see Chapter 3). On the other hand, the coexistence of different definitions applied in the same country but to different groups exemplifies the problems that, unfortunately, often occur with migration statistics.

4.4   Trends in international migration flows

International migration flows reported by the Czech Republic, Hungary, Poland, Slovakia and Slovenia have one predominant feature: the migration statistics either are or will be inconsistent over time. This results from the use of the permanent migration concept, which has recently been changed or will be soon. Therefore, the analysis of trends cannot be separated from the concept underlying the data and coverage of statistics in terms of different categories of migrants. A sudden deviation from a long-term trend should receive particular attention.

In the *Czech Republic*, there are two landmark dates that are reflected in the migration pattern (see Figure 4.1). After the dissolution of Czechoslovakia on 1 January 1993, the previously internal movements between the territories of the Czech Republic and Slovakia became international migration flows. The expected split of the common state evoked an increase in migration from Slovakia to the Czech Republic in 1992 (11 740) but soon after decreased (i.e. 7276 in 1993 and 4076 in 1994) and levelled off at about three thousand between 1996 and 2001. The patterns of emigration to Slovakia exhibited a sharp decrease in the volumes starting in 1994. During this year, only 56 persons emigrated, whereas the average level between 1980 and 1993 was around six thousand or more (e.g. 7232 in 1993). This decline is associated with the fact that, beginning in 1994, the Czech Republic and Slovakia no longer exchanged information on migration between their two countries.

The second important change that occurred in the Czech Republic took place in 2001. This change is associated with the change in migrant definition for foreigners, whereas before the concept of permanent migration was applied to both citizens and foreigners. The substantial increases in immigration and emigration starting in 2001 reflects mostly the change in definition and not a shift in migration trends.

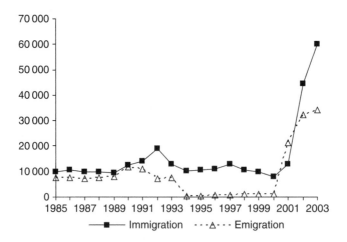

Figure 4.1   Immigration and emigration in the Czech Republic, 1985–2003.
Source: Statistical Yearbook of the Czech Republic (various years).

The sudden rise is visible especially for the emigration figure in 2001. According to the Czech Statistical Office this was a special case of parallel change of definition and implementation of the new Act on Residence of Foreigners. As a result, in 2001, a negative net migration was recorded.

The analysis of international migration trends in *Slovenia*, presented in Figure 4.2, is complicated by historical events. In addition to changes in migration definitions, there have been substantial changes in coverage. Up to 1991, statistics represent permanent migration of Socialist Federal Republic of Yugoslavia citizens. Data for 1992 to 1994 refer to permanent migration of Slovenian citizens. From 1995, migration flows of foreigners (i.e. permanent and temporary) are included in the statistics. From 1999 onwards, data on international migration of Slovenian citizens also include temporary moves. However, in 1999 there was a considerable decline in emigration of foreigners from 6003 in 1998 to 1643 in 1999, which is reflected in the overall emigration figures. This was a direct consequence of the clearing and verification of data by the Ministry of the Interior (i.e. the data provider).

In *Hungary*, there have been no substantial changes in the definition of migration. Since the first publication of international migration statistics in the 1992 volume of the Demographic Yearbook, the figures on international migration of foreigners have been produced from data on residence permits. The sharp increase in immigration figures in the late 1980s resulted from the liberalisation of international movements. Note, however, that there are many discrepancies between figures for total immigration and emigration flows available in various publications and databases (see Table 3.5 and discussion in Chapter 3).

International migration statistics in *Poland* are consistent over time and capture only registered changes of permanent place of residence. These migration flows have been relatively stable, following a short period of higher emigration rates in

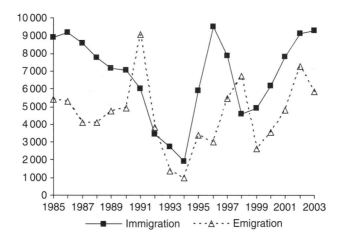

Figure 4.2   International immigration and emigration in Slovenia, 1985–2003.
Source: Statistical Yearbook of the Republic of Slovenia (various years).

the late 1980s. However, keep in mind that these statistics do not allow an adequate picture of the fluctuations in migration over time.

Finally, the concept of international migration underlying the official annual statistics in *Slovakia* has not really evolved over time and the concept of permanent migration has always been used. International migration data on non-nationals that are obtained from the register of foreigners follow a different definition, but they are not included in the official figures for overall migration.

4.5   Impact of migration definition on recorded volume of international migration flows

The development of inadequate migration concepts over time raises a more general issue of the impacts of changes in definitions on the recorded levels of international migration. This is very important for migration research, especially for comparative analyses. The variability of definitions applied in EU countries is described in Chapter 3. Double entry matrices allow one to identify differences in definitions (see Chapter 3), although differences may also result from various qualities of data provided. Valuable insights into the relation between the concept of migration and the volume of recorded migration flows can also be provided by examining a particular country's flows at different points in time or by examining data referring to the same year, but obtained from various sources with different definitions or collection methods.

The transition from the permanent (i.e. declared infinite duration of stay) to the long-term (i.e. actual stay exceeding one year[1]) migration concept, which took place in the Czech Republic for foreigners, is a good example of how a change in definition can affect the level of reported migration flows. It also emphasises the serious limitations of statistics on permanent migration. First, permanent migration accounts for a minor part of the broader category of migration that includes actual stay exceeding one year (see Figure 4.1). Flows between the Czech Republic and Poland since 2001 (see Figure 3.8 in Chapter 3) and between the Czech Republic and Slovakia since 2001 are remarkable examples. For instance, the 2002 and 2003 flows from Poland and Slovakia to the Czech Republic reported by the Czech Republic are on average 41 times higher than those reported by the sending countries. This conclusion can be strengthened by the comparison of the Slovak data on foreign migration derived from different sources. The 2002–2003 foreign permanent immigration and emigration data include only 16 % and 1 %, respectively, of the data contained in the register of foreigners.

The direct comparison of flows between Poland and Slovakia reported by the sending and receiving countries reveals another important feature of the statistics based on the concept of permanent place of residence, namely the underestimation of emigration flows. Here, the migration flows according to the receiving country

---

[1] Note that in the case of the Czech migration statistics a criterion of actual one-year duration of stay is not generally applicable to all persons (see Section 4.3).

are higher than those according to the sending country. Flows between Poland and Slovakia (Figure 3.4 in Chapter 3), between the Czech Republic and Poland until 2000 (Figure 3.8 in Chapter 3) and between the Czech Republic and Slovakia until 2000 are good illustrations of this relationship. The concept of permanent residence applied to the international migration statistics has additional serious drawbacks. Foreigners, who usually may be granted a permanent residence permit after several years of uninterrupted stay in the country, are registered as permanent residents and counted as immigrants with a considerable delay of an unspecified number of years. Thus, the comparison of data on flows between pairs of countries for a given year is questionable. The date of immigration of nationals and emigration of both nationals and non-nationals corresponds in principle to the date of their actual movement. Therefore, immigration and emigration data for foreigners are not compatible, which makes the total immigration and emigration flows incompatible as well. Besides, the very idea of permanency of movement imposes reliance on people's intentions about the length of their stay in the host country. Intentions are usually not stable, and in this case they cannot be verified before the death of the potential migrant.

The abandonment of the concept of permanent migration for foreigners in the Czech Republic should be recognised as a marked improvement. It has led to a significantly higher level of registered migration flows. The data are more comparable with the statistics produced by other countries in the European Union (see Tables 4.2 and 4.3). For the years 1996–2000 the Czech Republic's annual reported immigration flows from Germany, Austria, Denmark and the Netherlands accounted for, on average, 13 % of the reported flows by the sending countries (in the case of flows in the opposite direction, the Czech emigration data accounted for, on average, less than 4 % of the flows reported by the receiving countries). The decline in discrepancies between the two sources of data reported by the different countries occurred after the change in definition by the Czech Republic. Note that the differences also depend on the definitions of emigration used and reported by the origin countries. For example, the flows from Germany for 2002 and 2003, reported by the Czech Republic, still accounted for only 12 % of the corresponding flows reported by Germany. For flows between other countries using durations of three months (Austria) or six months (Denmark), the improvement of the Czech data are more apparent, from 10 % (1996–2000) to approximately 40 % (2002–2003). For the case of sending countries with longer duration of absence qualifying the movement of migration, the Czech Republic's immigration data for 2002 and 2003 are even higher than emigration data of those countries. After the changes were implemented, the level of emigration reported by the Czech Republic also improved, but the figures for 2002 and 2003 never exceed 45 % of the values published by the destination countries.

In comparing the migration flows between the Czech Republic and Germany, as reported by both countries (see Tables 4.2 and 4.3), one might wonder why such large differences remain. Might they result from a large share of short-term migrations between these two countries, which are counted only by Germany? A closer examination of the data reveals that it actually results from a large share of

Table 4.2 Receiving to sending country reported flow ratios: immigration to the Czech Republic, 1993–2003.[a]

Country of origin	Time criterion in definition of emigration in country of origin	1993	1994	1995	1996	1997	1998	1999	2000	2001	2002	2003
Germany	none	0.10	0.10	0.12	0.11	0.10	0.09	0.07	0.06	0.05	0.10	0.14
Austria	three months	...	...	...	0.13	0.10	0.09	0.09	0.09	0.08	0.54	0.41
Denmark	six months	0.16	0.11	0.09	0.12	0.09	0.11	0.08	0.14	0.06	0.36	0.36
Netherlands	eight out of 12 months	0.29	0.42	0.26	0.27	0.30	0.21	0.12	0.17	0.34	1.08	1.42
Sweden	one year	1.17	1.32	0.80	1.23	1.08	0.40	0.42	0.43	0.64	1.03	1.08
Poland	permanent	2.16	4.21	3.45	4.02	3.72	3.16	2.66	1.73	8.00	44.18	35.93
Slovakia	permanent	1.00	42.91	21.36	38.76	14.57	11.50	15.55	9.12	7.73	29.68	54.43

[a] ... Not available.

Table 4.3 Sending to receiving country reported flow ratios: emigration from the Czech Republic, 1993–2003.[a]

Country of desti- nation	Time criterion in definition of immigration in country of destination	1993	1994	1995	1996	1997	1998	1999	2000	2001	2002	2003
Germany	none	0.01	0.01	0.02	0.02	0.03	0.04	0.03	0.03	0.06	0.10	0.10
Austria	three months	...	...	...	0.04	0.04	0.09	0.05	0.06	0.14	0.35	0.26
Denmark	none/six months/three months[b]	0.02	0.00	0.02	0.01	0.05	0.03	0.02	0.01	0.09	0.28	0.20
Netherlands	four out of six months	0.01	0.03	0.00	0.02	0.02	0.04	0.04	0.04	0.19	0.40	0.45
Sweden	one year	0.04	0.04	0.03	0.06	0.05	0.03	0.05	0.03	0.22	0.38	0.35
Poland	permanent	0.14	0.05	0.25	0.41	0.22	0.40	0.15	0.43	11.83	32.85	22.61
Slovakia	permanent	1.00	0.02	0.09	0.21	0.30	0.46	0.39	0.33	8.80	19.30	28.10

[a] . . . Not available.
[b] Information referring to immigration of nationals/other EU citizens/non-EU citizens.

flows of Czech Republic citizens still using the old concept of permanent migration. According to German statistics almost 90 % of flow consists of Czech Republic citizens. When we take into account migration of foreigners only, the values were a lot more comparable, i.e. more than 60 % for immigration and more than 80 % for emigration, respectively, of the values reported by Germany – this is the author's own calculation based on data from Holá (2005). This example illustrates that application of diverse definitions for different groups in the population prevent straightforward conclusions that require additional investigations. In general, the application of a new definition only to foreigners distorts the picture of international migration. For example, the percentage of immigrants to the Czech Republic that were foreigners accounted for around 72 % between 1996 and 2000. Between 2002 and 2003, it was 96 %. For the corresponding emigration flows, the discrepancy was much greater, i.e. 15 % between 1996 and 2000, and 97 % between 2002 and 2003.

Finally, the statistics provided by the Czech Republic have some additional drawbacks. Emigration flows by country of next residence derived from the expiration of residence permits are assigned to the foreigner's country of citizenship. Of course, foreigners may not emigrate to the country of their citizenship. For instance, upon expiry of a residence permit that was granted to a Slovak citizen, the Czech Republic assumes that he or she migrated to Slovakia. However, that person could have migrated to Germany instead.

4.6   Conclusions

The liberalisation of international movements in the late 1980s and early 1990s has affected these five Central European countries greatly. There has been a considerable increase in international mobility. Unfortunately, the concept of permanent migration, dating from the communist era, captures only a minor part of the movements. Improvements in the recording of migration are clearly needed. However, this alone will not solve all the problems. Some incentives must also be created to encourage persons to report changes in their residence. Nationals of these countries are particularly prone not to report changes in residence. International migration of foreigners is better controlled, to some extent, through residence permits. Therefore, full transformation of statistics requires changes both to the system and to the behaviour of the migrants themselves. Only then can sound migration statistics be produced.

The gradual process towards providing internationally harmonised migration data leads to the undesirable side effect of inconsistency over time. It is of great importance for countries in transition to provide datasets, together with detailed metadata, to enable correct interpretation of the statistics. Currently, a considerable diversity of migration concepts are applied in the five Central European countries covered in this chapter. The differences exist not only between countries, but also within them. Despite the development of international migration statistics undertaken by some states, e.g. the Czech Republic, none of the countries produce statistics that comply completely with the UN definition to be adopted in the forthcoming EU Regulation; nor can they provide all of the tabulations. Thus, further improvements are necessary.

Acknowledgements

This chapter has been prepared using information collected in the THESIM project (*Towards Harmonised European Statistics on International Migration*), which was funded by the European Commission's Directorate-General for Research under the FP6 Programme. The author is grateful to the whole project team, especially Michel Poulain from the Université Catholique de Louvain who led the team and Dorota Kupiszewska from the Central European Forum for Migration and Population Research, for their invaluable cooperation. Special thanks are also due to the national experts from the Czech Republic, Hungary, Poland, Slovakia and Slovenia for their comments on country reports. Finally, both the Netherlands Interdisciplinary Demographic Institute in The Hague and the Central European Forum for Migration and Population Research in Warsaw provided support for this research.

References

European Commission. 2005. *Proposal for a Regulation of the Parliament and of the Council on Community statistics on migration and international protection*. Brussels. COM (2005) 375 final, 2005/0156 (COD). Available at http://eurlex.europa.eu/LexUriServ/site/en/com/2005/com2005_0375en01.pdf/

Holá B. 2005. Srovnatelnost statistiky zahraniční migrace (Comparability of external migration statistics). *Demografie* 47:177–187.

Kupiszewska D and Nowok B. 2006. Country reports for the Czech Republic, Hungary, Slovenia and the Slovak Republic. In *THESIM. Towards Harmonised European Statistics on International Migration*. Poulain M, Perrin N and Singleton A, eds, pp. 391–401, 529–539, 599–610, 611–619. Louvain-la-Neuve: UCL Presses Universitaires de Louvain.

Kupiszewska D, Nowok B and Kupiszewski M. 2006. Country report for Poland. In *THESIM. Towards Harmonised European Statistics on International Migration*. Poulain M, Perrin N and Singleton A, eds, pp. 577–589. Louvain-la-Neuve: UCL Presses Universitaires de Louvain.

United Nations. 1998. *Recommendations on statistics of international migration*, Revision 1. Statistical Papers, Series M, No. 58, UN Publication, Sales No. E.98.XVII.14. New York: United Nations.

5

Foreign migrants in Southern European countries: evaluation of recent data

Alessio Cangiano

ESRC Centre on Migration, Policy and Society, University of Oxford

5.1 Introduction

In Southern European countries (Italy, Spain, Greece and, by convention, Portugal), the measurement of both the level and characteristics of the foreign population is not an easy task. In addition to the usual problems affecting migration statistics in traditional immigration countries, the presence of a sizeable and dynamic irregular component is a major hindrance. Even for the estimation of the legal presence, administrative sources are in some cases not exhaustive, either because they capture different subsets of the whole universe of regular foreigners (as in the case of Italy) or because of very restrictive immigration laws not allowing immigrants to remain for very long in the country (as in the case of Greece). The analysis in this chapter focuses on the foreign population, as citizenship is the criterion most widely used in Southern European migration statistics. The foreign population in these countries significantly overlaps with the foreign-born population, since migration is a relatively recent phenomenon and most migrants still keep the citizenship of their country of birth. However, nationality is becoming

International Migration in Europe: Data, Models and Estimates   Edited by J. Raymer and F. Willekens
© 2008 John Wiley & Sons, Ltd

increasingly inadequate to identify migrants in the available statistics because of the rising number of naturalisations and size of second generations, as minors generally retain the citizenship of their parents.

The quantitative evolution of the foreign population in Southern Europe since the beginning of the 1990s is set out in this chapter. The approach is simple and based on the comparison and integration of several data sources. In Section 5.2, the type and quality of the information provided at national level by the most usually available migration statistics (residence permits, population registers and censuses) are reviewed. The weaknesses and strengths of each source are set out. For each of the four countries considered in this study, an illustration of how the available data sources can be integrated to improve the measurement of the stock of foreign nationals over time is provided. In Section 5.3, an attempt is made to reconstruct the quantitative evolution of the foreign population in Southern European countries since the beginning of the 1990s. The main patterns shared by all Southern European countries as well as some features specific to national contexts are pointed out.

In Sections 5.4 and 5.5, the analysis focuses on estimating the stock of undocumented migrants. Results of periodical regularisation programmes are used as *ex post* information on the number of unauthorised residents before the amnesty. Estimates provided by previous studies are also reviewed. A more analytical approach is possible for Spain, where the size of the undocumented migrant population can be estimated through a simple strategy based on the overlapping of three sources (residence permits, population registers and census). The degree of involvement of the main national groups in the last regularisation carried out in this country (2005) is assessed. Section 5.6 concludes the chapter and presents a few final remarks.

5.2   Overview of statistical sources

5.2.1   Italy

Three sources of data are currently available for determining the stock of foreigners living in Italy: residence permits, the population census, and the survey on municipal population registers. These sources all have different objectives and approaches to data collection.

The *residence permit* is the document allowing non-EU foreigners[1] aged 18 and over to reside legally in the country. It is granted by the 'Questura', a provincial body of the Ministry of the Interior. The files issued by the Ministry of the Interior are provided to the National Statistical Institute (Istat), which then revises them before publishing a series on the stock of foreign nationals residing legally in the country at the end of each year. Minors dependent on their parents are not

---

[1] In compliance with the Schengen Treaty, EU nationals do not need a residence permit to stay in the Italian territory. However, they may be required to obtain a residence permit if they want to register with the municipality or work for an Italian employer.

recorded by this source because they are not granted their own permit. For the adult population, this source provides the most reliable estimates of the stock of legal foreigners living in Italy. However, the evolution over time of statistics on residence permits is significantly affected by the implementation of regularisation schemes, as we shall see in more detail in the next section.

The survey on *municipal population registers* (*Anagrafi*) has been carried out by Istat since the beginning of the 1990s. This source is especially useful because minors are also included. As far as the adult population is concerned, foreigners recorded in the population register can be *roughly* considered as a subset of documented foreigners because the possession of a residence permit is a precondition for the registration in the municipality. A necessary condition to apply to the registries is a demonstration of continuous residence in the municipality.[2] Actually, especially in large municipalities, population registers are not updated with sufficient regularity. For this reason, it is possible to find foreign citizens who do not qualify as 'residents', either because they have not lived continuously in the same municipality or because they do not have a valid permit any more. Missed deregistrations are one of the main limitations of this source, leading to a possible overestimation of the number of foreigners, whose residential mobility is generally higher than that of natives. Another problem which affects the registration of the native population is that the legal residence in a municipality is related to electoral or taxation purposes and, more generally, linked to all rights and duties attached to this particular place. Therefore, voluntary misreporting of the place of residence is fairly common in Italy.

The *population census* collects information on a large number of characteristics of the foreign population every 10 years, such as citizenship, citizenship at birth and birthplace. The combination of these elements makes it possible to distinguish between the foreign and the migrant population,[3] whereas administrative sources only refer to the legal foreign population. The Italian census also includes foreigners according to their usual residence and their residence at the time of the census. Therefore, the census theoretically provides data on both legal and undocumented foreigners. In 2001, two different forms for the 'usual' and 'non-usual' residents were translated into six foreign languages in order to cope with an increasingly complex migration situation (Istat 2005a). The main limitations of this source, besides its periodical nature, are the well known problems of undercoverage. These are essentially of three kinds. First, and especially with regard to undocumented migrants, there is an issue of distrust towards census operations, to which control purposes are typically ascribed. Second, there may be language issues for the foreign residents, particularly recent arrivals. Third, there are many foreign workers living in one municipality and working in another. Problems can also derive from

---

[2] It is not clearly specified in the current legislation whether all residence permit holders are allowed to apply or if there is a minimum length of stay. According to some interpretations, permits whose duration is less than one year do not fulfil the 'continuous' aspect (Istat 2005b).

[3] Not all immigrants are foreign citizens (some are Italian by naturalisation) and not all foreigners are immigrants (some are born in Italy). For instance, the 2001 census counted 1335 thousand foreigners and 1447 thousand migrants (Istat 2005a).

the misinterpretation of the distinction between usual and non-usual residence. Given that the inclusion among the usual resident population depends on the self-declaration according to several requirements, a certain number of foreigners in 2001 declared themselves as residents even without holding a residence permit. This resulted in long and complex operations to check and correct these situations (Istat 2005a).

One of the main objectives of the Italian census, besides the enumeration of the legally resident population in each municipality, is to provide information on usual residence for the updating of population registers in municipalities. This operation should result in the highlighting of 'irregular' positions left erroneously in the municipalities' registries, which typically refer to foreign citizens who have left the country without communicating it to the municipality. At the same time, the municipal registers' data represent a reference benchmark for the census useful to cope with undercoverage problems. For every discrepancy municipalities have to verify which source must be corrected. These interventions are of course more complicated in large municipalities.

5.2.2   Spain

There are three main data sources in Spain which have helped to describe the evolution of the foreign population since the beginning of the 1990s: the register of foreigners legally resident in Spain obtained from residence permits, the population register (*Padròn Municipal*), and the 1991 and 2001 censuses.

In Spain, two different types of residence permits are issued: one for EU nationals and another for non-EU nationals. Statistics on the number of legal foreign residents are produced by the Ministry of the Interior. By definition, this source does not record unauthorised residents. Similar to Italy, the number of EU nationals is probably underestimated because, for this group, the need for a residence permit arises only because of work or other social needs. Unlike Italy, students are not included in the residence permits' database because they are granted a specific permit (*autorización de estancia por estudios*).

The population register records all inhabitants (usually) living in the Spanish municipalities. Since 1996, as a result of improved techniques for storage and accessibility of data, the local register legislation was modified to make it compulsory for them to computerise their registers. At the same time, the Spanish National Statistical Institute (INE) was given the responsibility of collecting, organising and assessing the data.[4] Unlike in Italy, registration is compulsory, regardless of the residence status in Spain. Thus, everybody who resides in the country is obliged to register with the municipality in which he or she (usually) resides, including illegal migrants and asylum seekers. In addition, migrants have a strong incentive to register because registration allows them to take advantage of welfare benefits (e.g.

---

[4] Monthly interchanges of information are passed between the municipalities and the INE. INE informs the municipalities of any discrepancies or double counts detected.

health and education). To register, foreigners must present a valid document (e.g. passport, residence permit, or birth certificate) and provide proof of their residence in the municipality. There is no time constraint for registering both immigration and emigration. For these reasons, this source should enable the whole foreign population living in Spain to be captured and estimates of the number of undocumented immigrants to be obtained by difference with the stock of legal residents. However, foreigners recorded in the population registers are likely to be overenumerated for a number of reasons related to the administrative nature of this source.[5] First, there is the issue of people who have left the country but did not notify the authorities. The second problem is related to the variety of documents that can be used to register with a municipality. [6] Finally, since the population register is used as the basis for allocating resources, municipalities may have an interest in swelling their populations to raise more funds (Villán Criado 2002; Jurado 2004).

A major characteristic of the 2001 census was that the information available in the population register was used as a reference for assessing the accuracy of the enumeration. Names and demographic data from the population registers were preprinted in the questionnaires, so that a partly completed form was delivered to the household. The census results were then used to clean and maintain the population register by adding persons found and by deleting those no longer living in the municipality. This operation of record linkage was particularly important for identifying foreign nationals, as a significant number of undocumented migrants were captured by the census. However, this ambitious operation was not exempt from difficulties. The problems of undercoverage in the census and overenumeration in the population register could not be completely solved (Villán Criado 2002).

5.2.3   Portugal

Similar to Italy and Spain, information on the dimension and features of the foreign population in Portugal can be obtained from both censuses and administrative registers. For regular foreigners, two different legal statuses are currently allowed by the authority in charge (i.e. *Serviço de Estrangeiros e Fronteiras*): the residence permit and the 'permanence' permit. The latter administrative status was introduced in January 2001, with documentation of a work contract being the main criterion. In practice, this new law prompted a policy of direct labour recruitment, with the new work permits representing a sort of labour visa obtained at destination (Peixoto 2002). The permanence permit was awarded for one year, with the possibility of renewal for another four years.[7] Throughout this process, it was important for the migrant to demonstrate the possession of a continuous work contract, which is also

---

[5] A (presumably small) number of migrants may not register because they do not need to do so (e.g. EU nationals), they are unaware of the process or they fear that enforcement authorities have access to the database. Of course, some applications may be rejected.

[6] Other problems include misspelling or poor transcription of names from different languages and incompleteness of the date of birth (frequent for Moroccans). These factors may cause problems of double counting. Also migrants may deliberately attempt to register with more than one municipality in order to present more than one application (Jurado 2004).

necessary for the application to be a permanent resident. The law was repealed in February 2003, replacing it with a system that encouraged the recruitment of foreigners living abroad. From a statistical point of view, the main problem with this source of data is that the *Serviço de Estrangeiros e Fronteiras* disseminates only flow statistics on the number of permanence permits issued in a given year and, more recently, on the number of renewals. No data on the stock of valid permanence permits are processed.

Statistics on foreigners recorded by censuses suffer basically from the usual problems of under-registration of persons not entitled of the right to reside in the country. However, it is worth mentioning that the Portuguese Statistical Institute (INE) adopted a restrictive definition for the status of resident, including only foreigners who had been living in Portugal for at least one year (Rosa *et al.* 2004).

5.2.4　Greece

Migration statistics in Greece are particularly poor. Data comparability is often hindered by discrepancies in the definitions. A major issue concerns Greek nationals with foreign citizenship (*homogeneis*, or ethnic Greeks). These migrants are subject to favourable regulation. A joint Ministerial Decision (Defence and Interior) in 1990 allowed them to remain in Greece without documentation. Other provisions defined the concept of 'repatriated Greeks' and established a rapid process for the granting of Greek citizenship (Kanellopoulos *et al.* 2005). As a consequence, ethnic Greeks are not included in the residence permits database and are indistinguishable from the native population when citizenship is used as a criterion (Baldwin-Edwards 2004b).

Throughout the 1990s *residence permits* were unsuitable for constructing the evolution of the foreign population in Greece. The main reason for this is that, when policy-makers became aware that a large number of migrants were flocking in from the neighbouring countries, they enacted a very restrictive immigration law, which made little provision for legal entries and stays. Thus, until the first regularisation programme (1998) documented migrants were only a small minority and, even after the regularisation, many foreigners quickly fell back into illegality because most of their granted authorisations had short durations or were affected by bureaucratic inefficiencies (Section 5.4). At the beginning of the 2000s, the Ministry of the Interior became responsible (for the first time) for immigration policy and the issuing of work and residence permits. When a second regularisation scheme was carried out in 2001, serious problems arose in the processing of the applications, mainly because of the heavy bureaucratic procedures required for legalisation in combination with the lack of experienced and well trained personnel. The Ministry was not able to provide data on residence permits until mid-2004. Moreover, because of the delays in entering the data into the main database (until recently), sporadically published data differed remarkably according to the date of

---

[7] The possession of a visa potentially can allow one to stay in Portugal for a long time due to extensions. In 2000, nearly 16 thousand visa extensions were issued (Rosa *et al.* 2004).

reference. As a result, the Greek Statistical Office started to produce the official statistics on the legal foreign population by combining data issued by the Ministry of the Interior and by the Ministry of Public Order (for EU citizens). Apart from the substantial undercoverage of EU nationals (Cypriots in particular) who may not need a residence permit (e.g. if they do not work), these data appear to provide a plausible assessment of the number of legal foreign residents in the country.[8]

Since reliable statistics on residence permits were not available during the 1990s, the only other option is to use information from the censuses and the 1998 regularisation. The 1991 census undercounted the foreign population, as only EU residents and those with legal status were included (Duquenne and Kaklamani 2004). The 2001 census, however, made an explicit attempt to capture a record of all immigrants (legal or undocumented) and appears to have been a fairly reliable measure of foreign population at that time, with less than 100 thousand people not counted (Baldwin-Edwards 2004b; Kanellopoulos *et al.* 2005). The main problem with the 2001 census was that migrants with a Greek ethnic background (*homogeneis*) were recorded as Greek nationals, with the implication that 150 thousand to 200 thousand migrants were not captured by the statistics on the foreign population.

5.3   The evolution of foreign populations

The research presented in this section strives to provide a comprehensive picture of the foreign populations in Southern European countries beginning in the early 1990s by integrating the available information obtained from different data sources of varying qualities.[9] The most recent trends are of particular interest.

5.3.1   Italy

In Italy, immigration from non-EU countries became important during the late 1980s. Following the first two regularisation programmes carried out in 1986 and 1990, the number of valid permits in 1991 was already above 600 thousand (Figure 5.1). A significant number was recorded in the municipal registers as well. However, only 356 thousand foreign residents were found by the 1991 census.

The evolution of the legal foreign presence in Italy has been largely affected by the chronology of the regularisation programmes. In the 1990s, the number of residence permits grew slowly until the third amnesty, carried out at the end of 1995. More than 250 thousand applications were submitted, giving rise to a sudden increase in the number of legal foreigners. A similar situation occurred three years later with over 253 thousand applications submitted for the fourth regularisation

---

[8] Efforts to harmonise the national legal framework with EU standards led to a new law concerning the entry and stay of third-country nationals which came into force in 2006. For the residence permit system, the main changes were the introduction of a single stay and work permit of two-year duration depending on local labour market supply and demand conditions, renewable for another two years.

[9] The analysis presented in this section is drawn from a paper presented at the EAPS–IRPPS Conference 'International Migration in Europe: New Trends, New Methods of Analysis' (Cangiano and Strozza 2004).

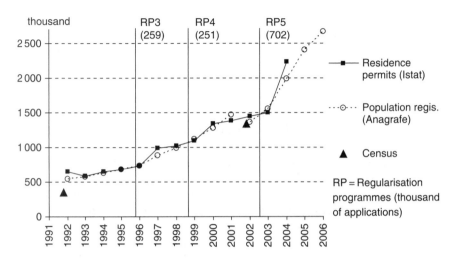

Figure 5.1 Foreign population in Italy (in thousands) according to different sources, 1 January 1991–2006.

programme. This trend is also reflected in the records of the municipal register and by the 2001 census.

The real turning point in the evolution of the legal foreign population in Italy took place with the most recent regularisation programme, following the new immigration law enacted in 2002 (the so-called *Bossi–Fini*). More than 700 thousand applications were submitted and more than 650 thousand new permits were issued, implying a 50 % increase in the legal foreign resident population. On 1 January 2004, the number of valid residence permits was around 2.2 million.

The results of the 2002 regularisation confirmed that only few migrants enter Italy holding a residence permit. A picture of the prevailing role of undocumented migrations can be obtained by comparing the cumulative number of new permits issued in the 1995, 1998 and 2002 regularisation programmes (i.e. 1.1 million) with the corresponding increase in the valid residence permits issued between 1991 and 2004 (i.e. 1.5 million). This comparison is, of course, only indicative as some may have applied for more than one regularisation scheme and no figures are available for migrants who have left Italy.

The stock of foreign residents registered with local municipalities reflects the evolution of the number of permits to stay. It is useful to recall that this source also records minors, who are not included in the residence permit statistics. Note that this source does not cover those who do not register with the municipality, as registration is not compulsory. Many immigrants tend to register only after one or more years of continuous legal residence. Also, over-enumeration is a problem in large municipalities (e.g. Rome) because of poor maintenance of registers (e.g. non-deletion of people who leave). All of these factors explain the smoother evolution in comparison to the residence permits (see Figure 5.1). The decrease observed in

2001 is mainly due to the revision of the municipal records, which followed the census operations (see Section 5.2.1).[10]

According to recent figures issued by Istat the number of foreigners recorded in the population register reached 2.7 million on 1 January 2006, with minors representing a significant share of the foreign population (i.e. 585 thousand were less than 18 years old) and contributing substantially to its growth. After the substantial increase that occurred in response to the 2002 regularisation, the foreign population is continuing to grow due to relatively high numbers of births (over 50 thousand in 2005) and because of more generous quotas for legal entry. Combining information on residence permit holders with those on minors registered in the *anagrafe*, it is clear that the total number of foreigners with legal status is over 2.8 million persons.[11]

5.3.2   Spain

The number of legal foreigners in Spain started to increase in the mid-1980s when the first regularisation programme took place. After the second amnesty in 1991, with more than 110 thousand immigrants applying for legal status, about 350 thousand legal foreign residents were recorded by the Ministry of the Interior (see Figure 5.2). A similar number was found by the census. During the 1990s, a fairly constant growth of residence permits was observed. In fact, a separate quota system for foreigners already working (illegally) in the country served basically as a *de facto* permanent regularisation from 1994 to 1999, excluding 1996 (Reyneri 2003). Despite the substantial success in reducing the illegal presence of migrants in this period, the number of migrants not holding a valid residence authorisation increased again by the end of the decade. Thus, in January 2000, when a new immigration act was passed providing for another regularisation scheme, nearly 250 thousand people applied. During the fifth amnesty in 2001, an even larger number of applications (351 thousand) were sent in. For this scheme, employment and strong social ties with Spain had to be proved. As a result of these campaigns, the stock of legal foreign residents grew very quickly, reaching 2.0 million at the beginning of 2005, with about 1.2 million persons added in the last five-year period. The 2001 census confirmed this trend with more than 1.5 million foreign nationals counted, of whom 300 thousand to 400 thousand were probably undocumented.

The time series of data obtained from the *Padròn Municipal* exhibited a growth in the overall foreign population in the early 2000s, which was more pronounced than the growth of legal foreigners. Despite the considerable participation in the 2000 and 2001 regularisation programmes, large and increasingly complex unauthorised migration flows gave rise to the formation of a new pool of undocumented foreign

---

[10] Not all municipalities have revised the list of residents after the census registration. When this operation was not carried out, aggregate estimates of the actual number of foreign residents were obtained by updating the census results with the net demographic balance and acquisition of citizenship.

[11] The latest data available from the two sources record 2238 631 foreigners aged 18 years and older holding a valid permit to stay at the beginning of 2005 (Istat provisional data) and 585 496 foreign minors registered in the *anagrafe* on 1 January 2006.

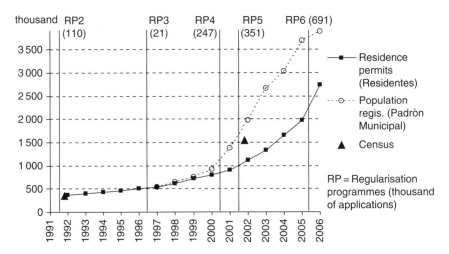

Figure 5.2   Foreign population in Spain (in thousands) according to different sources, 1 January 1991–2006.

residents. The difference between the number of foreigners recorded in the population registers and the stock of residence permits amounted to 1.7 million on 1 January 2005. Excluding minors and the citizens of other industrialised countries, the presence of 1.2 million undocumented migrants was assumed (Arango and Jachimowicz 2005). Hence, to reduce this amount to a more manageable one, a new regularisation programme took place between 7 February and 7 May 2005. This scheme was intended only for workers who had been residing in Spain since August 2004 or before and who had an employer willing to provide them with a work contract for at least six months. More than 690 thousand applications were gathered, almost equalling the Italian regularisation of 2002. Note that this programme left out a significant section of the undocumented residents, i.e. children, spouses, students, the self-employed and other ineligible workers. As a result of this new amnesty, the stock of legal foreigners underwent a further increase in size, reaching 2.7 million on 1 January 2006. According to the most recently available figures, the difference between the number of foreigners recorded in the population registers and that of residence permit holders has decreased but is still high at 1.1 million persons.

   In summary, the size of the immigrant population living in Spain has increased rapidly over a very short time period. The increase has been more dramatic than the corresponding trends in Italy. In the mid-1990s, around a half a million foreigners legally resided in Spain. By the mid-2000s, there were five times as many.

5.3.3   Portugal

In the early 1990s, the size of the foreign population in Portugal was around 110 thousand to 120 thousand persons, according to information obtained from the

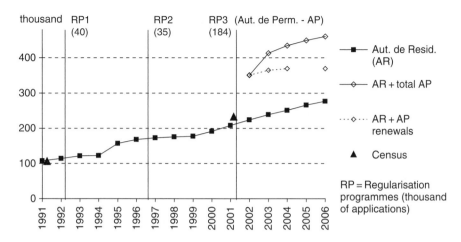

Figure 5.3 Foreign population in Portugal (in thousands) according to different sources, 1 January 1991–2006.

1991 census and from the number of residence permits (Figure 5.3). With the exception of some 40 thousand other European nationals, the stock of foreigners living in Portugal at that time could mostly be traced to the immigration flows that followed the process of decolonisation in Africa. In the mid-1970s, a number of Africans, especially from Cape Verde and Angola, moved to Portugal and established important migratory chains (Malheiros 2002).

After joining the European Union in 1986, Portugal became more attractive for non-EU migrants both as a destination and as a transit country. The foreign population in Portugal from overseas became larger and more varied as a result of the growing inflows from Africa and, to a lesser extent, from Brazil. The Portuguese government soon realised that a significant portion of migration was occurring through illegal channels and subsequently implemented its first regularisation programme in 1992. Nearly 40 thousand foreigners legalised their status under the programme, 72 % of whom originated in the PALOP states, i.e. African countries where Portuguese was spoken (Rosa *et al.* 2004). A second regularisation occurred in 1996 with 35 thousand applicants (61 % from the PALOP countries). Given the considerable bureaucratic delays which affected both procedures, their impact on the size of the legal foreign population is observable only some years later (Figure 5.3). The 2001 census confirmed the significant increase of immigrants by counting 233 thousand foreign nationals, a number which in all probability underestimated the actual number of foreigners.[12]

An important point in the evolution of the recorded immigrant population in Portugal occurred during the 2001–2003 programme, which granted a temporary

---

[12] This is likely to be the case for Cape Verdeans, most EU nationals (except French), and many Eastern Europeans.

legal status (i.e. permanence permit) to the illegal foreigners already working in the country. The possibility for migrant workers to regularise their status whenever engaged by an employer allowed them to fulfil specific needs of certain economic sectors in local labour markets (Peixoto 2002). Almost 184 thousand permits were issued during this period,[13] the majority of them to Ukrainians (64 300) and Brazilians (37 700).

Since statistics on the permanence permits record only those issued (or renewed) in a calendar year, and not the stock of permits valid at a given date, it is not possible to determine the total number of foreigners legally residing in the country after 2001. A rough estimate of the foreign population can be obtained by adding to the population of valid residence permits ('AR'), either the total number of permanence permits issued during the regularisation ('AR + total AP' in Figure 5.3), or the number of permanence permits renewed in a given year ('AR + AP renewals' in Figure 5.3). The former operation provides a maximum estimate since we are assuming that all people granted a permanence permit were still legally residing in the country in a given year. The latter is a minimum estimate because only people who renewed their permanence permit in that year were counted.[14] Both estimates are to some extent questionable: the former because some people who regularised through the permanence permit scheme might have left the country or fallen back into a status of irregularity; the latter because of administrative delays.[15] Given this premise, we can realistically assume that the number of legal foreign residents in Portugal on 1 January 2006 ranged between 370 thousand and 460 thousand persons (Figure 5.3).

5.3.4 Greece

Immigration to Greece increased substantially after the collapse of the communist Albanian government. The level of immigration from 1991 onwards is largely unknown. The 1991 census registered 167 thousand foreigners (Figure 5.4) in a total population of 10.3 million, with the foreign population representing only EU nationals and those with legal status. The number of foreigners with a valid residence permit in the early 1990s amounted to 250 thousand, with 60 thousand ethnic Greeks and 80 thousand EU or North American citizens. This means that the 'real' authorised migrants numbered only around 100 thousand. Available figures for the middle of the 1990s show a decrease of the legal residents (Figure 5.4), confirming the unreliability of this source in measuring the foreign population that was in fact rapidly increasing during this time.[16]

---

[13] A government decree in April 2004 included an article that opened the possibility of a further regularisation of non-EU foreign workers who could prove they were present in the Portuguese labour market before March 2003. There were around 40 thousand applications but only about three thousand foreigners had received work permits by the spring of 2005.

[14] AP renewal can be claimed within two months from the expiry date. Moreover, delays in processing the applications are frequent, meaning that persons may wait several months before their permit is renewed.

[15] The maximum estimate is most likely to be more reliable at the time the regularisation occurred or soon after. The minimum estimate is likely to reflect the true numbers of legal foreigners in the country during the most recent years.

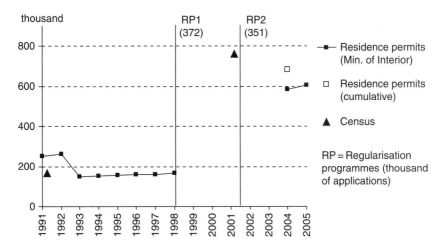

Figure 5.4 Foreign population in Greece (in thousands) according to different sources, 1 January 1991–2005.

By the mid-1990s it became clear that a large number of immigrants were residing illegally in Greece, despite the mass deportations. Thus, in 1998, the Greek government adopted its first regularisation programme. Over 370 thousand non-EU citizens (65 % of whom were Albanian) applied for and received the 'White Card' that entitled them to reside legally in Greece for some months. Applications were processed very slowly, so that by February 2000 only 107 thousand cards had been awarded (Fakiolas 2003). Ultimately over 90 % of applications were approved, but most of them were issued late and for one year only. To get a work permit, i.e. 'Green Card', immigrants had to make a second application, including proof of employment during their temporary stay. Because of the stricter requirements, only 220 thousand applied, with an overall renewal acceptance rate of 54 % (Baldwin-Edwards 2004a). The 150 thousand migrants who received a White Card but failed to obtain a Green Card fell quickly back into an irregular status (Reyneri 2003). There were also an estimated 150 thousand migrants who did not even apply for the White Card for various reasons, such as recent entry or fear of being identified by authorities (Cavounidis 2002; Fakiolas 2003). After the regularisation, no statistics are available on the number of valid permits issued by the Greek Manpower Employment Organisation (OAED), the authority in charge of the process. It is probable that, in addition to the unauthorised immigrants who applied for legal status, there were approximately another 200 thousand to 250 thousand illegal residents who did not submit an application (Cavounidis 2002).

---

[16] The immigration law of 1991 made it possible to deport immigrants found without legal authorisation to stay. Around 2.2 million persons were expelled to the neighbouring Balkan countries between 1992 and 2001, 85–90 % of whom were Albanians. No other European country has carried out such a large numbers of expulsions. The legality of such treatments has been questioned by the local Ombudsman's Office (Baldwin-Edwards 2004a).

The total foreign population at the time of the 1998 regularisation was likely to be around 600 thousand persons or more.

The first reliable information on the actual size of the foreign population in Greece was provided by the 2001 census. Thanks to advertising campaigns and public assurances of the confidentiality of the data, the national statistical office largely succeeded in counting the foreign population (legal and undocumented). At the time of the census, 762 thousand persons with non-Greek citizenship were enumerated. Important information was gained, for example, from the self-declared length of stay; nearly half of the foreigners declared residence in Greece for at least five years. However, a recent estimate suggests that around 100 thousand foreign immigrants (probably those residing in the country for a shorter period) did not participate in the census and should be added to the enumerated foreign population (Baldwin-Edwards 2004b). In addition, according to a special census carried out in 2000, over 200 thousand ethnic Greeks with foreign nationality were living in the country by that time, most of whom were recorded as Greek nationals in the 2001 general census. Immigrants of Greek descent with foreign nationality come mainly from the former USSR and Albania. The former are 'Pontians' whose ancestors migrated in the 19th and early 20th centuries from the Pontos region to the Caucasus and to the northern coast of the Black Sea. During Stalin's dictatorship, they were dispersed to various republics in the former Soviet Union. Some 150 thousand descendants of these Greek emigrants had repatriated, mainly from Georgia (79 thousand), Kazakhstan (31 thousand) and Russia (23 thousand). Ethnic Greeks with Albanian nationality come from the south of Albania, the region known to Greeks as 'Northern Epirus' – and were estimated to be 100 thousand individuals (Cavounidis 2002).

In 2001, the need for an actual immigration policy, rather than the exclusionary approach of the 1991 law, led to the enactment of a new immigration law with a last-minute amendment for a new regularisation procedure (i.e. 'Green Card II'). The 2001 regularisation was similar to the 1998 one with the main difference being that the work permit became a prerequisite for obtaining a residence permit. Over 350 thousand immigrants applied to OAED, but only 220 thousand met the eligibility criteria. Once again, no data on the number of residence permits issued were made available.

In 2004, the quality of migration statistics and their conformity with EU standards was analysed in a study using the Greek Ministry of the Interior's data on permits to stay (Baldwin-Edwards 2004b). Results from this study clearly demonstrated the unsuitability of these data to describe the evolution of the foreign population in Greece. A more reliable measure of legal foreign presence was argued to be the cumulative number of permits to stay issued in 2003 and early 2004 (683 thousand and 483 thousand, respectively). This figure, if compared with the 592 thousand adults counted in the 2001 census, would indicate a further increase of the immigrant population in the period following the census. Adding schoolchildren, EU nationals, and the estimated illegal population to this number results in 950 thousand foreign persons in the population (or about 8.4% of the total population). This number, albeit a rough estimate, excludes ethnic Greeks with *homogeneis* cards.

Recently, the Greek statistical office has started to make available official statistics on the legal foreign population based on the permits to stay issued by the Ministry of the Interior and by the Ministry of Public Order (for EU citizens). Apart from the substantial undercoverage of EU nationals and the obvious non-registration of undocumented migrants, these data seem to be consistent with the census figures, recording 604 thousand legal foreigners at 1 January 2005.

Finally, it is useful to stress some peculiarities of the Greek migration system in comparison to the other Southern European countries. First, the increase of the immigrant population in the last decade or so continued despite the large number of deportations. Second, Albanians represent about 60 % of the total foreign population. No similar level of dominance exists in the other three countries of Southern Europe. Greece's close proximity to Albania and other Central and Eastern European countries (e.g. Bulgaria and Romania) is an important factor in the Greek migration scenario (Cavounidis 2002). Third, Greece is the most sought-after country by refugees in Southern Europe, especially by Iraqis and Afghans arriving from Turkey, many of whom have intentions either to work in Greece or to transfer to other EU countries. Although not entitled to the right to work, many of these persons overstay and create networks for other potential migrants.

5.4 Regularisation programmes

By its nature, it is very difficult to provide accurate estimates of the illegal foreign presence in any country. The diversity of undocumented migration, which includes various categories of persons and movements, such as illegal entries, overstayers and rejected asylum seekers, makes the task even more difficult. As in most countries receiving immigrants, no direct figures exist on the size of illegal residents in Southern Europe. However, some information is available that can be used to provide rough estimates on the levels and trends over time using regularisation data. In this section, a brief history of the regularisation programmes carried out in Southern European countries is presented.

Applications for a regularisation programme provide useful information to evaluate the level and characteristics of immigrants residing illegally before the amnesty. However, data on applications do not capture the full extent of the illegal presence. Three main problems undermine the counting of the total irregular foreign population. First, and most obvious, not all undocumented foreigners participate in the regularisation. The eligibility criteria for the amnesty necessarily limit the involvement of some illegal residents. Also, a number of migrants may choose not to participate if, for example, they fear enforcement authorities' access to the database or because no employer is available to confirm their working condition or pay for their social security contributions. Second, there is the problem of persons who register simply for the opportunity to obtain legal status, regardless of their intentions to reside or work there. Finally, some migrants with residence permits may apply to obtain a different kind of residence permit. This is generally the case for immigrants who reside legally but are not entitled to the right to work. Despite all these issues, data obtained from the periodical regularisations carried out in Southern Europe can be used to point out at least some basic trends in the stock of irregular migrants.

In Spain, the first regularisation programme was launched in July 1985 and involved 44 thousand immigrants. At the end of 1986, Italy adopted its first programme to legalise the residence and work status of 119 thousand non-EU foreigners. Since then, four other programmes have been carried out in both countries. In order to clarify how regularisation data can be used retrospectively to measure the irregular foreign population, it is useful to point out some general characteristics of these programmes. First, some proof of residence in the country for a period of time prior to a given date is always required. In some cases, this implies proof of a relatively long and continuous period of stay (e.g. it was two years in the 2000 Spanish regularisation). Employment status is another important criterion used to define migrants' eligibility, particularly in Italy, where employed people were often the only category eligible to participate in the programmes (meaning that dependants and asylum seekers were ineligible). This was also the situation in the Spanish 1991 and 2005 regularisations. Finally, all programmes were announced as one-off procedures. Although this supposed exceptionality has been clearly belied by the periodicity with which the procedures were launched, usually every three or four years, this may in fact have encouraged future potential migrants to arrive from neighbouring countries.

As a general trend, the number of irregular migrants involved in the regularisation programmes has been increasing over time in both Italy and Spain (Table 5.1). In Italy, around 250 thousand people applied for each of the three regularisations carried out in the 1990s, and over 700 thousand irregular migrants were involved in the 2002 regularisation, which is the largest scheme ever to have taken place in Europe. In Spain, if the 1996 programme that was open only to participants in the 1991 regularisation with an expired residence permit is excluded, the upward trend is even more clear-cut: 110 thousand applications in 1991, 247 thousand in 2000, 351 thousand in 2001, and 691 thousand in 2005. These figures clearly show that the number of foreigners residing irregularly in the two countries has increased during the past two decades, despite the periodical reductions from the regularisations. However, this absolute increase in the irregular foreign population does not necessarily imply a larger incidence over time. In the most recent Italian and Spanish regularisations, the ratio between the number of applications and the size of the regular foreign population before the programme, referred as the 'regularisation ratio' (Table 5.1), was not significantly higher than in the past.[17] In the 2002 Italian regularisation 49 applications per 100 valid residence permits were recorded, a ratio similar to the 1990 programme, which had adopted less restrictive requirements. In contrast, a smaller incidence of the irregular component was found when the 1995 and 1998 regularisations took place. In Spain the regularisation

---

[17] This indicator should be interpreted with some caution, as the legal requirements to participate in the different programmes varied.

Table 5.1 Regularisations of foreigners in Southern Europe, 1985–2006.

Country	Year	Applications (in thousands)	Regularisation ratio[a]	Categories of applicants
Italy	1986	119	31	Employees
	1990	235	51	All
	1995	259	34	Employees and dependants
	1998	251	25	All workers
	2002	702	49	Employees
Spain	1985–86	44	18	All
	1991	110	33	All workers
	1996	21	4	Applicants for 1991 regular, with expired permit
	2000	247	28	All
	2001	351	39	All
	2005	691	33	Employees
Greece	1997–98	372[b]	...[d]	All workers
	2001	351	...	All workers
	2006	180[c]	...	All workers
Portugal	1992–93	40	35	All
	1996	35	21	All
	2001–03	184	88	Employees

[a] Applications for regularisation per 100 legal foreign residents before the programme.
[b] Number of white cards.
[c] Provisional data.
[d] ... Not available.
Source: Update on Cangiano and Strozza (2006: 15).

ratio was highest in 2001 (39 applications per 100 regular foreign residents), and decreased in 2005 despite near doubling in the number of applicants.

The adoption of regularisation programmes is a more recent practice in Greece and Portugal. The first legalisation in Portugal took place during 1992 to 1993, with around 40 thousand people applying for a residence permit. The second scheme was implemented only three years later and received a similar number of applications. Both programmes were open to all foreign residents regardless of their employment status; however, in 1996 there was a less restrictive criterion used to accommodate migrants from former Portuguese colonies (Papadopoulou 2005). Despite the relatively unrestrictive requirements, the ratio of applications to the size of the regular foreign population before the programmes was not very high (particularly in 1996), suggesting that illegality was widespread but not a prevalent condition for migrants living in Portugal in the early 1990s. In contrast, the 2001–2003 regularisation programme had a much greater impact on the size of the legal foreign population. This was mostly due to the new legal condition for 'temporary' workers that was created (i.e. the 'permanence permit', see Section 5.2.3) and because of the large time period the programme remained open. This meant that migrants who entered the country after the beginning of the procedure could also regularise their status. Overall, the number of irregular immigrants who applied for the programme almost equalled the size of the regular foreign population before the amnesty, i.e. 88 applications were presented for every 100 valid residence permits.[18]

Only foreign workers were allowed to participate in the two regularisations carried out in Greece during 1997 to 1998 and in 2001. These two programmes were more restrictive than the schemes implemented in the other Southern European countries. In fact, a two-stage procedure was adopted, with temporary authorisations issued in the first stage ('White Cards') and long-term residence permits ('Green Cards') granted only if strict requirements were fulfilled, including the payment of social security contributions and residence in the country for at least one year (Fakiolas 2003). Between 1997 and 1998, 372 thousand irregular migrants obtained a White Card, but over one-third of them were excluded from the second stage leading to the issue of a residence permit. In 2001 a similar number of applications (351 thousand) were presented, but no analytical data on the outcomes of the programme were ever disseminated. The lack of data on the stock of valid residence permits, which is related to the instability of the legal residence in this country, prevents one from constructing regularisation ratios. The exclusionary approach to legality throughout the 1990s resulted in the majority of non-EU foreigners residing and working without legal status during this time (Baldwin-Edwards 2002). More recently, in April 2006, a new and more selective programme to legalise migrant workers was put forward. Provisional figures indicate that the number of applications was significantly lower in comparison to the previous programmes (i.e. 180 thousand). Note that this procedure has been criticised by the Greek Ombudsman for the particularly strict requirements that applicants had to fulfil, which excluded a high number of irregular residents.

5.5   Estimates of irregular migrants in Italy, Spain and Greece

Given the importance of irregular migration in Southern Europe, several studies have tried to estimate the number of irregular foreigners living in these countries using various approaches, statistical techniques and assumptions. Most estimates referred to a specific point in time because the limited information made it difficult to propose reliable estimates for longer periods (Natale and Strozza 1997). The

---

[18] The 'continuous' nature of the last Portuguese regularisation makes comparisons with other programmes problematic, since data on the number of applications include both the stock of irregular foreign workers living in Portugal before the amnesty and a flow of new labour immigrants who entered the country between 2001 and 2003.

methods most commonly used were based on regularisation outcomes and typically referred to the date that foreigners had to prove residence in the country in order to participate in the amnesty. The assumptions focused mostly on the share of irregular migrants excluded from the amnesty and on the year of arrival in the country (Schoorl *et al.* 1996; Natale and Strozza 1997). Hypotheses regarding the development of the undocumented population after the regularisation were also put forward, including some estimates for periods following the amnesty (Strozza 2004). Other estimates are based on illegality ratios determined from sample surveys and using specific questions regarding the entry or residence status of the person being interviewed. In general, the results from surveys are likely to under-represent illegal foreigners, as questions related to the residence status are more likely to be either left blank or falsely answered, especially among those who do not comply with the current regulations. Some approaches rely on the comparison of different sources, some of which might capture a variable part of the irregular foreign population (e.g. censuses and population registers; see Section 5.5.2 on Spain). Finally, other more complex methods are used to estimate indirectly the stock or flow of irregular migrants using proxy information provided by judicial statistics (e.g. rejections at national borders, expulsions and detentions). The minimum values are usually those deriving from the comparison and juxtaposition of the sources, while the maximum ones are determined by indirect estimates and, in some cases, by the application of revised legalisation ratios which consider the share of illegal foreigners who do not benefit from the amnesties.

5.5.1 Italy

Blangiardo and Molina (2006) provide estimates of the illegal foreign presence in Italy using illegality ratios resulting from the ISMU survey conducted in Lombardy.[19] The estimates exhibit a 'roller-coaster' trend reflecting the chronology of the regularisation programmes (Figure 5.5). According to the authors, this trend indicates a twofold effect of the regularisation schemes: a so-called 'recall effect', that is, the increase of irregularity in the period immediately preceding the amnesties, and the reduction of irregularity immediately after a regularisation. These estimates suggest that in the last 15 years the lowest numbers of irregular migrants (150 thousand to 200 thousand persons) were found at the beginning of 1997 and 2000. A comprehensive review of the estimates published in the literature confirmed that this was one of the lowest levels ever experienced since the 1980s (Strozza 2004). In contrast, the maximum size of the illegal foreign population (750 thousand) would have been reached just before the 2002 programme. Since then, after the amnesty effect, the stock of irregular foreigners is estimated to have increased again very quickly, reaching 540 thousand people in mid-2005.

The Italian regularisations that occurred in the 1990s reduced the irregular presence. Studies that have examined the undocumented foreign presence in Italy from a

---

[19] The ISMU Foundation carries out an annual survey for Lombardy's regional observatory on immigration and integration, providing one of the most comprehensive sources of data on the foreign population in Italy.

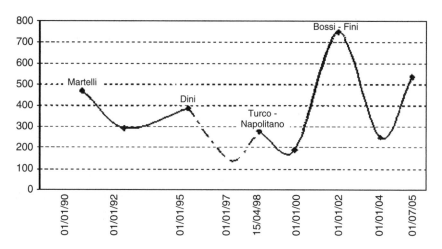

Figure 5.5   Estimates of the irregular foreign population in Italy (in thousands), 1990–2005.
Notes: The names refer to the politicians who promoted the corresponding regularisation/amnesty laws.
Source: Blangiardo and Molina (2006). Reproduced with permission from Edizioni della Fondazione Giovanni Agnelli

longitudinal perspective found that the majority of migrants (80–90 %) participating in regularisation programmes had not applied in earlier programmes (Carfagna 2003). Survey data have showed that migrants who have been granted a permit are unlikely to fall back into illegality (Blangiardo 2005).

5.5.2   Spain

During the 1990s, the number of irregular foreigners in Spain was lower than in Italy. In the late 1990s, the estimated number of undocumented immigrants was between 150 thousand and 200 thousand (Icduygu and Unalan 2001). One of the reasons for this lower level was a relatively light regulation of residence permits, which basically enabled migrants who were irregularly employed (e.g. as cleaners, personal carers or agricultural workers) to apply for a residence and work permit. The quota system issued 20 thousand to 30 thousand residence permits each year between 1994 and 1999 (Reyneri 2003). A significant number of applicants were recorded in the 2000 and 2001 regularisations. Despite high participation in these two programmes, the number of illegal residents continued to increase in the 2000s, perhaps attracted by the prospect of a new regularisation[20] or because of difficulties in renewing residence permits issued in earlier amnesties (Pérez 2003).

---

[20] A number of immigrants were caught entering the country with photocopied Spanish registration forms. The suspicion arose that family members or other immigrants already settled in Spain could have registered foreigners still living abroad while awaiting the regularisation programme (Arango and Jachimowicz 2005).

For the past few years, estimates of the illegal foreign presence in Spain can be obtained by comparing census and administrative data, which allow one to estimate roughly the number of undocumented residents as a residual. In particular, the following two simple strategies are possible:

1. The population of illegal residents can be estimated on the day of the census (1 November 2001) by subtracting the number of valid residence permits from the number of foreigners recorded by the census. Illegal residents often escape the census enumeration, but the 2001 census is believed to have captured a sizeable portion of the undocumented residents (see Figure 5.2). This procedure may be used to provide a *minimum estimate* of the illegal presence.

2. Foreigners recorded in the population register can be assumed to represent the total population of foreign residents. The population of illegal residents is estimated as a residual between the *empadrònados* and the number of valid residence permits. Considering that the population register's records could outnumber the number of foreigners actually living in the country, this approach is assumed to provide a *maximum estimate* of the undocumented presence, though probably with a lower error range than the minimum estimate.

Estimates of undocumented residents and their share of the total foreign population for the main national groups are presented in Table 5.2. The breakdown by nationality allows one to identify different levels of illegality for the different groups and some issues of measurement affecting the two approaches described above. At the end of 2001, the overall number of non-EU foreigners residing illegally in Spain can be estimated to be between 400 thousand and 700 thousand individuals. Ecuadorian and Colombians, owing to recent arrivals, were by far the two groups with the greatest number of illegal residents. Moroccans exhibited much lower values, but were, by far, the most sizeable community in the 2000 regularisation.

Only three years after the 2001 regularisation, the number of undocumented foreigners had more than doubled. At the beginning of 2005, just before the launching of the latest regularisation programme, Ecuadorians represented the largest number of illegal residents (i.e. 270 thousand). However, the number of illegal Romanians increased more rapidly with nearly 200 thousand additional undocumented individuals in comparison to the level in 2001. Conversely, following their participation in the 2001 regularisation, the size of the Colombian illegal presence was lower than three years before.

Aside from the levels of illegal migrants, it is worth while also to consider the relative share of the different national groups over time. As expected, the proportion of illegal residents is much higher among the migrant groups who arrived more recently. In 2001, illegality ratios were above 60–70 % for all Latin Americans and not much lower for Eastern Europeans. Over the last few years these shares have been decreasing significantly for Ecuadorians and especially for Colombians, because of their involvement in the 2001 regularisation and because more recent

Table 5.2 Undocumented foreigners[a] (in thousands) and illegality ratios[b] (in per cent) for the main immigrant groups in Spain, as at 1 November 2001 and 1 January 2005.

	Undocumented foreigners			Illegality ratios		
	2001		2005	2001		2005
	min.	max.	max.	min.	max.	max.
All sending countries[c]	402	686	1 427	35	48	49
Morocco	10	73	116	4	24	23
Ecuador	129	175	270	60	67	55
Romania	32	42	231	56	63	74
Colombia	109	142	131	69	75	49
Argentina	27	36	96	57	64	63
Bolivia	8	10	85	70	75	88
Ukraine	12	17	38	58	65	58

[a] Minimum estimate = census data − permits to stay at 1st January 2002; maximum estimate = 'Empadrònados' − permits to stay.
[b] Proportion of illegal foreigners in the whole foreign population.
[c] European Union, North America and other industrialised countries are not considered.

immigration flows have been smaller. In contrast, the proportion of illegal residents has been rising for Romanians and especially for Bolivians, who are currently immigrating at high levels.

Finally, in order to assess the involvement of the different immigrant groups in the most recent regularisation programme, a simple procedure set out in Table 5.3 is applied to construct the regularisation rates, i.e. the ratio of regularised individuals to the number of illegal residents eligible for application (column (5)). Given that the programme was open only to migrant workers, the number of potential applicants (column (4)) has been estimated by multiplying the pool of illegal residents of working age (column (2)) by the activity rates found at 2001 census (column (3)). According to these estimates, the level of participation in the amnesty was relatively high for all immigrant groups, with the exception of Argentines. The results also confirm the (probably high) underestimation of the Moroccan illegal presence, finding the number of potential applicants to be significantly lower than the number of applications actually accepted.

5.5.3 Greece

The proportion of irregular migrants in Greece was much higher than in other Southern European countries throughout the 1990s. This stems from a number of factors primarily related to the more restrictive approach applied to regulate migration flows and to some peculiarities of the Greek system. Unlike Italy or Spain,

Table 5.3   Estimate of the regularisation rates[a] for the main immigrant groups in Spain, 2005.

	Illegal resid. (1)	Illegal resid., 17–64 (2)	Act. rate (%) (3)	Potential applicants (4) = (2) * (3)	Regularised[b] (5)	Regularisation rates (6) = (5)/(4) * 100
All sending countries[c]	1 427	1 099	79	866	632	73
Morocco	116	75	76	57	77	136
Ecuador	270	198	86	170	136	80
Romania	231	194	85	165	108	66
Colombia	131	104	78	81	55	68
Argentina	96	68	71	49	23	46
Bolivia	85	72	81	59	43	74
Ukraine	38	33	83	28	21	77

[a] Regularised foreigners per 100 illegal active foreigners.
[b] Number of applications accepted.
[c] Citizens of the European Union, North America and other industrialised countries are not considered.

where frequent regularisation schemes periodically reduced the stock of irregular foreign residents, in Greece these type of programmes (i.e. in 1998 and 2001) only temporarily affected the level of the illegality (Cavounidis 2002; Fakiolas 2003). Furthermore, most of the migrants enter the country illegally, whereas in the other Southern European countries, the majority of irregular residents are overstayers. Even those deported are often found to re-enter the country because of the high demand for flexible labour (Baldwin-Edwards 2004a). Finally, for migrants since 1990, persons claiming Greek ethnicity have been allowed to remain undocumented in Greece (Baldwin-Edwards 2004b).

The evolution of the irregular foreign population in Greece since 2002 is difficult to document. However, the reduction in the stock of illegal residents which followed the 2001 regularisation was mirrored by the rapid decline in the number of expulsions (i.e. around 45 thousand per year between 2002 and 2004 in comparison to over 200 thousand per year during the 1990s), by the significant increase in the number of residence permits, and by the rising number of foreigners included in the main social fund (550 thousand persons). Trade union representatives and officials from the Ministry of Public Order estimated the stock of irregular migrants before the last regularisation at around 400 thousand persons, but this figure was deemed to be an overestimation (Kanellopoulos *et al.* 2005). A relatively small number of applications (180 thousand) were presented for the 2006 regularisation, but no information on the actual involvement of the undocumented migrant population is available yet.

5.6   Final remarks

Over the last few years, the statistical information available to researchers for the study of international migrations in Southern Europe has significantly improved. The substantial success of the 2000 round of censuses in catching the foreign populations (Greece and Spain), the availability of new administrative sources able to quantify the total foreign population (Spain), the issuing of a new residence permits' database (Greece), and the extensive datasets on the regularisation proce- dures (Italy) all contributed to a better understanding of the migratory process in these countries. Nevertheless, some limits still remain, as none of these sources of data are exempt from problems of coverage or maintenance of the files. The differences in adopted definitions can sometimes also be an issue, e.g. the case of 'ethnic' Greeks holding foreign passports.

The measurement of foreign populations in Southern Europe is hampered by the presence of many illegal residents. The illegal population varies in size across space and time and takes different forms. It is something that all Southern European countries experience, despite frequent regularisation programmes and an increasing use of longer-term residence permits. The data collected by the regularisation programmes are useful for providing retrospective information on the irregular presence before the amnesties. However, a number of migrants do not partici- pate in the schemes. Furthermore, data comparability over time and across coun- tries is affected by the varying requirements that participants are asked to fulfil. Despite all these problems, this chapter has demonstrated how combining infor- mation obtained from multiple data sources may be used to estimate the size of irregular foreign residents, such as in Spain, where the juxtaposition of the total foreign population recorded by the population register with the residence permit database allows one to assess the evolution of the irregular foreign population over time.

References

Arango J and Jachimowicz M. 2005. *Regularizing immigrants in Spain: a new approach.* MPI/Migration Information Source. Available at www.migrationinformation.org.

Baldwin-Edwards M. 2002. Semi-reluctant hosts: Southern Europe's ambivalent response to immigration. *Studi Emigrazione/Migration Studies* 145:27–47.

Baldwin-Edwards M. 2004a. Immigration into Greece, 1990–2003: a Southern European paradigm? Presented at the *UNECE European Population Forum*, Geneva.

Baldwin-Edwards M, ed. 2004b. *Statistical data on immigrants in Greece: an analytic study of available data and recommendations for conformity with European Union standards.* Final Report for the Hellenic Migration Policy Institute, Mediterranean Migration Obser- vatory, Panteion University, Athens.

Blangiardo GC. 2005. I processi di immigrazione: dall'illegalità alla regolarizzazione. In *L'incidenza economica dell'immigrazione.* Livi Bacci M, ed., Quaderni Cesifin, no. 20. Turin: Giappichelli.

Blangiardo GC and Molina S. 2006. Immigrazione e presenza straniera. In *Generazioni, famiglie, migrazioni fondazioni*. Fondazione Giovanni Agnelli and Gruppo di Coordinamento per la Demografia, eds., Pensando all' Italia di Domani. Turin: Fondazione Giovanni Agnelli.

Blangiardo GC and Tanturri ML. 2006. How many and who? An up-date picture of the foreign migrants in Italy. Presented at the *European Population Conference*, Liverpool.

Cangiano A and Strozza S. 2004. Foreign immigration in Southern European receiving countries. Presented at the *Conference 'International Migration in Europe: New Trends, New Methods of Analysis'*, EAPS and IRPPS–CNR, Rome.

Cangiano A and Strozza S 2006. Le procedure straordinarie di regolarizzazione: regole e risultati delle diverse tornate. In *Il Mezzogiorno dopo la grande regolarizzazione: Vecchi e nuovi volti della presenza migratoria*. Strozza S and Zucchetti E, eds., pp.13–40. Milan: Franco Angeli.

Carfagna M. 2003. I sommersi e i sanati. Le regolarizzazioni degli immigrati in Italia. In *Stranieri in Italia. Assimilati ed esclusi*. Sciortino G and Colombo A, eds. Bologna: Il Mulino.

Cavounidis J. 2002. Migration in Southern Europe and the case of Greece. *International Migration* 40(1):45–70.

Duquenne MN and Kaklamani S. 2004. Dimensions et caractéristiques dominantes de l'immigration économique en Grèce. Presented at the *13th AIDELF Conference, Les Migrations Internationales: Observation, Analyse et Perspectives*, Budapest.

Fakiolas R. 2003. Regularising undocumented immigrants in Greece: procedures and effects. *Journal of Ethnic and Migration Studies* 29(3):535–561.

Icduygu A and Unalan T. 2001. Tides between Mediterranean shores: undocumented migration in the South of Europe. Presented at the *XXIV IUSSP General Population Conference*, Salvador de Bahia.

Istat. 2005a. *The measurement of immigrants stock: immigrants, foreigners and naturalized citizens according to Italy's 2001 population census*. Working Paper 20. UNECE/Eurostat Seminar on Migration Statistics, Geneva.

Istat. 2005b. *Comparison and integration among different sources for determining the legal foreign population stock in Italy*. Working Paper 11. UNECE/Eurostat Seminar on Migration Statistics, Geneva.

Jurado A. 2004. Diferencias entre Censo de población y Padrón municipal. *Indice* 3:12–13.

Kanellopoulos CN, Gregou M and Petralias A. 2005. *Illegally resident third country nationals in Greece: State approaches towards them, their profile and social situation*. Athens: Centre of Planning and Economic Research.

Malheiros J. 2002. *Portugal seeks balance of emigration, immigration*. MPI/Migration Information Source. Available at www.migrationinformation.org

Natale M and Strozza S. 1997. *Gli immigrati stranieri in Italia. Quanti sono, chi sono, come vivono?* Bari: Cacucci Editore.

Papadopoulou A. 2005. *Exploring the asylum-migration nexus: a case study of transit migrants in Europe*. Global Migration Perspectives no. 23. Global Commission on International Migration, Geneva.

Peixoto J. 2002. Strong market weak state: the case of recent foreign immigration in Portugal. *Journal of Ethnic and Migration Studies* 28(3):483–497.

Pérez NO. 2003. *Spain: forging an immigration policy*. MPI/Migration Information Source. Available at www.migrationinformation.org

Reyneri E. 2003. Immigration and the underground economy in new receiving South European countries. *International Review of Sociology* 13(1):117–143.

Rosa MJV, de Seabra H and Santos T. 2004. *Contributos dos 'imigrantes' na demografia portuguesa. O papel das populações de nacionalidade estrangeira, estrangeira.* Estudos e Documentos do Observatório da Imigração, no. 4. Available at www.oi.acime.gov.pt

Schoorl J, Brujin B, Kuper EJ and Heering L. 1996. Migration from African and Eastern Mediterranean countries to Western Europe. Presented at the *Mediterranean Conference on Population, Migration and Development,* Council of Europe, Palma di Maiorca, 15–17 October.

Strozza S. 2004. Estimates of the illegal foreigners in Italy: a review of the literature. *International Migration Review* 38(1):309–331.

Villán Criado I. 2002. El Padrón Continuo como instrumento para el conocimiento sobre la inmigración en España. *Revista Fuentes Estadísticas,* no. 69, November–December.

Part II
Models

6

Models of migration: observations and judgements

Frans Willekens

Netherlands Interdisciplinary Demographic Institute, The Hague

6.1 Introduction

The monitoring of international migration in Europe calls for an adequate representation of migration processes and adequate data on the processes.[1] Models are abstract representations of some portion of the real world. They are important for our understanding of the world. When the founding father of econometrics, Jan Tinbergen, received the Nobel Prize in 1969, he delivered the Nobel Prize Lecture. In it, he asserted that models force us to present an internally consistent theory and to confront that theory with reality (Tinbergen 1981:17). Around the same time, Britton Harris, a founding father of quantitative urban planning research, viewed properly constructed models as theories (Harris 1983). More recently, Tom Burch has expressed the view that models are the central element of scientific knowledge (Burch 1999, 2003). In his view, a model represents theory when it starts from empirical observations and arrives at an abstract and therefore general description of the real world (Burch 2003:269). Burch argues that models do not have to be true to be useful. Sufficient realism to the purpose at hand supports understanding,

---

[1] For a discussion, see e.g. Willekens (1994) and Chapters 2 to 5 of this book.

International Migration in Europe: Data, Models and Estimates   Edited by J. Raymer and F. Willekens
© 2008 John Wiley & Sons, Ltd

explanation, prediction and policy guidance. The models that are reviewed in this chapter are abstract but they acknowledge that some people are more likely to migrate than other people and that the likelihood of migration and the directions of migration are determined by characteristics of potential migrants, by contextual factors at the place of origin (e.g. push factors) and at the place of destination (e.g. pull factors) and intervening factors. The models also acknowledge the fact that the occurrence and timing of migration can never be predicted with certainty, which points to the role of chance. Finally, the models provide a vehicle for integrating systematic factors and random factors in the prediction of migration.

The aim of this chapter is to provide a perspective on the modelling of migration that distinguishes between systematic and random factors and that accommodates several migration measures and several data types, including statistical and judgemental data. Migration is a renewable or repeatable event. Renewable events are characterised by quantum and tempo. The *quantum* or level of migration is measured by the number of migrants during a period of time, the share of migrants in a population, or the rate of migration. These measures are related in some way. The *tempo* refers to the timing of migration, more particularly the ages at which migration occurs and the interval between successive migrations. The migration models are rooted in survival models that have been developed for time-to-event data, where the event is a transition from one state of existence to another. Migration models are transition models. The dependent variable is a count, a rate or a probability.

This chapter consists of six sections. Section 6.2 is a non-technical discussion of issues in migration modelling. The issues include the distinction between migrations and migrants, the spatial pattern of migration and the age structure of migrants. The section also introduces the use of regression models for predicting migration flows from incomplete data. Section 6.3 reviews the essentials of probability theory. Section 6.4 is the main part of the chapter. It is a review of migration models from the perspective of probability theory. Different models are associated with different data types. Harmonisation of migration data and comparison of migration levels in space and time require that one data type can be converted into another type. Probability models serve that function. Some models predict the number of migrants during a given interval, others predict the probability that a randomly observed individual is a migrant, while still other models predict the rate of migration. The parameters of the probability models are generally estimated from observations on migration. Empirical evidence on migration is often lacking, incomplete or of questionable quality. Parameters estimated from inadequate data may be misleading. To obtain accurate parameter estimates, hard evidence may be augmented by soft evidence such as expert opinions and judgements. Section 6.5 discusses migration modelling in the absence of adequate data. The method, which is widely used in migration analysis, is presented as a special case of the EM (expectation–maximisation) algorithm. The EM algorithm is the most widespread statistical method for model estimation when data are incomplete. The chapter demonstrates that the modelling of migration can benefit from recent developments in modelling of life events and life histories. Section 6.6 concludes the chapter.

6.2   Data types and data structure

Migration is an event and the person who migrates is a migrant. The event must be properly defined. Any migration involves relocation but not all relocations are migrations. Migration is generally defined as a change of usual residence (address) beyond administrative boundaries. Relocations that do not involve a change of address do not qualify as migrations. They may be travel, commuting or temporary change of residence. These spatial movements may need to be considered because they may lead to a migration. The administrative unit considered in the definition of migration may be a village, a town, a district or a country. The general definition of the event does not change with the spatial unit. The definition of migration often involves a temporal dimension – an intended duration of stay or an actual duration of residence. Sometimes relocation beyond an administrative boundary involves a change of address but does not meet the duration criterion and is therefore not viewed as a migration. The migration models reviewed in this chapter focus on the spatial dimension of migration and disregard the time dimension. Any relocation across administrative boundaries involving a change of address is a migration irrespective of the intended or actual duration of stay at the destination. If the address is changed for a very small duration, the event is seen as a migration. Migration is said to be an event in continuous time. The person relocating beyond administrative boundaries is a migrant.

A major problem in migration studies is the confusion that exists between concept and measurement. Although the concept of migration is unambiguous, for practical reasons migration is often defined in terms of its measurement. For instance, relocation (a change of address) is sometimes measured by comparing the addresses at two points in time, a fixed or variable number of years apart (one year, five years, lifetime). If the addresses are in different administrative units, a migration is said to have occurred although the event of migration has not been recorded. Several authors therefore make a distinction between migrant data that result from a comparison of addresses at two points in time and migration data. Different lengths of the time interval have occupied researchers for years (see e.g. Long and Boertlein 1981; Kitsul and Philipov 1981; Courgeau 1982; Rogers *et al.* 2003b). As early as 1973, Courgeau (1973) distinguished between migration and migrant. A migration is an event and a migrant is a person.

Events occur in continuous time, i.e. an event may occur at any time. Although an event occurs in continuous time it is generally not feasible to record the exact time or date of the event. The month or year of occurrence is recorded instead, i.e. the date of the event is recorded in discrete time. The measurement approach that focuses on events and records the timing of events in continuous or discrete time is referred to as the event-based approach (Willekens 2001). A different approach is to measure events indirectly by comparing the places of residences at two points in time. For instance, in the United States census, migrations are measured by comparing the place of residence at census night and the place of residence five years prior to the census. This approach is the status-based approach. The status-based measurement of migration distinguishes time intervals and therefore takes place in discrete time. The distinction between continuous time and discrete time and the distinction between the event-based approach and the status-based approach are essential. In the literature on migration, the measurement of events in continuous or discrete time has been referred to as the movement approach and the measurement of a migration by comparing places of residence at two points in time as the transition approach (Ledent 1980; Rees and Willekens 1986). Following Courgeau (1973), the event-based type of data is often referred to as migration data and the status-based type of data as migrant data. Other authors distinguish between direct transitions (event-based) and discrete-time transitions (status-based).

The different ways of measuring migration lead to different data types. The distinction between the event-based approach and the status-based approach has been accepted in the literature as a basis for a typology of data types associated with the measurement of migration (see e.g. Rees and Willekens 1986; Bell *et al.* 2002:437; Duke-Williams and Blake 1999; Kupiszewski and Kupiszewska 2003). The authors also discuss the sources of the differences. Typically, event data are associated with population registers and (discrete-time) transition data with population censuses.

Events have characteristics and persons have characteristics and these should be kept separate. Characteristics of a migration include the origin and destination of migration, and the reason for migration. Examples include rural–urban migration, international migration by country of origin and country of destination, marriage migration, family reunion, job-related migration and forced migration. Note that the destination is the current place of residence, which is also a characteristic of the person who migrates, i.e. the migrant. Age, sex, level of education, marital status, employment status, country of birth and country of residence at a given time are characteristics of migrants. In the present chapter and in life history analysis in general, age is treated differently from the other characteristics. Age is a duration variable, which measures the time elapsed since a reference event or event origin. Any event can be selected as the event origin (e.g. birth, marriage, last migration). If the reference event is birth, the duration variable is age. If, on the other hand, the reference event is the last migration, the duration variable is the duration of current residence. The personal characteristics except age are referred to as covariates.

The modelling framework presented in this chapter is rooted in the multivariate analysis of time-to-event data, also known as multistate survival analysis. In multistate analyses a personal attribute such as the place of residence is denoted as a state and the variable that identifies the state is the state variable. The collection of all possible states is the state space. In mathematics and engineering, multistate models are known as state-space models. A change of attribute, i.e. a change of state, is a transition. In this chapter two key concepts are distinguished. The first is *state occupancy*. It is the state occupied at a given point in time (e.g. at a given exact age). The second is the *state transition*. It refers to a change in state occupied. Transitions may be expressed in continuous time or discrete time (Andersen *et al.* 1993:93). Transitions in continuous time are referred to as *direct transitions* or events (Rajulton 1999:5). *Discrete-time transitions* are identified by comparing states occupied at two consecutive points in time (t), e.g. the region of residence at t_1 and the region of residence at t_2.

Characteristics of migrations and migrants are specified at the individual level. An individual occupies an address and has a set of attributes. At the population level, the distribution of addresses and attributes is the subject of study. With a distribution is associated a *data structure*: age structure, covariate structure, motivational structure and spatial structure. The covariate structure relates to the attributes of the migrants, e.g. country or region of birth and sex, but also employment status and marital status. The motivational structure relates to the reasons for migration. The spatial structure relates to the origins and destinations of migrations. In the study of migration, the spatial structure is of particular relevance. The origins and destinations of migration flows define a spatial pattern. The spatial distribution of potential destination locations, their attributes and the interlinkages that connect them shape migration flows in ways which accord them spatial structure that is likely to affect the directions of subsequent migrations (Rogers *et al.* 2001). When residents of a given area are more likely to move to a particular destination rather than to another destination, a spatial dependence exists. Spatial dependences generate spatial structure. The modelling framework proposed in this chapter captures different spatial dependences and spatial structures. Based on a review of the recent literature on spatial structure, Bell *et al.* (2002:436) argued that four broad dimensions of spatial structure could be recognised, relating to (1) migration intensities, (2) migration distance, (3) migration connectivity and (4) the effect of migration on the redistribution of the population. Recently, Rogers *et al.* (2001, 2002a,b, 2003a) proposed statistical models to capture the spatial structure of migration flows or a migration system. The approach is adopted in this chapter. The proposed framework encompasses different types of spatial dependence. Consider the following examples. Spatial dependence is absent if the destination is not affected by the origin of the migrant. The spatial focus, which is the concentration of migration in a few flows, is another example. The spatial focus is measured using a variety of indices to capture the extent to which migration flows between regions are concentrated or dispersed (Plane and Mulligan 1997; Rogers and Sweeney 1998; Rogers and Raymer 1998). The study of spatial focusing also includes the spatial dominance exerted in varying degrees by destinations on origins (Pooler 1992).

Each structure calls for a different modelling approach that may be logically integrated in the comprehensive framework. The age structure is modelled by models of age (duration) dependence. The Rogers–Castro model migration schedule is an example (see Chapter 8). The covariate structure is described by transition rate models and logistic regression models. The spatial structure is captured by spatial interaction models. Conventional spatial interaction models capture the effects of distance on the level and direction of migration flows (see e.g. Mueser 1989; Sen and Smith 1995). Most spatial interaction models today include other variables. Underlying the various spatial interaction models of migration has been a recognition that the decision to move both shapes and is shaped by the population geography within which the movement takes place. The spatial structure changes over time and, although the spatial structure is often remarkably stable even in periods of socio-economic change, models of spatial structure should be able to capture continuity and change. In the unlikely case that the four structures are independent, the structures can be modelled separately. Different dependences or interaction effects are identified and integrated into the framework. For instance, the age structure of migration may differ by origin and/or destination and/or covariate and/or reason for migration.

A unified perspective on the modelling of migration flows has four significant advantages. First, it provides a single, comprehensive framework for the *analysis of data* on migration. Data may be of different types. Second, it provides a framework for the *harmonisation of migration statistics*. Migration data are obtained in many different ways. The estimation of comparable indicators of levels and trends of migration (and direction of migration) and the comparative analysis of patterns of migration require comparable data or techniques for converting one data type into another, e.g. migrant data into migration data.

Third, it provides a framework for the *prediction (estimation) of missing data on migration*. The prediction of a missing value or a set of missing values on the basis of available data is similar to the imputation of missing values. Imputation is receiving much attention in the literature. Methods for statistical data imputation may be divided into two broad groups: model imputation and donor imputation. In model imputation, the imputed values are directly derived from a data model, i.e. a statistical or demographic model of the data. Common data models take the format of regression models. The regression model is estimated from the available data that may be augmented with qualitative information on migration (expert opinion, judgemental data). In donor imputation, the imputed values are derived from a set of observed values (donors). The imputed value is based on information in the closest valid record (nearest neighbour matching characteristics that are not missing). Post-imputation edits (post-editing) make sure that the nearest neighbour is close enough to be used as a donor. All available auxiliary information is used to assure a best estimate or imputation. The estimation of missing migration data may benefit from the literature on imputation. Donor imputation is not considered in this chapter.

A fourth advantage of a unified perspective is that seemingly different migration models may be grouped into classes of models. For instance, the gravity model, the entropy model and the log-linear model of interregional migration have a common structure that links them to the family of generalised linear models (Willekens 1983). The migrant pool model, which is much used in regional population projections, is a special case of the origin–destination migration flow model. The migrant pool model[2] is a migration flow model with the origin–destination interaction removed.

---

[2] A survey of European official subnational population projection practices in the 1990s by van Imhoff *et al.* (1994) revealed that one variant in common use is the migration pool model. In the migration pool model, migration is projected in two stages. The first stage is the projection of the number of outmigrations from each region. The migrants are placed in a common pool. In the second stage, the migrations in the pool are distributed over the possible destinations. For a recent illustration of the two-stage modelling of migration, see van Wissen *et al.* (2005) and Stillwell (2005).

Several authors have assessed the impact of choice of migration model in the context of regional population projections (Kupiszewski and Kupiszewska 2003; Wilson and Bell 2004; Stillwell 2005; van Wissen *et al.* 2005).

The framework for the modelling of migration is still incomplete. Four limitations are singled out. First, it does not include a measure of the reliability of the estimates or the degree of confidence that one may attach to the estimates. Second, it does not cover the indirect estimation of migration from data on *populations* at two points in time and natural increase during the period. Third, it is of no use to estimate undocumented migration if undocumented migrants are not included in the aggregate data or if qualitative information (e.g. expert opinion, educated guess) is lacking. Fourth, the framework does not yet include a unified approach to model temporal changes in migration flows. In the 1980s, Plane and Rogerson (1986) and Jackson *et al.* (1990) proposed causative matrices to link matrices of migration rates from one time period to another and to extrapolate migration rates using a geometric regression based on two data points. The causative method was recently discussed by De Mesnard (2004) in the context of structural changes in input–output analysis. The method is not considered in this chapter.

6.3   Probability models: generalities

Observations on migration are manifestations of a migration process. People migrate at different times in different directions for different reasons. The migration process that is revealed to the observer depends on the observation, i.e. the measurement of the migration process. The true migration process is often referred to as the under-lying process. *Data models* describe the observations while *process models* describe the underlying process. Our aim is to model processes rather than observations on the processes.

Migration is the result of several factors. They include personal factors and contextual or situational factors. The latter include push, pull and intervening factors. Migration is however also an outcome of chance, i.e. random factors. As a consequence, an observation on migration is in part a result of chance. To separate the effects of systematic factors from the effects of random factors, the chance process is made explicit, which is done by specifying a probability model rather than a deterministic model. A probability model is a mathematical representation of a random phenomenon (see e.g. Taylor and Karlin 1998). The variable that describes an observation or a set of observations is a *random variable* and the distri-bution of the random variable reflects the random distribution of the observations. Many authors do not use random variables but work in terms of variates. A variate is a generalisation of the idea of a random variable. It has similar probabilistic properties but is defined without reference to a particular probabilistic experiment (Evans *et al.* 2000:4). A *variate* is the set of all random variables that obey a given probabilistic law. A multivariate is a vector or set of elements, each of which is a variate. Different types of observations (data types) are represented by different variates. Counts are represented by Poisson variates. Like counts, they can take on only nonnegative integer values. The Poisson model is the probability model for counts. A response variable that has only two categories (yes, no; success, failure) is a binary response variable. A single observation may be viewed as the outcome of a Bernoulli trial, and a sequence of observations as a sequence of Bernoulli trials. The number of successes in a number of trials is a binomial variate. The number of failures before the first success is a geometric variate, and the number of failures before the kth success is a negative binomial variate. A variate may take on several values (a finite number in the case of a discrete variate and an infinite number in the case of a continuous variate). In this chapter we consider discrete variates. With each value of the discrete variable may be associated a probability, and the distribution of the values is a probability distribution. The distribution of observations (empirical distribution) is approximated by a theoretical distribution, such as the Poisson distribution and the binomial distribution.

The probability distributions describe the random phenomenon. Theoretical distributions are fully characterised by one parameter or a few parameters. For instance, the normal distribution has two parameters (mean and variance). The Poisson distribution has a single parameter (mean and variance are equal). The values of the parameters of the distributions depend on the systematic factors such as personal characteristics and situational variables. In order to predict the parameter value from the factors (predictors), the parameter may need to be transformed to assure that the predicted value is within the range of acceptable values. The Poisson distribution has a single parameter, which must be nonnegative. To assure that the value is nonnegative irrespective of the values of the systematic factors or predictors, the log (logarithmic) transformation is used. The binomial distribution has a single parameter too – the probability of success – generally denoted by p for probability. Its value must be between 0 and 1, and to assure that the requirement is met, a logit transformation is often used to relate the parameter to the systematic factors. The logit transformation imposes a logistic distribution onto p. That means that, as the predictors vary from $-\infty$ to $+\infty$, the parameter p varies between 0 and 1 following a cumulative logistic distribution. Alternative transformations include the probit (cumulative normal distribution) and the log-log transformation (cumulative extreme value distribution).

Now we consider the Poisson model and the logit model in more detail. The Poisson variate describes the number of migrations during an interval of unit length (e.g. year, month). The number of migrations is not restricted in any way. Subjects in a (sample) population may experience an event more than once during the unit interval. The probability of observing n migrations during the interval is given by the Poisson model

$$\Pr\{N = n\} = \frac{\lambda^n}{n!} \exp(-\lambda), \tag{6.1}$$

where N is the variate denoting the number of migrations, n is the observed number of migrations during a unit interval, and λ is the expected number of migrations during the unit interval. The latter is the parameter of the Poisson model. Note that λ is not the same as the migration rate. The migration rate is the expected number of migrations during the unit interval per person, i.e. the number of migrations per person-year or person-month. It is obtained by dividing the number of migrations during the interval by the total time spent during the interval by all subjects combined. The migration rate will be denoted by μ.

The parameter of the Poisson model may be made dependent on covariates:

$$E[N] = \lambda = \exp(\beta_0 + \beta_1 Z_1 + \beta_2 Z_2 + \cdots). \tag{6.2}$$

The model may be written as a log-linear model:

$$\ln \lambda = \beta_0 + \beta_1 Z_1 + \beta_2 Z_2 + \cdots. \tag{6.3}$$

In principle, Z_p can be any covariate. In conventional log-linear analysis, all covariates are discrete or categorical. The observations on event occurrences may therefore be arranged in a contingency table. The covariates refer to rows, columns, layers and combinations of these (to represent interaction effects). Log-linear models of age and spatial structures of migration flows are studied by Rogers *et al.* (2003a) among others.

The migration rate associated with the parameter of the Poisson model is the expected number of migrations divided by the person-years lived during the interval (assuming that the unit of time is a year). Let the person-years be denoted by *PY*. The migration rate is $\mu = \lambda/PY$. The person-years are assumed to be known and independent of the number of migrations. Hence, *PY* is not a variate.

The Poisson variate describes the number of migrations during an interval. It is applicable when event data are available (event-based approach). Migration is often measured by comparing the places of residence at two consecutive points in time. In that status-based approach, a migrant is an individual whose residence at time t is different from the residence at $t - 1$. Multiple migrations during the interval are not recorded and the number of migrants can never be larger than the number of subjects in the (sample) population. If an individual whose residence at time t differs from that at $t-1$ is coded 1 and an individual whose residence is the same (although he may have changed residence several times during the interval) is coded 0, then the number of migrants in a (sample) population is a binomial variate. The probability of observing n migrants among a (sample) population of m individuals is the binomial distribution with parameter p and index m:

$$\Pr\{N_m = n\} = \frac{m!}{n!(m-n)!} p^n (1-p)^{m-n}. \tag{6.4}$$

The parameter p is the probability of being a migrant. The logit model relates p to predictors while assuring that the predicted value of p is between 0 and 1. The logit model is

$$\text{logit}(p) = \ln\left(\frac{p}{1-p}\right) = \beta_0 + \beta_1 Z_1 + \beta_2 Z_2 + \cdots, \tag{6.5}$$

where Z_p can be any covariate and $p/(1-p)$ is the odds of being a migrant.

In many applications, m is the total number of migrations out of a given region (of origin) and n is the number of migrations that end in a given destination. In that case, $m - n$ is the number of outmigrations that end in another destination. The expected number of migrations during an interval that end in a given destination rather than in any of the other destinations is λp. The probability of observing n migrations to the given destination during a unit interval is given by the Poisson distribution with parameter λp with $\lambda = \mu PY$. If the destination choice is independent of the decision to outmigrate, then μ (and λ) and p can be studied separately (see e.g. Hachen 1988). If the decision to outmigrate is not independent of the destination μ and p must be studied simultaneously.

If the possible number of destinations exceeds two, then the distribution of the outmigrations over the destinations, say three, is given by the multinomial distribution

$$\Pr\{N_1 = n_1, N_2 = n_2, N_3 = n_3\} = \frac{m!}{n_1! n_2! n_3!}\, p_1^{n_1} p_2^{n_2} p_3^{n_3}, \tag{6.6}$$

which may be written in general as

$$\Pr\{N_1 = n_1, N_2 = n_2, \ldots\} = \frac{m!}{\prod_{i=1}^{I} n_i!} \prod_{i=1}^{I} p_i^{n_i}, \tag{6.7}$$

where I is the number of possible destinations, n_i is the number of individuals ending in i and with $\sum p_i = 1$ and $\sum N_i = \sum n_i = m$.

6.4   Probability models of migration

This section reviews models of state occupancies and state transitions. State occupancy is expressed in terms of the probability that an individual selected at random from a (sample) population occupies a given state. It is approximated by the proportion of the (sample) population in a given state. State transitions during an interval (in continuous time or discrete time) are expressed in terms of three risk indicators: counts, probabilities (proportions) and rates. Probabilities relate transitions to the *risk set* (population at risk) at the beginning of an interval. The risk set accounts for attrition during the interval for reasons unrelated to the transition being studied (censoring), which is the migration from the current place of residence to a given destination. Rates relate transitions to *exposure*. Probabilities are obtained as the ratio of the number of events (in continuous time or discrete time) to the risk set; rates are obtained as the ratio of the number of events (in continuous time) to the exposure time. Counts refer to the number of events that occur during a given interval. It is the numerator in probability and rate measures.

In developing probability models of migration and migrants, the occurrence of an event (migration) is assumed to be the result of an underlying random mechanism. The occurrence of a migration depends on both personal attributes (systematic factors) and chance. Our approach is to model the random mechanism by specifying a probability model. A problem is that a migration is not necessarily associated with a single random mechanism. Different mechanisms may result in the same event of migration. Hence different probability models may describe the level and direction of migration. If an event is observed, it should be possible to identify the set of plausible mechanisms and to identify the mechanism that *most likely* produced the event. To determine the most likely mechanisms and the model that describes that mechanism, the maximum likelihood method is applied. The method identifies the 'best model', i.e. the model that has the greatest probability of predicting the observations on event occurrence. That model describes the random mechanism that most likely underlies the event.

6.4.1 State probabilities

Let S denote the state space: $S = \{1, 2, 3, \ldots, i, \ldots, I\}$. The state space contains I possible geographical areas or regions. At a given age, an individual resides in one area and one area only. In other words, the states are mutually exclusive. Let $Y_k(x)$ be a polytomous random variable denoting the state occupied by individual k at exact age x. The probability that an individual resides in state i is the *state probability*. The probability that individual k resides in state i at exact age x is $_k\pi_i(x) = Pr\{Y_k(x) = i\}$. If all individuals are independent and identical, then $_k\pi_i(x) = \pi_i(x)$ for all k. If individuals differ in a few characteristics only or if a few characteristics suffice to predict the state occupied at age x, then $_k\pi_i(x) = \pi_i(x, Z)$, where Z represents a specific combination of characteristics or covariates. The probability that individual k occupies state i at exact age x depends on the covariates only, and individuals with the same covariates have the same state probability.

The state occupied at x may be denoted differently. Let $Y_{ki}(x)$ be an indicator variable (binary) which is 1 if individual k occupies state i at x and 0 otherwise. The state probability is the probability that the random variable takes on the value of 1.

Consider a sample of m individuals. We do not consider covariates, implying that all individuals are identical. Covariates are introduced below. In addition, age is omitted for convenience. The number of individuals observed in state i is

$$N_i = \sum_{k=1}^{m} Y_{ki}. \tag{6.8}$$

The probability of observing N_1 individuals in state 1, N_2 in state 2, N_3 in state 3, etc., is given by the multinomial distribution

$$Pr\{N_1 = n_1, N_2 = n_2, \ldots\} = \frac{m!}{\prod_{i=1}^{I} n_i!} \prod_{i=1}^{I} \pi_i^{n_i}, \tag{6.9}$$

where n_i is the observed number of individuals in i and with $\sum \pi_i = 1$ and $\sum N_i = \sum n_i = m$. The most likely values of the parameters π_i, given the data, are obtained by maximising the likelihood that the model predicts the data, which is the maximum likelihood method. The values of π_i $(i = 1, 2, \ldots, I)$ that maximize the above multinomial distribution is $\hat{\pi}_i = n_i/m$.

The expected number of individuals occupying state i is $E[N_i] = \pi_i m$ and the variance is $\mathrm{Var}[N_i] = \pi_i(1 - \pi_i)m$. The probability that an individual is found in state i is the expected value of $Y_i : \pi_i = E[Y_i]$. The variance of Y_i is

$$\mathrm{Var}[Y_i] = \mathrm{Var}[N_i/m] = \mathrm{Var}[N_i]/m^2 = [\pi_i(1 - \pi_i)]/m.$$

The variance declines with increasing sample size.

Now we introduce covariates. They are denoted by Z ($Z = \{Z_1, Z_2, Z_3, \dots\}$), and Z_p may represent a single attribute or a combination of attributes (to denote interaction effects). The state probability $\pi_i(Z)$ that an individual with covariates Z occupies state i is given by the logit equation

$$\mathrm{logit}(\pi_i) = \ln\left(\frac{\pi_i}{1 - \pi_i}\right) = \eta_i = \beta_{i0} + \beta_{i1}Z_1 + \beta_{i2}Z_2 + \beta_{i3}Z_3 + \cdots, \qquad (6.10)$$

where $\pi_i/(1 - \pi_i)$ is the odds of occupying state i. The logit transformation assures that the state probabilities lie between 0 and 1, and that their sum is equal to unity. The value of η may range from $-\infty$ to $+\infty$, but the value of π_i stays within 0 and 1. To obtain the probabilities, the logit scale is converted into the probability scale

$$\pi_i = \frac{\exp(\eta_i)}{\exp(\eta_1) + \exp(\eta_2) + \cdots + 1 + \cdots} = \frac{\exp(\eta_i)}{\sum_{j=1}^{J} \exp(\eta_j)}, \qquad (6.11)$$

where the 1 is associated with the reference category. The model is the multinomial logistic regression model.

6.4.2   Transition probabilities

The state occupied at a given age generally depends on the states occupied at previous ages, in addition to personal attributes at the given age. Hence the probability of being in state j at $x+1$ (or more generally y) depends on the states occupied at previous ages. It is often assumed that only the most recent state occupancy is relevant:

$$\Pr\{Y(x+1) = j \,|\, Y(x), Y(x-1), \cdots ; Z\} = \Pr\{Y(x+1) = j \,|\, Y(x); Z\}. \qquad (6.12)$$

If the state occupied at x is i, then

$$\Pr\{Y(x+1) = j \,|\, Y(x) = i\} = p_{ij}(x). \qquad (6.13)$$

Here $p_{ij}(x)$ is the probability that an individual who resides in state i at x resides in state j at $x+1$. It is the discrete-time transition probability. The interval can be of any length but is generally one or five years. This model is suited for describing migrant data, i.e. data that infer migration by recording the places of residence at two consecutive points in time.

The status dependence may also be written as

$$\text{logit}[\pi_j(x+1)] = \beta_{j0}(x) + \beta_{j1}(x)Y_i(x), \tag{6.14}$$

where $Y_i = 1$ if state i is occupied at x and 0 otherwise. Hence the transition probability may be written as

$$p_{ij}(x) = \frac{\exp[\beta_{j0}(x) + \beta_{j1}(x)Y_i(x)]}{\sum_{r=1}^{I} \exp[\beta_{j0}(x) + \beta_{j1}(x)Y_r(x)]}. \tag{6.15}$$

The transition probabilities may depend on covariates in a way that is similar to that of state probabilities. Transition probabilities out of a given state i that depend on covariates may be estimated using multinomial logistic regression software.

6.4.3 Transition rates

The transition probabilities discussed in Section 6.4.2 are defined for discrete time intervals. They depend on the number of persons at risk at the beginning of the interval, which is generally known as the risk set. The probabilities are not directly related to the *duration* that individuals in i are at risk of migrating to j. Since the event of migration (direct transition) may occur at any time during the interval from x to y (with $y = x+1$, for instance), the transition probability is defined for very small intervals. The probability that an individual in i transfers to j during an infinitesimally small interval following x is the instantaneous rate of transition:

$$\mu_{ij}(x) = \lim_{(y-x)\to 0}\left(\frac{p_{ij}(x,y)}{y-x}\right) \quad \text{for } i \text{ not equal to } j. \tag{6.16}$$

The instantaneous rate of transition is also known as the transition intensity and the force of transition. The term $\mu_{ii}(x)$ is defined such that $\sum_j \mu_{ij}(x) = 0$. It is sometimes referred to as the intensity of passage because it relates to the transition from i to any other state different from i. Schoen (1988:65) refers to it as the 'force of retention'.

The intensities are the basic parameters of a continuous-time multistate process. Under the restrictive Markov assumption, the probability that an individual leaves a state depends only on the state occupied and the individual's age. It is independent of other characteristics.

The matrix of instantaneous rates with off-diagonal elements $-\mu_{ij}(x)$ and with $\mu_{ii}(x)$ on the diagonal is known as the *generator* of the stochastic process $\{Y_k(x); x \geqslant 0\}$ (Çinlar 1975:256). The matrix is denoted by $\boldsymbol{\mu}(x)$, and has the following configuration:

$$\boldsymbol{\mu}(x) = \begin{bmatrix} \mu_{11}(x) & -\mu_{21}(x) & \cdots & -\mu_{I1}(x) \\ -\mu_{12}(x) & \mu_{22}(x) & \cdots & -\mu_{I2}(x) \\ \vdots & \vdots & \ddots & \vdots \\ -\mu_{1I}(x) & -\mu_{2I}(x) & \cdots & \mu_{II}(x) \end{bmatrix}. \tag{6.17}$$

Note that

$$\lim_{(y-x)\to 0}\left(\frac{\mathbf{P}(x,\,y)-\mathbf{I}}{y-x}\right)=-\boldsymbol{\mu}(x). \tag{6.18}$$

The matrix of discrete-time transition probabilities is:

$$\mathbf{P}(x,\,y)=\begin{bmatrix} p_{11}(x,\,y) & p_{21}(x,\,y) & \cdots & p_{N1}(x,\,y) \\ p_{12}(x,\,y) & p_{22}(x,\,y) & \cdots & p_{N2}(x,\,y) \\ \vdots & \vdots & \ddots & \vdots \\ p_{1N}(x,\,y) & p_{2N}(x,\,y) & \cdots & p_{NN}(x,\,y) \end{bmatrix}. \tag{6.19}$$

An element of $\mathbf{P}(x,\,y)$, namely $p_{ij}(x,\,y)$, denotes the probability that an individual who is in state i at exact age x is in state j at exact age y. The Markovian assumption implies the following relationship between $\mathbf{P}(x,\,x+v)$ and $\mathbf{P}(x+v,\,y)$:

$$\mathbf{P}(x,\,y)=\mathbf{P}(x,\,x+v)\cdot\mathbf{P}(x+v,\,y). \tag{6.20}$$

Subtraction of $\mathbf{P}(x+v,\,y)$ from both sides of the equation yields

$$\frac{\mathbf{P}(x,\,y)-\mathbf{P}(x+v,\,y)}{v}=\frac{[\mathbf{P}(x,\,x+v)-\mathbf{I}]\mathbf{P}(x+v,\,y)}{v} \tag{6.21}$$

and

$$\lim_{v\to 0}\frac{\mathbf{P}(x,\,y)-\mathbf{P}(x+v,\,y)}{v}=\lim_{v\to 0}\frac{[\mathbf{P}(x,\,x+v)-\mathbf{I}]\,\mathbf{P}(x+v,\,y)}{v} \tag{6.22}$$

or

$$\frac{\mathrm{d}\mathbf{P}(x)}{\mathrm{d}x}=-\boldsymbol{\mu}(x)\mathbf{P}(x). \tag{6.23}$$

Recall that

$$\lim_{(y-x)\to 0}\frac{\mathbf{P}(x,\,y)-\mathbf{I}}{y-x}=-\boldsymbol{\mu}(x). \tag{6.24}$$

Multiplying both sides with the vector of state probabilities at age x, $\mathbf{P}(x)$, leads to

$$\frac{\mathrm{d}\mathbf{P}(x)}{\mathrm{d}x}=-\boldsymbol{\mu}(x)\mathbf{P}(x), \tag{6.25}$$

where $\mathbf{P}(x)$ is a vector of state probabilities.

The model is a system of differential equations. In multistate demography, two avenues are followed to solve the system. Both introduce age intervals (Rogers and Willekens 1986:370ff.). The first avenue postulates a piecewise constant intensity function, $\mu(t)=\mu(x)$ in the interval from x to $y\,(x\le t<y)$. This implies an

exponential distribution of demographic events within each age interval. The model that results is referred to as the *exponential model*. The second avenue postulates a piecewise linear survival function. A piecewise linear survival function is obtained when demographic events are uniformly distributed within the age intervals. The model that results is referred to as the *linear model*. The first avenue is followed by van Imhoff (1990) and van Imhoff and Keilman (1991) among others; the second by Willekens and Drewe (1984) among others. The state occupancies and the sojourn times must be estimated simultaneously from the population at the beginning of the interval and the events during the interval.

To solve the system of differential equations, it may be replaced by a system of integral equations:

$$\mathbf{P}(x, y) = \mathbf{I} - \int_0^{y-x} \boldsymbol{\mu}(x + t) \mathbf{P}(x, x + t)\, dt. \tag{6.26}$$

To derive an expression involving transition rates during the interval from x to y, we write

$$\mathbf{P}(x, y) = \mathbf{I} - \left[\int_0^{y-x} \boldsymbol{\mu}(x + t) \mathbf{P}(x, x + t)\, dt \right] \left[\int_0^{y-x} \mathbf{P}(x, x + t)\, dt \right]^{-1}$$
$$\times \left[\int_0^{y-x} \mathbf{P}(x, x + t)\, dt \right] \tag{6.27}$$

and

$$\mathbf{P}(x, y) = \mathbf{I} - \mathbf{M}(x, y)\, \mathbf{L}(x, y), \tag{6.28}$$

where $\mathbf{M}(x, y)$ is the matrix, with elements $m_{ij}(x, y)$, of average transition rates during the interval from x to y and

$$\mathbf{L}(x, y) = \int_0^{y-x} \mathbf{P}(x, x + t)\, dt$$

is the sojourn time spent in the different states between ages x and y per person in a state occupied at age x.

6.4.3.1   Exponential model

The transition intensities $\mu(x)$ are assumed to remain constant during the age interval from x to y and to be equal to the model transition rates $\mathbf{M}(x, y)$. It is furthermore assumed that they can be estimated by empirical occurrence–exposure rates for that age interval. This assumption is consistent with the general assumption in demography that life-table rates are equal to empirical rates. In this chapter no separate notation is used for model rates and empirical rates. The matrix of transition probabilities between x and y is

$$\mathbf{P}(y, x) = \exp\left[-(y - x)\mathbf{M}(y, x) \right], \tag{6.29}$$

where $\mathbf{M}(x, y)$ is the matrix of empirical occurrence–exposure rates or transition rates for the age interval from x to y and $\mu_{ij}(t) = m_{ij}(x, y)$ for $x \leq t < y$ and $\mu(t) = \mathbf{M}(x, y)$ for $x \leq t < y$.

A number of methods exist to determine the value of $\exp[-\mathbf{M}]$ (see e.g. Director and Rohrer 1972:431ff.; Aoki 1976:387; Strang 1980:206). For example, the Taylor series expansion may be used for $\exp(\mathbf{A})$,

$$\exp(\mathbf{A}) = \mathbf{I} + \mathbf{A} + \frac{1}{2!} \mathbf{A}^2 + \frac{1}{3!} \mathbf{A}^3 + \cdots . \tag{6.30}$$

Hence one obtains

$$\exp[-(y - x)\mathbf{M}(x, y)] = \mathbf{I} - (y - x)\mathbf{M}(x, y) + \frac{(y - x)^2}{2!} [\mathbf{M}(x, y)]^2$$
$$- \frac{(y - x)^3}{3!} [\mathbf{M}(x, y)]^3 + \cdots \tag{6.31}$$

(see also Schoen 1988:72).

The transition rates $\mathbf{M}(x, y)$ are estimated from the data. The transition rate $m_{ij}(x, y)$ is equal to the number of moves (or direct transitions) from i to j during the interval from x to y, divided by the exposure in the state i:

$$m_{ij}(x, y) = \frac{n_{ij}(x, y)}{L_i(x, y)}, \tag{6.32}$$

where $n_{ij}(x, y)$ is the observed number of moves from i to j during the interval and $L_i(x, y)$ is the duration in i exposed to the risk of moving to j. It is the sojourn time in i during the (x, y) interval. Exposure is measured in person-months or person-years. In the case of two states, the rate equation may be written as

$$\begin{bmatrix} m_{11}(x, y) & -m_{21}(x, y) \\ -m_{12}(x, y) & m_{22}(x, y) \end{bmatrix} = \begin{bmatrix} n_{11}(x, y) & -n_{21}(x, y) \\ -n_{12}(x, y) & n_{22}(x, y) \end{bmatrix} = \begin{bmatrix} L_1(x, y) & 0 \\ 0 & L_2(x, y) \end{bmatrix}^{-1}, \tag{6.33}$$

where $m_{11}(x, y) = m_{12}(x, y)$ and $m_{22}(x, y) = m_{21}(x, y)$. In matrix notation:

$$\mathbf{M}(x, y) = \mathbf{n}(x, y) [\mathbf{L}(x, y)]^{-1} . \tag{6.34}$$

Let $\overline{\mathbf{L}}(x, y)$ be the vector of sojourn times containing the diagonal elements of $\mathbf{L}(x, y)$ and let $\mathbf{K}(x)$ be a vector with the state occupancies at age x by surviving cohort members as its elements:

$$\mathbf{K}(x) = \begin{bmatrix} K_1(x) \\ K_2(x) \end{bmatrix}, \tag{6.35}$$

with $K_i(x)$ the number of cohort members in state i at exact age x. The vector of sojourn times by all cohort members in the various states is obtained by the following equation:

$$\overline{L}(x, y) = \left[\int_0^{y-x} P(x, x+t)\,dt \right] K(x). \tag{6.36}$$

Since the transition intensities are constant in the interval from x to y, the equation may be written as follows:

$$\overline{L}(x, y) = \left[\int_0^{y-x} \exp[-t\,M(x, y)]\,dt \right] K(x). \tag{6.37}$$

Integration yields

$$-[M(x, y)]^{-1} \,|\exp[-t\,M(x, y)]\,|_0^{y-x}, \tag{6.38}$$

which is equal to

$$[M(x, y)]^{-1} \{I - \exp[-(y-x)\,M(x, y)]\}. \tag{6.39}$$

Hence the sojourn times in the various states during the (x, y) interval are given by

$$\overline{L}(x, y) = [M(x, y)]^{-1} \{I - \exp[-(y-x)\,M(x, y)]\}\, K(x). \tag{6.40}$$

6.4.3.2 Linear model

To solve $P(x, y) = I - M(x, y)\,L(x, y)$, one may introduce an approximation of $L(x, y)$. A simple approximation is that $P(x, x+t)$ is linear in the interval $x \le x+t < y$. Hence $L(x, y)$ may be approximated by a linear integration:

$$L(x, y) = \int_0^{y-x} P(x, x+t)\,dt \approx \frac{y-x}{2}[I + P(x, y)]. \tag{6.41}$$

Introducing this expression in the equation to be solved gives

$$P(x, y) = I - \frac{y-x}{2}M(x, y)\,[I + P(x, y)], \tag{6.42}$$

$$P(x, y) = I - \frac{y-x}{2}M(x, y) - \frac{y-x}{2}M(x, y)\,P(x, y), \tag{6.43}$$

$$P(x, y) + \frac{y-x}{2}M(x, y)\,P(x, y) = I - \frac{y-x}{2}M(x, y) \tag{6.44}$$

$$\mathbf{P}(x, y) = \left[\mathbf{I} + \frac{y - x}{2} \mathbf{M}(x, y) \right]^{-1} \left[\mathbf{I} - \frac{y - x}{2} \mathbf{M}(x, y) \right]. \tag{6.45}$$

The linear approximation implies the assumption that the events are uniformly distributed over the interval. The assumption is adequate when the transition rates are small or the interval is short. It can be shown that the linear model is an approximation to the exponential model that retains the first three terms of the Taylor series expansion.

The instantaneous rates of transition $\mu_{ij}(x)$ may be written as the product of two terms, a rate and a probability. The first is the instantaneous rate of leaving state of origin i irrespective of destination, and the second is the conditional probability of selecting j as the destination provided the state of origin is left (i.e. upon leaving i). The first term, the exit rate, determines the timing of the transition, while the second, the destination probability, determines the destination (new attribute). The exit rate is

$$\mu_{i+}(x) = \mu_i(x) = \sum_{j \neq i} \mu_{ij}(x).$$

Note that $\mu_i(x) = \mu_{ii}(x)$. The transition rate is

$$\mu_{ij}(x) = \mu_{i+}(x)\, \xi_{ij}(x),$$

where $\xi_{ij}(x)$ is the probability that an individual who leaves i selects j as the destination. It is the conditional probability of a *direct transition* from i to j. Note that the above expression is that of a competing risk model or a transition rate model with multiple destinations (Blossfeld and Rohwer 2002). In the terminology of competing risks, the first term is the rate of the event and the second term (destination) indicates the type of the event. If the occurrence of the event and the type of event are unrelated, the two terms may be estimated and studied separately (Hachen 1988:29; Sen and Smith 1995:372). The first term is studied using a transition rate (hazard rate) model; the second using a logit model or a logistic regression model.

In the migration literature, the first term $\mu_i(x)$ is known as the generation component and the second $\xi_{ij}(x)$ as the distribution component (Rogers *et al.* 2002b). It is often assumed that the factors that influence the level of outmigration differ from those that influence the choice of destination, and consequently the generation and the distribution components are modelled separately. If, in addition, the choice of destination is independent of the origin, the model that results is the migrant pool model.

The discrete-time transition probabilities are related to the transition intensities and the transition intensities are related to the probabilities of direct transition in an interesting way. The off-diagonal elements of $\mathbf{M}(x, y)$ may be replaced by $-m_{i+}(x, y)\xi_{ij}(x)$, where $m_{i+}(x, y)$ is the rate of leaving i, which is assumed to be constant in the interval from x to y. The diagonal elements are $m_{i+}(x, y)$. The μ matrix may be written as

$$
\begin{bmatrix}
\mu_{11}(x) & -\mu_{21}(x) & \cdots & -\mu_{I1}(x) \\
-\mu_{12}(x) & \mu_{22}(x) & \cdots & -\mu_{I2}(x) \\
\vdots & \vdots & \ddots & \vdots \\
-\mu_{1I}(x) & -\mu_{2I}(x) & \cdots & \mu_{II}(x)
\end{bmatrix}
=
\begin{bmatrix}
\xi_{11}(x) & -\xi_{21}(x) & \cdots & -\xi_{I1}(x) \\
-\xi_{12}(x) & \xi_{22}(x) & \cdots & -\xi_{I2}(x) \\
\vdots & \vdots & \vdots & \vdots \\
-\xi_{1I}(x) & -\xi_{2I}(x) & \cdots & \xi_{II}(x)
\end{bmatrix}
$$

$$
\times
\begin{bmatrix}
\mu_{1+}(x) & 0 & \cdots & 0 \\
0 & \mu_{2+}(x) & \cdots & 0 \\
\vdots & \vdots & \ddots & \vdots \\
0 & 0 & \cdots & \mu_{I+}(x)
\end{bmatrix}
\tag{6.46}
$$

Now we introduce covariates. As above, they are denoted by $Z(Z = \{Z_1, Z_2, Z_3, \dots \})$. The exit rate is modelled using a transition rate model for a single event (leaving the state of origin). The elementary transition rate model is the basic exponential model, with the rate being independent of age (Blossfeld and Rohwer 2002):

$$
m_i = \exp\left(\beta_{i0} + \beta_{i1}Z_1 + \beta_{i2}Z_2 + \cdots\right).
\tag{6.47}
$$

The model may be written as a log-linear model

$$
\ln m_i = \beta_{i0} + \beta_{i1}Z_1 + \beta_{i2}Z_2 + \cdots
\tag{6.48}
$$

The model is also known as the log-rate model (see e.g. Yamaguchi 1991: Chapter 4).

The age dependence may be introduced in two ways: nonparametric and parametric. In the first approach, the population is stratified by age and a transition rate is estimated for each age separately. In the parametric approach, age dependence is represented by a model. A common model is the Gompertz model, which imposes onto the transition rate an exponential change with duration. The Gompertz model has two parameters and each may be made dependent on covariates – for detailed treatment, see Blossfeld and Rohwer (2002). Other parametric models of duration dependence may be used. In migration studies, the model migration schedule is a common representation of the age dependence of the migration rate. Each parameter of the model may be related to covariates. In practice, only one or a selection of parameters is assumed to depend on covariates. The computer package for transition data analysis (TDA) developed by Rohwer and used by Blossfeld and Rohwer (2002) has a facility for user-defined rate models (Rohwer and Pötter 1999: Section 6.17.5). The program may be downloaded from Professor G. Rohwer's homepage at http://www.stat.ruhr-uni-bochum.de/ . The manual can be downloaded from the same site.

In some cases, the researcher is not interested in the age dependence of migration rates, but in the effect of covariates on the migration level. Rather than omitting age altogether, as in the basic exponential model, the migration rate is allowed to vary with age but the effect of the covariates on the migration rate does not vary with age. The transition rate model that results is a Cox proportional hazard model. It is written as

$$m_i(x) = m_{i0}(x) \exp [\beta_{i0} + \beta_{i1} Z_1 + \beta_{i2} Z_2 + \cdots], \qquad (6.49)$$

where $m_{i0}(x)$ is the baseline hazard. It is the set of age-specific migration rates for the reference category. Note that if the age dependence (age structure) of migration is independent of the dependence on covariates (motivational structure), the baseline hazard may be represented by a parametric model and the two components may be estimated separately.

This brief discussion illustrates that transition rate models are ideally suited to impose age structures onto migration data. The same applies for spatial structures and motivational structures. Spatial interaction models may be used for that purpose, but also Poisson regression models (log-linear models) with offset. The migration rates in the model presented in this section depend on origin, destination and age, and they may vary over time. Several authors have found that spatial patterns of migration are stable in time (Baydar 1983; Rogers and Raymer 2001; Berentsen and Cromley 2005). It implies that the origin–destination interaction does not vary in time. Higher-order interaction effects may be omitted if empirical evidence shows that they are not significant. Van Imhoff *et al.* (1997) and van der Gaag *et al.* (2000) used a series of empirical tests to determine which combinations of the four dimensions of interregional migration (origin, destination, age and sex) are needed to describe the migration patterns adequately. Many of the simplifications that result from omitting higher-order interaction effects are studied in the context of demographic projection models.

6.4.4   From transition probabilities to transition rates

In this section, we assume that migration is measured in discrete time. Examples include the census (based on the residence at time of census and five years prior to the census). From that information, the approximate transition rates can be derived. The problem is equivalent to one in which we are given $\mathbf{P}(x, y)$ and $\mathbf{M}(x, y)$ is required. The derivation starts with the exponential expression in Equation (6.29). The exponential expression may be approximated by the linear model:

$$\mathbf{P}(x, y) = \left[\mathbf{I} + \tfrac{1}{2}\mathbf{M}(x, y)\right]^{-1} \left[\mathbf{I} - \tfrac{1}{2}\mathbf{M}(x, y)\right]. \qquad (6.50)$$

The approximation is adequate when the transition rates are small or the interval is short.

The derivation of the rate of migration during an interval from information in regions of residence at two consecutive points in time is known as the inverse problem: transition rates are derived from transition probabilities (Singer and Spilerman 1979). The transition rates may be written as:

$$\mathbf{M}(x, y) = \frac{y - x}{2} [\mathbf{I} - \mathbf{P}(x, y)][\mathbf{I} + \mathbf{P}(x, y]^{-1}. \qquad (6.51)$$

This inverse relation may be used to infer transition probabilities for intervals that are different from the measurement intervals. For instance, if changes of address are recorded over a period of five years, the inverse relation may be used to infer the average migration rates $\mathbf{M}(x,y)$ and the transition probabilities over a one-year period. The expression is

$$\mathbf{P}(x, x+1) = \exp[-\mathbf{M}(x, x+1)], \qquad (6.52)$$

where $\mathbf{M}(x, x+1)$ is estimated from $\mathbf{P}(x, y)$ using the inverse method. The method assumes that migration rates are constant during the (x, y) interval and that the linearity is an adequate approximation of the exponential model.

6.5   Incomplete data

6.5.1   Adding statistical data

In the previous section, it is assumed that the data are adequate to estimate the parameters of the probability models that are specified. The method applied is the maximum likelihood method. In this section the assumption is relaxed. Some data may be missing.

If data are missing, the strategy consists of two steps. The first is to *predict* the missing data and the second step is to *estimate* the parameters of the model assuming that the data are complete. This two-step procedure is the EM (expectation–maximisation) algorithm (McLachlan and Krishnan 1997). Suppose that we are interested in migration by origin and destination (N_{ij}), but the data are limited to departures and arrivals by region $(n_{i+}$ and $n_{+j})$. The model is

$$E[N_{ij}] = \lambda_{ij} = \alpha_i \beta_j. \qquad (6.53)$$

In the first step, the expected value of N_{ij} is determined assuming values for the parameters (*expectation*). Assume $\alpha_i = 1$ and $\beta_j = 1$. Hence $E[N_{ij}] = 1$. In the second step, the parameters of the model are estimated by maximising the probability that the model predicts the data (maximisation). If the observations are independent, the probability model is the Poisson model

$$\Pr\{N_{ij} = n_{ij}\} = \frac{\lambda_{ij}^{n_{ij}}}{n_{ij}!} \exp(-\lambda_{ij}) \qquad \text{with } E[N_{ij}] = \lambda_{ij} = \alpha_i \beta_j. \qquad (6.54)$$

The maximisation of the probability is equivalent to maximising the log-likelihood:

$$l = \sum_{ij} \left[n_{ij} \ln(\alpha_i \beta_j) - \alpha_i \beta_j \right]. \tag{6.55}$$

The first-order conditions result in the following equations:

$$\hat{\alpha}_i = \frac{n_{i+}}{\sum_j \hat{\beta}_j} \quad \text{and} \quad \hat{\beta}_j = \frac{n_{+j}}{\sum_i \hat{\alpha}_i}. \tag{6.56}$$

The EM algorithm results in the well known expression,

$$\lambda_{ij} = \frac{n_{i+}}{n_{++}} n_{+j},$$

which may be written as $\lambda_{ij} = n_{i+} p_j$, where n_{i+} is the number of migrations originating in i and p_j is the probability that a migration ends in j. The destination probability applies to all migrations in the pool of migrations and is independent of the origin.

The method just described imposes a spatial structure onto migration flows (Willekens 1999). The spatial structure implies that destinations are independent of regions of origin. Suppose we have reason to believe that origin matters, i.e. the choice of destination is influenced by the origin. For instance, migrants are more likely to settle in a region not too far from the region of origin. If distance matters, the prediction of λ_{ij} may be improved by the addition of a distance factor to the model:

$$E\left[N_{ij} \right] = \lambda_{ij} = \alpha_i \beta_j \gamma_{ij}, \tag{6.57}$$

where the distance factor γ_{ij} may be a function of the friction between i and j, i.e. the higher the friction, the lower the value of γ_{ij}. An expression that satisfies the relation between distance factor and spatial friction is $\gamma_{ij} = \exp\left(-c_{ij}\right)$, where c_{ij} is a measure of friction between i and j. The model that results is the gravity model

$$E\left[N_{ij} \right] = \lambda_{ij} = k \, a_i \beta_j \exp\left(-c_{ij}\right), \tag{6.58}$$

where k is a scaling factor. Note that the model is a log-linear model,

$$\ln \lambda_{ij} = u + u_i^A + u_j^B + u_{ij}^{AB}, \tag{6.59}$$

where u_i^A is the effect associated with origin (variable A) i, u_j^B is the effect associated with destination (variable B) j, and u_{ij}^{AB} is the interaction effect between origin and destination. The model is the conventional log-linear model (see e.g. Agresti 1996: Chapter 6). The gravity model and other spatial interaction models may be represented as log-linear models (Willekens 1980, 1983; Rogers *et al.* 2003a). Although the translation opens new perspectives and provides a thorough statistical basis for the modelling of spatial interaction, the log-linear specification of spatial interaction models has not caught on – see e.g. the review of spatial interaction modelling by Roy and Thill (2004).

Interaction effects between origin and destination may be derived from different data sources. One source is a historical migration matrix. Snickars and Weibull (1977), working on internal migration, found that migration tables of some period in the past provide much better estimates of accessibility than any distance measure. Since the publication of that article, authors started to use historical migration matrices instead of distance measures to represent the spatial interaction. Later, authors did not limit the *a priori* information to a single year in the past, but use trends over several years (see e.g. Jörnsten *et al.* 1990). The historical migration may be accurate measurements of the 'revealed' residential preferences. For a prediction of migration flows involving historical migration matrices, see Rogers *et al.* (2003a:64ff.). Table 6.1 shows one of the results. The prediction problem is to determine the interregional migration flows in the United States during the period 1980–1985 from information on the number of residents in the different regions in 1985, the number of residents in 1980 by region, and the flows during the period 1975–1980. The model is

$$E\left[N_{ij}\right] = \lambda_{ij} = \alpha_i^* \beta_j^* m_{ij}^0, \qquad (6.60)$$

where m_{ij}^0 is an element of the 1975–1980 migration matrix and λ_{ij} is the predicted number of (i, j) migrants during the period 1980–1985. The historical matrix serves as an initial guess of migrations between origins and destinations. Note that the model is equivalent to the biproportional adjustment method or RAS model that is commonly used in input–output analysis.[3] The 1975–1980 migrant flow, the observed 1980–1985 flow and the predicted 1980–1985 flow are shown in Table 6.1(a)–(c). The marginal totals of the predicted migration flows are consistent with the data (1980–1985). The origin–destination interactions exhibited by the predicted values are identical to the interactions exhibited by the 1975–1980 flows. The interaction can be interpreted in terms of odds and odds ratios. The odds that, in 1975–1980, a migrant from the Midwest selects the South rather than the West is $1845/1269 = 1.454$. The odds for a migrant from the Northeast is $1800/753 = 2.390$. A migrant from the Northeast prefers the South over the West considerably more than a migrant from the Midwest. The odds ratio is $2.390/1.454 = 1.644$. The odds of selecting the South rather than the West are 64 % higher for a migrant from the Northeast than for a migrant from the Midwest. The migration flow that is predicted for the period 1980–1985 exhibits the same odds ratio: $[1614/632]/[1977/1272] = 1.644$. Main effects and interaction effects that are not included in the contemporary data are 'borrowed' from the historical data (Willekens 1982). For detailed

---

[3] In the input–output literature, the method has become known as RAS because of the notation used by Sir Richard Stone (1961), who developed the method. The method projects a matrix **A** to give it column and row sums of another matrix; the result is **RAS**, where **R** and **S** are diagonal matrices. For a history of method development, see Lahr (2004).

Table 6.1 US interregional migration flows (in thousands): 1975–1980 and 1980–1985.

Region of origin	Region of destination				
	Northeast	Midwest	South	West	Total
(a) 1975–1980 data					
Northeast	43 123	462	1 800	753	46 138
Midwest	350	51 136	1 845	1 269	54 600
South	695	1 082	67 095	1 141	70 013
West	287	677	1 120	37 902	39 986
Total	44 455	53 357	71 860	41 065	210 737
(b) 1980–1985 data					
Northeast	44 845	379	1 387	473	47 084
Midwest	326	52 311	1 954	1 144	55 735
South	651	855	68 742	1 024	71 272
West	237	669	1 085	40 028	42 019
Total	46 059	54 214	73 168	42 669	216 110
(c) 1980–1985 flows predicted based on marginal totals and 1975–1980 matrix					
Northeast	44 445	393	1 614	632	47 084
Midwest	431	52 055	1 977	1 272	55 735
South	814	1 047	68 324	1 087	71 272
West	369	719	1 253	39 678	42 019
Total	46 059	54 214	73 168	42 669	216 110
(d) 1980–1985 flows predicted based on West survey and 1975–1980 matrix					
Northeast	21 181	272	1 037	473	22 963
Midwest	247	43 135	1 526	1 144	46 052
South	488	909	55 235	1 024	57 656
West	237	669	1 085	40 028	42 019
Total	22 153	44 985	58 883	42 669	168 690
(e) 1980–1985 flows predicted based on West survey, 1975–1980 matrix and judgemental data					
Northeast	40 243	516	2 365	899	44 023
Midwest	296	51 762	2 197	1 373	55 628
South	488	909	66 282	1 024	68 703
West	237	669	1 302	40 028	42 236
Total	41 264	53 856	72 146	43 324	210 590

Source: Rogers *et al.* (2003a).

discussions on how to impose spatial structures and age structures onto migration flows, the reader is referred to Rogers *et al.* (2002a, 2003a).

6.5.2 Adding judgemental data

Combining data from several sources implies the use of prior information on the association between the cross-classified variables. In Section 6.5.1, the prior information is derived from other statistical data, such as a historical migration matrix. The interaction effects that are not contained in the contemporary data are 'borrowed' from the more detailed, but outdated, source. The estimation algorithm makes use of this prior information only when more reliable (recent) information is lacking. The estimation problem may be given a Bayesian interpretation, making the use of prior information more explicit (Albert and Gupta 1983; Congdon 2001).

The prior beliefs on the degrees of association between the variables are not restricted to information contained in statistical data. The prior information may also be derived from established theory or from expert judgements. Most experience to date with the quantification and utilisation of expert knowledge is in the field of risk analysis and artificial intelligence. The combination of statistical information and judgemental knowledge is common practice in forecasting to improve forecasts. Today it is generally accepted that judgemental and statistical methods each have unique strengths that they can bring to the forecasting process (Armstrong and Collopy 1998). Expert judgements are being used for the forecasting of internal and international migration (for an early review, see Willekens 1994:28ff.). Harker (1986) combined subjective judgements of experts in the migration field with quantitative data such as physical distance and wage and employment rates to make predictions of future migration patterns in the United States. His method of combining judgemental and statistical data is based on Saaty's analytic hierarchy process (AHP). The AHP is a method by which subjective weights are assigned to a set of objects or alternatives. Cook *et al.* (1984) use the AHP method in combination with time series analysis to make predictions of intra-urban migration. The AHP is used to correct the results of a statistically based forecast. George and Perreault (1992:93–95) report that Canada uses a consensus approach based on opinions of experts and/or administrators and that the Netherlands considers discussions with experts on international migration. The uncertain opinions of experts are used often because no other reliable data exist. Expert opinion can be a very useful source of data. But, as Cooke stresses in his book *Experts in uncertainty*, proper use of this source requires new techniques (Cooke 1991:3). Particularly, the estimation and forecasting of international migration may benefit from new techniques developed in the context of judgemental forecasting, since international migration depends on many factors that are difficult to incorporate into a formal model (van de Kaa 1993:87–88):

> The flow of migrants between countries is, normally, regulated by charters, covenants, treaties and similar agreements between (groups of) states, and by such rules and practices as individual countries choose to apply . . . Thus the way international migration is shaped and is likely to develop, depends to a great extent on the nature of the relations between the countries involved.

What is relevant to forecasting, i.e. the estimation of missing information pertaining to the future, may be relevant to the estimation of missing information in general. Expert opinion may be viewed as data (Cooke 1991:80). It turns out that the use of judgemental data in the modelling of migration is not at all a new idea. Knudsen (1992) discusses the method of how to include *a priori* information in the context of generalised linear models. He suggests including *a priori* information into the model by treating this information as a covariate having a known parameter value of unity. Note that this approach is similar to the method discussed in Section 6.4.1. In order to understand and monitor migration flows, expert knowledge on the causal structure of migration and on any other feature of migration should be used in combination with statistical data. The question of how to use expert knowledge properly remains largely open.

Rogers *et al.* (2003a) use judgemental data in combination with historical statistical data to predict migration flows between the four regions of the United States (Northeast, Midwest, South and West). The estimates in Table 6.1 were obtained assuming that the arrivals and departures are given for the period 1980–1985 and that, in addition, the full migration matrix for the period 1975–1980 is given. Now we assume that the information at hand is restricted to the historical migration matrix (1975–1980) and a migration survey carried out in the West. The survey gives the arrivals in the West during the period 1980–1985 by region of origin, and the departures from the West during the same period by region of destination. The data are shown in the fourth row and fourth column of the 1980–1985 migration flows in Table 6.1. The 1980–1985 migration flows are predicted using the survey data and the historical matrix of 1975–1980. Table 6.1(d) shows the results. The migrations from the West and to the West are equal to the input data. The other migration flows are obtained assuming the interaction patterns observed in the 1975–1980 period.

Suppose that experts indicate that the attractiveness of the West diminished in the early 1980s and that the South became more attractive. In addition, assume that other studies showed an increased propensity to leave the Northeast and the Midwest. That information can be incorporated into the parameter values of the log-linear model – that is, into the odds. Suppose the odds that a migrant selects the South rather than the West is 20 % higher than revealed by the migration pattern to the West between 1980 and 1985, and suppose that the odds that a migrant into the West originated in the Northeast (rather than in the West) is 9 % higher, and that in the Midwest it is 20 % higher than revealed by the patterns of arrival in the West. The parameters of the log-linear model with offset would then change accordingly. Table 6.1(e) shows the predictions. The predictions are considerably improved, except for the migration from the Northeast to the West, which is highly overpredicted.

6.6   Conclusion

This chapter presents a unified perspective on the modelling of migration. The level and direction of migration may be represented by different data types. Three types are distinguished: counts, rates and probabilities. The different representations of migration may be reduced to one of these types or a combination. Different types of migration data may be modelled using techniques that have been developed for the analysis of life history data. The core consists of a (multistate) transition model in discrete time and continuous time. Transitions recorded in continuous time are direct transitions or events (migrations). The migration model is a transition rate model. Transitions recorded in discrete time refer to migrants who are identified by comparing the place of residence of a person at two consecutive points in time. The appropriate model is a transition probability model, which is a multinomial logistic regression model.

A major conclusion of the chapter is that models that have been developed to study life histories are perfectly suited for the study of migration. They allow the analysis of data of different types and the conversion of one data type into another. They also allow the treatment of incomplete data. Initially migration models were spatial interaction models. Today, they are increasingly considered as applications of multistate event history models.

References

Agresti A. 1996. *An introduction to categorical data analysis.* New York: John Wiley & Sons, Inc.

Albert JH and Gupta AK. 1983. Estimation in contingency tables using prior information. *Journal of the Royal Statistical Society B* 45(1):60–69.

Andersen PJ, Borgan O, Gill RD and Keiding N. 1993. *Statistical models based on counting processes.* New York: Springer.

Aoki M. 1976. *Optimal control and system theory in dynamic economic analysis.* New York: North-Holland.

Armstrong JS and Collopy F. 1998. Integration of statistical methods for judgment for time series forecasting: principles and empirical research. In *Forecasting with judgment.* Wright G and Goodwin P, eds, pp. 269–293. New York: John Wiley & Sons, Inc. Available at http://design.cwru.edu/researchArticles/ForecastIntegration.pdf (accessed 7 September 2005).

Baydar N. 1983. *Analysis of the temporal stability of migration in the context of multiregional forecasting.* Working Paper no. 38. Netherlands Interuniversity Demographic Institute, Voorburg.

Bell M, Blake M, Boyle P, Duke-Williams O, Rees P, Stillwell J and Hugo G. 2002. Cross-national comparison of internal migration: issues and measures. *Journal of the Royal Statistical Society* 165(3):435–464.

Berentsen WH and Cromley RG. 2005. Interstate migration flows in Germany since unification: temporal and spatial patterns. *Eurasian Geography and Economics* 46(3):185–201.

Blossfeld HP and Rohwer G. 2002. *Techniques of event history modeling. New approaches to causal analysis,* 2nd edn. Mawah, NJ: Lawrence Erlbaum.

Burch TK. 1999. *Computer modelling of theory: explanation for the 21st century.* Discussion Paper no. 99–4. Population Studies Center, University of Western Ontario, London, Ontario.

Burch TK. 2003. Demography in a new key. A theory of population theory. *Demographic Research* 9(11):261–284.

Çinlar E. 1975. *Introduction to stochastic processes.* Englewood Cliffs, NJ: Prentice-Hall.

Congdon P. 2001. *Bayesian statistical modelling.* Chichester: John Wiley & Sons, Ltd.

Cook T, Falchi P and Mariano R. 1984. An urban allocation model combining time series and analytic hierarchy methods. *Management Science* 30:198–208.

Cooke RM. 1991. *Experts in uncertainty. Opinion and subjective probability in science.* Oxford: Oxford University Press.

Courgeau D. 1973. Migrants et migrations. *Population* 28:95–128.

Courgeau D. 1982. Comparaison des migrations internes en France et aux Etats-Unis. *Population* 6:1184–1188.

De Mesnard L. 2004. Biproportional methods of structural change analysis: a typological survey. *Economic Systems Research* (16(2):205–230.

Director SW and Rohrer RA. 1972.*Introduction to system theory.* New York: McGraw-Hill.

Duke-Williams O and Blake M. 1999. Database fusion for the comparative study of migration data. Presented at the *Fourth International Conference of Geocomputation,* Fredricksburg, Virginia, 25–28 July. Available at http://www.geovista.psu.edu/sites/geocomp99/Gc99/068/gc_068.htm (accessed 10 September 2005).

Evans M, Hasting N and Peacock B. 2000. *Statistical distributions,* 3rd edn. New York: John Wiley & Sons, Inc.

George MV and Perreault J. 1992. Methods of external migration projections and forecasts. In *National population forecasting in industrialized countries.* Keilman N and Cruijsen H, eds, pp. 87–103. Amsterdam: Swets and Zeitlinger.

Hachen DS. 1988. The competing risk model. *Sociological methods and research,* 17(1): 21–54. Reprinted in *Readings in population research methodology.* Bogue DJ, Arriaga EE and Anderton DL, eds, pp. 21.85–21.101. Chicago: Social Development Center; and New York: UNFPA.

Harker PT. 1986. The use of expert judgments in predicting interregional migration patterns: an analytic hierarchy approach. *Geographical Analysis* 18(1):62–80.

Harris B. 1983. Positive and normative aspects of modelling large-scale social systems. In *Systems analysis in urban policy-making and planning.* Batty M and Hutchinson B, eds, pp. 475–490. New York: Plenum.

Jackson RW, Rogerson P, Plane D and O'Huallachain B. 1990. A causative matrix approach to interpreting structural change. *Economic Systems Research* 2(3):259–269.

Jörnsten KO, Larsson T, Lundgren JT and Migdalas A. 1990. An entropy model with variable target. *Environment and Planning A* 22(4):493–506.

Kitsul P and Philipov D. 1981. The one year/five year migration problem. In *Advances in multiregional demography.* Rogers A, ed. Research Report RR-81-6, pp. 1–33. Laxenburg: International Institute for Applied Systems Analysis.

Knudsen DC. 1992. Generalizing Poisson regression: including a priori information using the method of offsets. *Professional Geographer* 44(2):202–208.

Kupiszewski M and Kupiszewska D. 2003. *Internal migration component in subnational population projections in member states of the European Union.* CEFMR Working Paper 2/2003. Central European Forum for Migration Research, Warsaw.

Lahr M. 2004. Biproportional techniques in input–output analysis: table updating and structural analysis. *Economic Systems Research* 16(22):115–134.

Ledent J. 1980. Multistate life tables: movement versus transition perspectives. *Environment and Planning A* 12:533–562

Long JF and Boertlein CG. 1981. *Using migration measures having different intervals.* Washington, DC: US Bureau of the Census.

McLachlan GJ and Krishnan T. 1997. *The EM algorithm and extensions*. New York: John Wiley & Sons, Inc.

Mueser P. 1989. The spatial structure of migration: an analysis of flows between states in the USA over three decades. *Regional Studies* 23(3):185–200.

Plane DA and Mulligan GF. 1997. Measuring spatial focusing in a migration system. *Demography* 34:251–262.

Plane DA and Rogerson PA. 1986. Dynamic flow modeling with inter-regional dependency effects: an application to structural change in the US migration system. *Demography* 23(1):91–104.

Pooler J. 1992. Spatial uncertainty and spatial dominance in interaction modelling: a theoretical perspective on spatial competition. *Environment and Planning A* 24(7):5–1008.

Rajulton F. 1999. LIFEHIST: Analysis of life histories: a state-space approach. Presented at the *Workshop on Longitudinal Research in Social Science: A Canadian Focus*. Windermere Manor, London, Ontario, 25–27 October .

Rees P and Willekens FJ. 1986. Data and accounts. In *Migration and settlement: A multiregional comparative study*, Rogers A and Willekens FJ, eds, pp. 19–58. Dordrecht: Reidel.

Rogers A and Raymer J. 1998. The spatial focus of US interstate migration flows. *International Journal of Population Geography* 4:63–80.

Rogers A and Raymer J. 2001. Immigration and the regional demographics of the elderly population in the United States. *The Journals of Gerontology Series B: Psychological Sciences and Social Sciences* 56(1):S44–S55.

Rogers A and Sweeney S. 1998. Measuring the spatial focus of migration patterns. *Professional Geographer* 50(2):232–242.

Rogers A and Willekens FJ. 1986. A short course on multiregional demography. In *Migration and settlement: A multiregional comparative study*. Rogers A and Willekens FJ, eds, pp. 355–384. Dordrecht: Reidel.

Rogers A, Willekens FJ and Raymer J. 2001. Modeling interregional migration flows: continuity and change. *Mathematical Population Studies* 9:231–263.

Rogers A, Willekens FJ and Raymer J. 2002a. Capturing the age and spatial structures of migration. *Environment and Planning A* 34:341–359.

Rogers A, Willekens FJ, Little K and Raymer J. 2002b. Describing migration spatial structure. *Papers in Regional Science: Journal of the Regional Science Association International* 81(1):29–48.

Rogers A, Willekens FJ and Raymer J. 2003a. Imposing age and spatial structures on inadequate migration flow data sets. *The Professional Geographer* 55(1):56–69.

Rogers A, Raymer J and Newbold KB. 2003b. Reconciling and translating migration data collected over time intervals of different widths. *The Annals of Regional Science* 37(4):581–601.

Rohwer G and Pötter U. 1999. *TDA user's manual*. Ruhr-Universität Bochum, Fakultät für Sozialwissenschaften, Bochum, Germany.

Roy JR and Thill JC. 2004. Spatial interaction modelling. *Papers in Regional Science* 83(1):339–361.

Schoen R. 1988. *Modeling multigroup populations*. New York: Plenum Press.

Sen A and Smith T. 1995. *Gravity models of spatial interaction behavior*. Berlin: Springer.

Singer B and Spilerman S. 1979. Mathematical representations of development. theories. In *Longitudinal research in the study of behavior and development*. Nesselroade JR and Baltes PB, eds, pp. 155–177. New York: Academic Press.

Snickars F and Weibull JW. 1977. A minimum information principle. Theory and practice. *Regional Science and Urban Economics* 7:137–168.

Stillwell J. 2005. Inter-regional migration modeling: a review and assessment. Presented at the *45th Congress of the European Regional Science Association*, Amsterdam, 23–27 August.

Stone R. 1961. *Input–Output and national accounts*. Organization for Economic Cooperation, Paris

Strang G. 1980. *Linear algebra and its applications*, 2nd edn. New York: Academic Press.

Taylor HE and Karlin S. 1998. *An introduction to stochastic modeling*, 3rd edn. San Diego: Academic Press.

Tinbergen J. 1981. The use of models: experiences and prospects. *American Economic Review* 71(6):17–29.

van de Kaa DJ. 1993. European migration at the end of history. *European Review* 1(1):87–108.

van der Gaag NE, van Imhoff E and van Wissen L. 2000. Internal migration scenarios and regional population projections for the European Union. *International Journal of Population Geography* 6:1–19.

van Imhoff E. 1990. The exponential multidimensional demographic projection model. *Mathematical Population Studies* 2(3):171–182

van Imhoff E and Keilman N. 1991. *LIPRO 2.0: An application of a dynamic demographic projection model to household structure in the Netherlands*. Amsterdam: Sweets and Zeitlinger.

van Imhoff E, van Wissen L and Spiess K. 1994. *Regional population projection in the countries of the European Economic Area*. NIDI & CBGS Publication no. 31. Amsterdam: Swets & Zeitlinger.

van Imhoff E, van der Gaag N, van Wissen L and Rees P. 1997. The selection of internal migration models for European regions. *International Journal of Population Geography* 3:137–159.

van Wissen LJG, van der Gaag N, Rees P and Stillwell J. 2005. In search of a modelling strategy for projecting internal migration in European countries: demographic versus economic–geographical approaches. Presented at the *45th Congress of the European Regional Science Association*, Amsterdam, 23–27 August. Available at http://www.feweb.vu.nl/ersa2005/final_chapters/787.pdf

Willekens FJ. 1980. Entropy, multiproportional adjustment and the analysis of contingency tables. *Systemi Urbani* 2:171–201.

Willekens FJ. 1982. Multidimensional population analysis with incomplete data. In *Multidimensional mathematical demography*. Land KC and Rogers A, eds, pp. 43–112. New York: Academic Press. Reprinted in *Readings in population research methodology*. Bogue DJ, Arriaga EE and Anderton DL, eds, pp. 22.49–22.82. Chicago: Social Development Center; and New York: UNFPA.

Willekens FJ. 1983. Log-linear modelling of spatial interaction. *Papers of the Regional Science Association* 52:187–205.

Willekens FJ. 1994. Monitoring international migration in Europe: towards a statistical data base combining data from different sources. *European Journal of Population* 10(1):1–42.

Willekens FJ. 1999. Modeling approaches to the indirect estimation of migration flows: from entropy to EM. *Mathematical Population Studies* 7(3):239–278.

Willekens FJ. 2001. Theoretical and technical orientations toward longitudinal research in the social sciences. *Canadian Journal of Population* 28(2):189–217. Revised version of The life course: models and analysis. In *Population issues: An interdisciplinary focus.* van Wissen LJG and Dykstra PA, eds, pp. 23–51. New York: Kluwer Academic/ Plenum.

Willekens FJ and Drewe P. 1984. A multiregional model for regional demographic projection. In *Demographic research and spatial policy.* ter Heide H and Willekens FJ, eds, pp. 309– 334. London: Academic Press.

Wilson T and Bell M. 2004. Comparative empirical evaluations of internal migration models in subnational population projections. *Journal of Population Research* 21(2):127–160.

Yamaguchi K. 1991. *Event history analysis.* Newbury Park, CA: Sage.

7

Bayesian estimation of migration flows

Matthew J. Brierley*, Jonathan J. Forster*, John W. McDonald† and Peter W. F. Smith‡

School of Mathematics, University of Southampton
†*Centre for Longitudinal Studies, Institute of Education, University of London*
‡*Division of Social Statistics, School of Social Sciences, University of Southampton*

7.1 Introduction

In this chapter, we explore a Bayesian modelling framework for generating estimates of place-to-place migration flows. The framework is designed to offer a flexible approach which is capable of dealing with flows of varying quality and missingness and providing reliable estimates that have meaningful measures of precision. More specifically, the motivation for this work stems from a paper by Raymer (2007), particularly his analysis and estimation of international migration between countries in Northern Europe during 1999–2000. The modelling framework assumes that the marginal totals of these migration flow tables have already been constructed. The modelling framework developed is capable of performing in one direct step, through a Markov chain Monte Carlo (MCMC) approach, a similar analysis to the more *ad hoc* sequential approach implemented by Raymer (2007).

International Migration in Europe: Data, Models and Estimates Edited by J. Raymer and F. Willekens
© 2008 John Wiley & Sons, Ltd

This chapter first discusses the main issues. The new methodology is introduced in Section 7.2. In Section 7.3, the flows in a small migration dataset are analysed to test the modelling framework, which is extended in Sections 7.4 and 7.5 by adding noise to the observed data and by including contiguity in the model. The Northern European migration dataset is analysed in Section 7.6. Here, the new approach and generated estimates are compared and contrasted with those in Raymer (2007). The chapter ends with a conclusion. The aim is to develop a modelling process that generates estimates for an origin-by-destination flow table given either a corresponding table of reported values, or a subtable of the true flow table.

7.1.1   Using log-linear models to explain migration flows

It is natural to use *log-linear models* to capture the spatial patterns of migration (see also Chapters 6 and 10), modelling the origin–destination flows (y_{ij}) as a function of four components. Consider the saturated log-linear model, which may be expressed as

$$\log y_{ij} = \mu + \alpha_i + \beta_j + \gamma_{ij}, \qquad i \neq j. \tag{7.1}$$

This model is convenient because of the natural interpretation of the parameters: μ can be thought of as a reflection of the overall level of migration present in the system; $\boldsymbol{\alpha}$ is a vector representing a *pushing* term for each region, and expresses the relative likelihood of any individual outmigrating from that particular region; $\boldsymbol{\beta}$ is a vector representing a *pulling* term for the region, and expresses the relative likelihood of a person inmigrating to the particular region; and γ_{ij} represents an interaction factor between origin i and destination j. Regarding the interaction factor, if two regions neighbour each other, we may expect an increased number of migrants to destination j from origin i and vice versa, so that both γ_{ij} and γ_{ji} may be large. Conversely, if two regions lie far apart, then we may expect a reduced flow between the two regions. Finally, constraints are usually applied to the model parameters to ensure identifiability. Throughout the modelling procedure, where a log-linear model is used, we set $\alpha_1 = \beta_1 = 0$ (and, when used, $\gamma_{i1} = 0$, $\gamma_{1j} = 0$), which are corner-point constraints.

7.1.2   Quasi-independence

One important log-linear model often used at some stage of predicting migration flows is the *quasi-independence model*. The full independence model is inappropriate, since typically we do not model the diagonal entries. This relatively simple model can often account for much of the variation observed in an origin–destination flow table. The quasi-independence model can be expressed as

$$\log y_{ij} = \mu + \alpha_i + \beta_j, \qquad i \neq j. \tag{7.2}$$

Table 7.1 US interregional migration, 1985–1990: observed and quasi-independent fitted values (in parentheses).

Residence in 1980	Residence in 1985				
	Northeast	Midwest	South	West	Total
Northeast		100 (126.6)	366 (312.9)	124 (150.5)	590
Midwest	87 (117.4)		515 (531.1)	302 (255.5)	904
South	172 (133.2)	225 (243.8)		270 (290.0)	667
West	63 (71.4)	176 (130.6)	286 (323.0)		525
Total	322	501	1 167	696	2 686

Source: Agresti (1990).

The μ, α and β parameters have the same interpretations as before. This model gives us the migration flows we would expect to observe if there were no interaction effects between any of the regions in the migration flow table. For example, Table 7.1 shows the observed and quasi-independent fitted values for the 1980–1985 interregional migration flows in the United States, obtained from Table 10.2 in Agresti (1990). The parameter estimates for the quasi-independence fit are: $\hat{\mu} = 4.24$, $\hat{\alpha} = (0, 0.53, 0.66, 0.03)^T$ and $\hat{\beta} = (0, 0.60, 1.51, 0.78)^T$. The greatest relative difference between the observed values and quasi-independent fitted values is 34 % for the migration flow between the Midwest and the Northeast. However, with additional information we would wish to improve upon these estimates. For example, it appears that the fitted values for both flows from West to Midwest and from Midwest to West are much lower than the true values; if we had some external information or prior knowledge that we could expect an interaction between these two regions, we could incorporate it into a modelling procedure to improve the estimates.

7.1.3   A brief overview of Raymer's approach

The examples included in Raymer (2007) assume that the margins of an origin–destination flow table are either known or have previously been estimated. The quasi-independent fitted values based on these margins form *expected migration flows*. Flows where no reported data are available are estimated based on the quasi-independence fitted values, but with an adjustment, based upon whether or not the two regions share a common border. The adjustment is derived by examining discrepancies between available observed flows and quasi-independence fitted values across contiguous and non-contiguous regions. The disadvantage of this approach is that the observed flows, and those estimated using the process described above, are not designed to provide values which match the margins of the true migration flow table. To achieve this, some *ad hoc* further adjustment (scaling of flows) is required. In addition, it may be difficult to discern what level of confidence we should have in the estimated flows, since only point estimates are produced. The aim, therefore, is ultimately to perform a similar analysis to Raymer, using quasi-independent fitted values as a guide and including a correction for the increased migration flows that we may expect to observe in contiguous regions.

7.2   A modelling framework

The framework outlined here is designed to predict the true values of an origin–destination flow matrix, with associated measures of precision. We take the following as a starting point:

1. The margins of the true origin–destination migration flow matrix are known.

2. We have either some reported values or some known subsection of the true migration flow matrix.

Of course, assumption 1, that the exact margins of the origin–destination flow matrix are known, is unlikely to be true in practice. However, we assume that some prior analysis has been conducted which has resulted in acceptably accurate estimates of the margins. This may involve *double usage* of the data, i.e. the observed data are used in the process of margin estimation, which are then used for prediction of the cell values in combination with the observed data. In practice, the margin estimation is often computed by independent methods. For example, net migration flows may be computed for a particular time period as a residual by subtracting natural increase (i.e. births − deaths) from the total population change. Clearly, assumption 2 is necessary, since, without any reported values or known migration flows, there is little we can do other than simply fit a quasi-independence model to the margins.

The initial model considered forms a framework from which more complex models can be easily developed. For the model, we take as a starting point a matrix $\mathbf{Z}$ of reported migration flows and a matrix $\mathbf{Y}$ of true migration flows with unknown entries, but fixed margins:

$$\mathbf{Z} = \begin{pmatrix} 0 & z_{12} & z_{13} & \cdots & z_{1n} \\ z_{21} & 0 & z_{23} & \cdots & z_{2n} \\ z_{31} & z_{32} & 0 & \cdots & z_{3n} \\ \vdots & \vdots & \vdots & \ddots & \vdots \\ z_{n1} & z_{n2} & z_{n3} & \cdots & 0 \end{pmatrix}, \quad \mathbf{Y} = \begin{pmatrix} 0 & y_{12} & y_{13} & \cdots & y_{1n} \\ y_{21} & 0 & y_{23} & \cdots & y_{2n} \\ y_{31} & y_{32} & 0 & \cdots & y_{3n} \\ \vdots & \vdots & \vdots & \ddots & \vdots \\ y_{n1} & y_{n2} & y_{n3} & \cdots & 0 \end{pmatrix} \begin{pmatrix} y_{1+} \\ y_{2+} \\ y_{3+} \\ \vdots \\ y_{n+} \end{pmatrix}.$$

$$\begin{pmatrix} y_{+1} & y_{+2} & y_{+3} & \cdots & y_{+n} \end{pmatrix} \qquad (7.3)$$

Initially we assume that the matrix $\mathbf{Z}$ is complete, and that the relationship between $\mathbf{Y}$ and $\mathbf{Z}$ is

$$\log z_{ij} \sim N(\log y_{ij}, \sigma^2), \qquad i \neq j, \tag{7.4}$$

where the $\log z_{ij}$ are independent, with common variance σ^2.

We choose the log-normal distribution here, for computational tractability. A Poisson model for the observed counts might be considered more appropriate. However, for the large counts observed in migration tables, we expect a normal approximation to be adequate. Here, we are making two implicit assumptions. First, that each of the reported values is of the same accuracy. If this assumption is not reasonable, then we could allow for heterogeneity by using σ_i^2, σ_j^2 or σ_{ij}^2 accordingly. In Section 7.6.3, we relax the common variance assumption above to reflect more accurately the mean–variance relationship in a Poisson model. Second, we assume that the reported values are *unbiased*. Typically, we may anticipate migration estimates to suffer from undercounting, as people are missed by registration systems, or do not properly complete surveys.

The matrix $\mathbf{Y}$ is the actual unknown origin–destination migration matrix. The column and row sums of the matrix are assumed to be known, i.e. we know y_{i+} for all i and we know y_{+j} for all j. We assume that the true migration flows, $\log y_{ij}$, follow a *prior distribution* which is centred on quasi-independence:

$$\log y_{ij} \sim N(\mu + \alpha_i + \beta_j, \tau^2), \qquad i \neq j. \tag{7.5}$$

In Section 7.5 we extend the model by including a contiguity parameter. Thus the idea behind the initial model is to produce estimates which best reflect the observed values given that they must also satisfy known margins, and given that quasi-independent fitted values often provide reasonable estimates of the true migration flows.

7.2.1   The initial model

The full hierarchical model is given below; we introduce normal prior distributions for the parameters α, β and μ, and inverse gamma distributions (IG) for σ^2 and τ^2. These distributions are conditionally conjugate, i.e. the prior and posterior densities have the same algebraic form, which makes computation more straightforward:

$$\log z_{ij} \sim N\left(\log y_{ij}, \sigma^2\right), \qquad i \neq j;$$

$$\log y_{ij} \sim N\left(\mu + \alpha_i + \beta_j, \tau^2\right) \times I\left[\sum_j y_{ij} = y_{i+}, \sum_i y_{ij} = y_{+j}\right], \qquad i \neq j;$$

$$\mu \sim N\left(0, \sigma_\mu^2\right); \qquad \alpha_i \sim N\left(0, \sigma_\alpha^2\right), \qquad i = 2, \ldots, n;$$

$$\beta_j \sim N\left(0, \sigma_\beta^2\right), \qquad j = 2, \ldots, n; \qquad \alpha_1 = \beta_1 = 0; \tag{7.6}$$

$$\sigma^2 \sim \text{IG}\left(a, b\right); \qquad \tau^2 \sim \text{IG}\left(c, d\right).$$

Here σ_μ^2, σ_α^2, σ_β^2, a, b, c and d are all required to be specified as initial conditions for the model, and $I[\cdot, \cdot]$ denotes the indicator function, which is 1 if the y_{ij} have the required marginal totals. The values for σ_μ^2, σ_α^2 and σ_β^2 are chosen to be large enough that the quasi-independent fitted parameter values (given in Section 7.1.2) are well supported. Initially, we assume that we have no real information as to the accuracy of the observed data, or the suitability of a quasi-independent fit, and set diffuse priors for σ^2 and τ^2, namely, $\sigma_\mu^2 = 5$, $\sigma_\alpha^2 = \sigma_\beta^2 = 2$ and $a = b = c = d = 0.001$.

7.2.2   The model

The full initial model can be expressed as:

$$
\begin{aligned}
f &\left(\mathbf{Z}, \mathbf{Y}, \boldsymbol{\alpha}, \boldsymbol{\beta}, \mu, \sigma^2, \tau^2 \right) \\
&= f\left(\mathbf{Z} \mid \mathbf{Y}, \sigma^2 \right) f\left(\mathbf{Y} \mid \boldsymbol{\alpha}, \boldsymbol{\beta}, \mu, \tau^2 \right) f\left(\boldsymbol{\alpha} \right) f\left(\boldsymbol{\beta} \right) f\left(\mu \right) f\left(\sigma^2 \right) \\
&\quad \times f\left(\tau^2 \right) f\left(\{ y_{i+}, y_{+j} \} \mid \mathbf{Y} \right) \\
&\propto \left(\sigma^2 \right)^{-n(n-1)/2} \exp\left[-\frac{1}{2\sigma^2} \sum_{ij} \left(\log z_{ij} - \log y_{ij} \right)^2 \right] \\
&\quad \times \left(\tau^2 \right)^{-n(n-1)/2} \exp\left[-\frac{1}{2\tau^2} \sum_{ij} \left(\log y_{ij} - \mu - \alpha_i - \beta_j \right)^2 \right] \\
&\quad \times \exp\left(-\frac{1}{2\sigma_\alpha^2} \sum_i \alpha_i^2 - \frac{1}{2\sigma_\beta^2} \sum_j \beta_j^2 - \frac{1}{2\sigma_\mu^2} \mu^2 \right) \\
&\quad \times \left(\sigma^2 \right)^{-(a+1)} \exp\left(-b/\sigma^2 \right) \left(\tau^2 \right)^{-(c+1)} \exp\left(-d/\tau^2 \right) \\
&\quad \times I\left[\sum_j y_{ij} = y_{i+}, \sum_i y_{ij} = y_{+j} \right].
\end{aligned}
\tag{7.7}
$$

7.2.3   Markov chain Monte Carlo techniques

Unfortunately, the model is too complex for direct computations, so we are forced to consider other methods to learn about the posterior distribution of the migration flow matrix $\mathbf{Y}$. *Markov chain Monte Carlo* (MCMC) techniques offer the best way of achieving this. Gamerman (1997) provides a thorough overview of the theory behind MCMC methods, and summarises the thinking behind the approach.

To construct a Markov chain with a limiting distribution we use a hybrid of *Gibbs sampling* and the *Metropolis–Hastings algorithm*. Gibbs sampling is based on a Markov chain where the dependence of each observation on its predecessor is governed by the conditional distributions that arise from the model. The constraint that the margins of the $\mathbf{Y}$ are fixed means that a pure Gibbs sampler is inappropriate for our model. Metropolis–Hastings algorithms are based on constructing a Markov chain where the dependence of each observation on its predecessor is split into two parts, a proposal and an acceptance probability p. The proposal distribution is essentially arbitrary, and the acceptance probability makes sure the appropriate limiting distribution of the chain is maintained by stochastically rejecting certain proposed moves of the chain. Therefore, we employ a Metropolis–Hastings approach to generate the proposed **Y** matrices, and use a Gibbs sampler for each of the remaining model parameters.

7.2.4   The updating equations

Both the Gibbs sampling and Metropolis–Hastings sampling processes require computation of the posterior conditional distributions of each of the parameters considered in the model. The priors set for the initial model have been chosen to make this task as simple as possible, and the results are as follows:

$$
f\left(\mathbf{Y} \mid \mathbf{Z}, \boldsymbol{\alpha}, \boldsymbol{\beta}, \mu, \sigma^2, \tau^2\right)
$$

$$
\propto I\left[\sum_j y_{ij} = y_{i+}, \sum_i y_{ij} = y_{+j}\right] \prod_{ij} f\left(z_{ij} \mid y_{ij}, \sigma^2\right) f\left(y_{ij} \mid \alpha_i, \beta_j, \mu, \tau^2\right)
$$

$$
\propto I\left[\sum_j y_{ij} = y_{i+}, \sum_i y_{ij} = y_{+j}\right]
$$

$$
\times \prod_{ij} \exp\left[-\frac{1}{2\sigma^2}\left(\log z_{ij} - \log y_{ij}\right)^2\right] \exp\left[-\frac{1}{2\tau^2}\left(\log y_{ij} - \mu - \alpha_i - \beta_j\right)^2\right]
$$

$$
\propto I\left[\sum_j y_{ij} = y_{i+}, \sum_i y_{ij} = y_{+j}\right]
$$

$$
\times \prod_{ij} \exp\left\{-\frac{1}{2}\left(\frac{1}{\sigma^2} + \frac{1}{\tau^2}\right)\left[\log y_{ij} - \left(\frac{\sigma^2\left(\mu + \alpha_i + \beta_j\right) + \tau^2 \log z_{ij}}{\tau^2 + \sigma^2}\right)\right]^2\right\}.
$$

$$(7.8)$$

This is the density of a set of independent normally distributed log y_{ij}, conditioned to satisfy the marginal constraints. Note that the indicator function $I[\cdot, \cdot]$ assures that the y_{ij} have the required marginal totals.

For each α_i we require the conditional distribution given the other components of $\boldsymbol{\alpha}$, which we denote by $\boldsymbol{\alpha}_{\backslash i}$,

$$
\alpha_i \mid \mathbf{Y}, \mu, \boldsymbol{\alpha}_{\backslash i}, \boldsymbol{\beta}, \tau^2 \sim N\left(\frac{\sigma_\alpha^2 \sum_j (\log y_{ij} - \beta_j - \mu)}{\sigma_\alpha^2(n-1) + \tau^2}, \frac{1}{(n-1)/\tau^2 + 1/\sigma_\alpha^2}\right). \quad (7.9)
$$

For each β_j we require the conditional distribution given the other components of $\boldsymbol{\beta}$, which we denote by $\boldsymbol{\beta}_{\backslash j}$,

$$\beta_j \,|\, \mu, \boldsymbol{\alpha}, \boldsymbol{\beta}_{\backslash j}, \tau^2 \sim N\left(\frac{\sigma_\beta^2 \sum_i (\log y_{ij} - \alpha_i - \mu)}{\sigma_\beta^2(n-1) + \tau^2}, \frac{1}{(n-1)/\tau^2 + 1/\sigma_\beta^2}\right). \quad (7.10)$$

For μ, we obtain

$$\mu \,|\, \mathbf{Y}, \boldsymbol{\alpha}, \boldsymbol{\beta}, \tau^2 \sim N\left(\frac{\sigma_\mu^2 \sum_{ij} (\log y_{ij} - \alpha_i - \beta_j)}{n\sigma^2(n-1) + \tau^2}, \frac{\sigma_\mu^2 \tau^2}{n\sigma_\mu^2(n-1) + \tau^2}\right). \quad (7.11)$$

Finally, for the variance parameters σ^2 and τ^2 we obtain

$$\sigma^2 \,|\, \mathbf{Y}, \mathbf{Z} \sim \text{IG}\left(a + \frac{n(n-1)}{2}, b + \frac{\sum_{ij}(\log z_{ij} - \log y_{ij})^2}{2}\right) \quad (7.12)$$

and

$$\tau^2 \,|\, \mathbf{Y}, \mu, \boldsymbol{\alpha}, \boldsymbol{\beta} \sim \text{IG}\left(c + \frac{n(n-1)}{2}, d + \frac{\sum_{ij}(\log y_{ij} - \alpha_i - \beta_j - \mu)^2}{2}\right). \quad (7.13)$$

7.2.5   Generating the proposals

Since the margins of the migration flow matrix $\mathbf{Y}$ are fixed, each of the new proposals $\mathbf{Y}'$ generated in the MCMC process must always meet these marginal constraints or the acceptance probability will automatically be zero. The following procedure was chosen to ensure the suitability of the proposals. First, two (different) rows and columns are selected, such that the four points of intersection all lie off the diagonal of the current $\mathbf{Y}$. Then, a random number, ε, is added to or subtracted from these entries, in such a way that the overall row and column sums are unaltered:

$$\mathbf{Y}' = \mathbf{Y} + \begin{pmatrix} 0 & 0 & 0 & 0 & 0 & 0 \\ 0 & 0 & +\varepsilon & 0 & -\varepsilon & 0 \\ 0 & 0 & 0 & 0 & 0 & 0 \\ 0 & 0 & 0 & 0 & 0 & 0 \\ 0 & 0 & 0 & -\varepsilon & +\varepsilon & 0 \\ 0 & 0 & 0 & 0 & 0 & 0 \\ 0 & 0 & 0 & 0 & 0 & 0 \end{pmatrix}. \quad (7.14)$$

Here, ε is generated from a uniform distribution, where the lower and upper bounds are determined by the selected rows and columns to ensure that none of the entries in the proposal can be negative. For example, if the four entries of $\mathbf{Y}$ selected are a, b, c and d such that the collapsed matrix of selected entries is

$$\begin{pmatrix} a & b \\ c & d \end{pmatrix},$$

then $\varepsilon \sim \text{Unif}(-\min\{a, d\}, \min\{b, c\})$.

Using the Metropolis–Hastings algorithm, the proposed new $\mathbf{Y}$ is then accepted with probability

$$p = \min\left\{1, \frac{f(\mathbf{Y}')}{f(\mathbf{Y})}\frac{q(\mathbf{Y}\mid\mathbf{Y}')}{q(\mathbf{Y}'\mid\mathbf{Y})}\right\} = \min\left\{1, \frac{f(\mathbf{Y}')}{f(\mathbf{Y})}\right\}, \qquad (7.15)$$

since the uniform proposal density q is symmetric in that

$$\begin{aligned}
q(\mathbf{Y}'\mid\mathbf{Y}) &= \frac{1}{\min\{b, c\} + \min\{a, d\}} \\
&= \frac{1}{\min\{b-\varepsilon, c-\varepsilon\} + \min\{a+\varepsilon, d+\varepsilon\}} \\
&= q(\mathbf{Y}\mid\mathbf{Y}'). \qquad (7.16)
\end{aligned}$$

Note that, although the true migration flows must be integers, we do not insist that ε and hence the proposals are integers. This relaxation makes construction of an irreducible Markov chain to sample from the posterior distribution much more straightforward. Also, since we assume that the marginal totals of the true migration flow matrix $\mathbf{Y}$ are known, the iterative proportional fitting (IPF) algorithm can be employed to generate an initial $\mathbf{Y}$ matrix for the MCMC simulation.

7.3   Results of the initial simulation

A program was written in S-PLUS to perform the MCMC simulation outlined in the previous section. Initially, the data from Table 7.1 were used for the $\mathbf{Z}$ matrix of reported values, and the true margins of this table ($\mathbf{Z}$) were used as the known margins of $\mathbf{Y}$. The advantages of beginning with this small dataset are twofold: first, in a practical sense, it is easier to discover errors with the S-PLUS code and ensure that the MCMC process is working correctly; and second, even in this scenario, it is not intuitively obvious what the expected results are. It is not clear if the sampled densities of each of the y_{ij} will favour the observed values or the quasi-independent fitted values. Naturally, we anticipate the densities to peak somewhere in between these two values, but the relative location and the shape of the densities are not at all trivial. By working on this smaller dataset, an understanding of the behaviour of the Markov chain can be gained prior to working on larger, more complex datasets.

An IPF procedure was applied to the known margins to generate the quasi-independent fit which was used as the initial value for $\mathbf{Y}$. The priors that were set for the model were as in Section 7.2.1, and the MCMC simulation was run for 10 thousand iterations. That is, starting from initial $\mathbf{Y}$ values satisfying quasi-independence and calculated using IPF, we successively update the parameter values for $\mathbf{Y}$, $\boldsymbol{\alpha}$, $\boldsymbol{\beta}$, μ, σ^2 and τ^2 using the distributions derived in Sections 7.2.4 and 7.2.5. The limiting distribution is the joint posterior distribution over all unknowns.

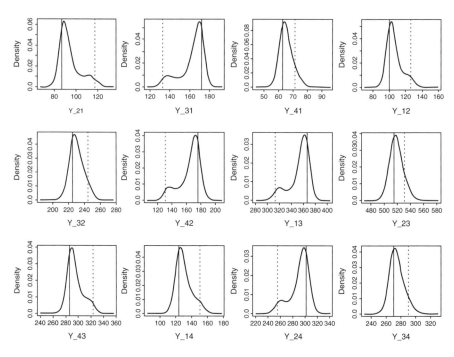

Figure 7.1   The sampled densities for each of the y_{ij} (full line = observed values, dashed line = quasi-independence fitted values).

Figure 7.1 shows the densities for each of the simulated entries of the true migration flow matrix **Y**. The densities are all maximised close to the observed values and the density at the observed value is always much greater than the density at the quasi-independent fitted value. The quasi-independent values plotted here are from the quasi-independent fit of the **Y** margins, and not from the sampled values of μ, α and β. The densities are mostly unimodal, but some have a secondary peak close to the quasi-independent fitted values. In this case, we would wish for the model to show a preference for the observed values; since they perfectly match the true margins, it makes sense that we should have a great deal of confidence in them, and we would want them to take precedence over the quasi-independent fitted values.

The sampled values of σ^2 and τ^2 are also interesting since they reflect the posterior estimates of the variance of the observations and deviance from the quasi-independent fitted values. Figure 7.2 is a scatterplot showing that they are highly correlated. In the MCMC simulation, σ^2 tends to hover very close to zero, whilst τ^2 tends to be larger and more variable. However, there are short periods where this is reversed, and τ^2 becomes very close to zero whilst σ^2 increases and becomes much more variable. This pattern continues throughout, rather than some balancing equilibrium being reached, and this is the same phenomenon that causes the densities to be bimodal, rather than unimodal. The means of the sampled values are 0.010

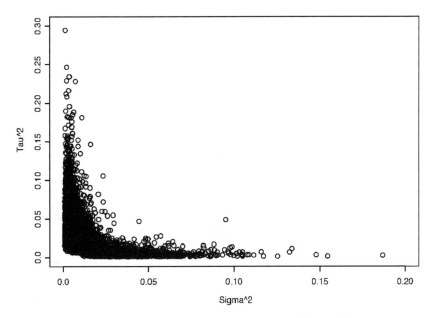

Figure 7.2   The relationship between σ^2 and τ^2.

and 0.040 for σ^2 and τ^2, respectively, showing that there is less variance between the observed values and the simulated values, than between the quasi-independent values and the simulated values. The remaining parameters in the model are less interesting for interpretation. The MCMC sample means are $\mu = 4.14$, $\boldsymbol{\alpha} = (0, 0.56, 0.75, 0.11)^{\mathrm{T}}$ and $\boldsymbol{\beta} = (0, 0.64, 1.56, 0.81)^{\mathrm{T}}$; comparing these to the quasi-independent parameters in Section 7.1.2 shows that they are very similar.

7.4   Adding noise to the observations

In the previous section, it was noted that the estimated densities for the true migration flows were pulled towards the observed values and away from the quasi-independent fitted values. In this section we investigate the effect of adding some noise to the reported values, and compare the resulting densities of the simulated y_{ij}. Again, we assume that the true migration values are those from Table 7.1. Alternative $\mathbf{Z}$ matrices are generated using a fixed value for σ^2. The MCMC process is then rerun with the same priors and the results are compared.

7.4.1   Generating observations with $\sigma = 0.03$

For the example presented here, we fixed σ as 0.03. Each z_{ij} was then randomly generated in turn from

$$z_{ij} \sim N(\log y_{ij}, 0.03^2).$$ (7.17)

The new **Z** matrix obtained was

$$\mathbf{Z} = \begin{pmatrix} 0 & 98 & 355 & 120 \\ 90 & 0 & 499 & 306 \\ 167 & 229 & 0 & 261 \\ 64 & 168 & 289 & 0 \end{pmatrix} \begin{pmatrix} 573 \\ 895 \\ 657 \\ 521 \end{pmatrix}.$$
$$\begin{pmatrix} 321 & 495 & 1143 & 687 \end{pmatrix}$$

Comparing this with the original **Z** (the true migration flow values) shows that the differences are slight. The greatest absolute difference between the new and previous flows is 16 for z_{23}. The relative differences for the flows are mostly around 3 %, and the relative differences for the margins are slightly smaller.

Figure 7.3 shows the estimated densities under this new scenario. Although the densities are still maximised around the observed values, the quasi-independent fitted values are much more prominent and the bimodality much more pronounced. In some cases, such as for y_{31} and y_{42} we have the strange situation of the density at the observed values and the quasi-independent fitted values being much greater than some of the region in between. This reflects that the system is not settling into some unimodal distribution, but oscillates between generating sampled values

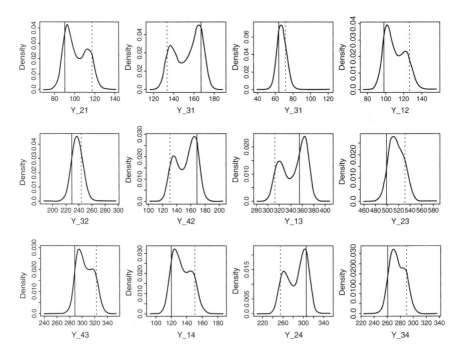

Figure 7.3 The sampled densities for each of the y_{ij} with observation matrix generated with $\sigma = 0.03$ (full line = observed values, dashed line = quasi-independence fitted values).

Table 7.2 Means of the simulated US interregional 1980–1985 migration flow estimates.

Residence in 1980	Residence in 1985			
	Northeast	Midwest	South	West
Northeast		111	346	134
Midwest	102		516	287
South	153	238		276
West	68	153	305	

close to the observed values, and then close to the quasi-independent values. One problem with bimodal densities is that it is unclear how we should provide point estimates of the migration flows from such densities. The mean of the sampled values is often the preferred estimator but in some cases the density at the mean may be much lower than for other values. For example, Table 7.2 shows the mean of the sampled values for y_{13} is 346 and the density plot shows that the density here is rather low. One attractive property of using the mean is that, since each of the sampled $\mathbf{Y}$ matrices has fixed margins, the means are guaranteed to satisfy the true margins. In contrast, taking the value where the density is maximised is not very attractive since it would not reflect the nature of the results and the resulting estimates will typically not meet the specified margins.

The means of the σ^2 and τ^2 values generated were 0.016 and 0.024, respectively. Comparing these to the original scenario in Section 7.3, we see that σ^2 is higher and τ^2 is lower, which is consistent with the increased prominence of the quasi-independent values in the sampled densities. Comparing the mean for σ^2 with the actual value used to generate the observed values shows that the estimate is much higher than the true value, suggesting that the modelling process may be giving less credibility to the observed values than we would like.

7.4.2 The priors for σ^2 and τ^2

Thus far, we have used uninformative priors for σ^2 and τ^2, reflecting that we have no prior beliefs either on the quality of the reported data, or on how well the data can be explained by a quasi-independent fit. Indeed, it is unlikely that we will ever have a large amount of information on how well the data can be expected to conform to a quasi-independent fit. It is much more likely, however, that we have some prior beliefs as to the quality of the reported data. Consulted experts should be able to give some indication of their beliefs regarding the accuracy of reported data values. For example, if it is believed that the reported flows are correct to within 15 %, a prior may be set accordingly. In Section 7.4.1 it was noted that, when σ^2 was fixed and a corresponding $\mathbf{Z}$ matrix was generated, the posterior estimate for σ^2 was higher than the actual value used. We now set priors with means equal to the true value of σ used, and rerun the MCMC to compare the results.

The prior was chosen to support the actual value of σ (0.03) used to generate the data. An IG(16, 0.0144) distribution was used as the prior for σ^2. Figure 7.4 shows the resulting posterior density estimates. There is no sign of the bimodality present when the prior for σ^2 was uninformative; with the new prior, the densities are now unimodal and the observed values dominate. Although it is unlikely that our observations would be expected to be so accurate, if we did, then we would want the estimates to be strongly favouring the observed values. The densities are not all maximised at the observed values though. Indeed, since the margins of the observed matrix no longer match the true margins, all of the observed values cannot be correct. In particular, the densities for region 3 are centred slightly away from the observed values. The most extreme example is y_{13}, where the density is very low at the observed value. The mean of the sampled σ^2 values now is 0.0009, and the mean of the τ^2 values is 0.0372. However, it seems that the quasi-independent fitted values are not very influential in this model.

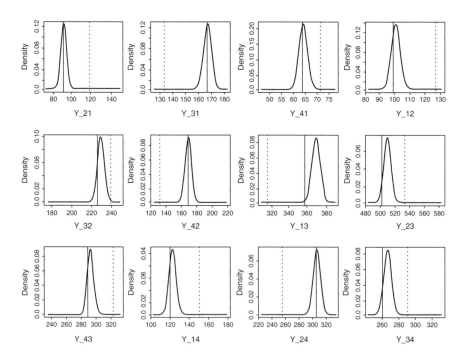

Figure 7.4   The sampled densities for each of the y_{ij} under the new prior for σ^2 (full line = observed values, dashed line = quasi-independence fitted values).

7.5   Introducing a contiguity parameter

Thus far, we have been assuming that the underlying structure of the true migration flow matrix is quasi-independent. In reality, this approach is likely to be oversimplistic; countries or regions which share borders, common languages or are closer in socio-economic or historical ways are likely to have greater migration flows than the quasi-independence model predicts, since there will be an interaction term which the model lacks. In this section we follow Raymer's (2007) approach by adding a fixed parameter for contiguity into the modelling process. It is straightforward to incorporate this directly into the modelling framework already set out. The new model becomes

$$\log y_{ij} = \mu + \alpha_i + \beta_j + \gamma d_{ij}, \quad i \neq j, \tag{7.18}$$

where $d_{ij} = 1$ if i,j share a border and 0 otherwise.

An artificial dataset was constructed to test the augmented model. We consider a system of four regions. Using $\boldsymbol{\alpha} = (0, 0.5, 1, 1.5)^T$, $\boldsymbol{\beta} = (0, 0.8, 0.8, 1.2)^T$ and $\mu = 4.0$ and assuming that only regions 3 and 4 share a border, with $\gamma = 1$, a migration flow matrix was generated as

$$\mathbf{Y} = \begin{pmatrix} 0 & 122 & 122 & 181 \\ 90 & 0 & 200 & 299 \\ 148 & 330 & 0 & 1339 \\ 245 & 545 & 1480 & 0 \end{pmatrix},$$

where, for example, $y_{34} = \exp(4 + 1 + 1.2 + 1) = 1339$. This artificial dataset may be thought of as a system of two islands (regions 1 and 2) and two neighbouring regions (3 and 4). The S-PLUS code was updated to include the new contiguity parameter, and an $N(0,2)$ prior was set for γ.

In the first scenario, we assume that we have perfect reported values and so $\mathbf{Z}$ was set to equal $\mathbf{Y}$ given above. For the initial value of $\mathbf{Y}$ in the MCMC process a quasi-independent fit of the true margins was computed. After running the process for three thousand iterations, the means of each of the sampled y_{ij} were equal to the true values when rounded to the nearest integer. The average values were $\sigma^2 = 0.003$, $\tau^2 = 0.004$ and $\gamma = 1.008$, reflecting that the process has managed to identify the true parameters used to generate the dataset.

7.5.1   Introducing some noise into the reported values

Here the earlier process of fixing σ^2 and using the known true migration flow table to generate some noisy observations is repeated to learn how the results change when the margins of the **Z** matrix do not match the true margins. As before, we fix $\sigma = 0.03$ to generate **Z**:

$$\mathbf{Z} = \begin{pmatrix} 0 & 117 & 128 & 180 \\ 88 & 0 & 192 & 292 \\ 157 & 330 & 0 & 1280 \\ 242 & 549 & 1433 & 0 \end{pmatrix}.$$

Figure 7.5 shows the sampled densities when **Z** was used for the reported values. For the two largest flows between regions 3 and 4 (i.e. the neighbouring regions), the densities are maximised close to the true values and virtually zero at the observed values. For the remaining 10 flows, the observed values are much closer to the true values, i.e. some of the densities are maximised closer to the observed values, some at the true values, and some almost exactly in between. For each flow, the density is unimodal and there are no signs of the bimodality that we observed in previous models. The means of the sampled values for σ^2 and τ^2 were both 0.004 and the sampled mean for γ was 0.955. Overall, the model seems to be working well here; the estimate for the distance parameter is very close to the true value used, and the predicted migration flows from this model appear sensible.

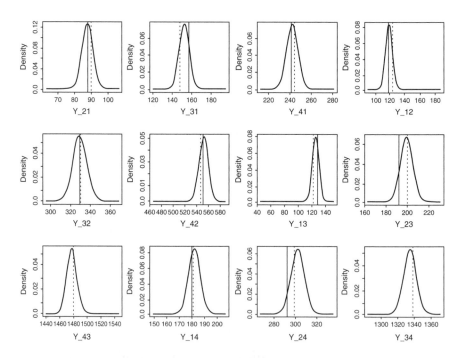

Figure 7.5   The sampled densities for each of the y_{ij} when the MCMC process was run with observation matrix **Z** (full line = observed values, dashed line = true values). Note that some observed flows are outside the plotting range.

There are many ways in which it may be interesting to continue the analysis of the behaviour of this modelling approach. For example, we may build more complex datasets with more regions sharing borders to see how capable the model is at finding the parameters used to construct the datasets. Also, we could investigate setting more informative priors on the accuracy of the reported values.

7.5.2 Where some of the y_{ij} are assumed known

Another situation that may occur is one where some reported values in the origin–destination flow matrix do not need to be predicted, i.e. they are already considered to be accurate. Here, the focus turns to estimating the remaining migration flows. If we have a scenario where there are some observations deemed perfect and no reported values for the other flows, we may dispense with the $\mathbf{Z}$ matrix altogether in the modelling approach.

Owing to the process in which the proposals are generated (Section 7.2.5), it is necessary that the unknown region of the overall migration flow table is at least 4 × 4. Therefore the artificial migration flow matrix created in Section 7.5.2 is expanded to include a fifth region, using $\boldsymbol{\alpha} = (0, 0.5, 1, 1.5, 1)^{\mathrm{T}}$, $\boldsymbol{\beta} = (0, 0.8, 0.8, 1.2, 0.8)^{\mathrm{T}}$ and $\mu = 4.0$. Again, we assume that the only regions sharing a border are regions 3 and 4 and that $\gamma = 1.0$. The augmented migration flow matrix now becomes

$$\mathbf{Y} = \begin{pmatrix} 0 & 122 & 122 & 181 & 122 \\ 90 & 0 & 200 & 299 & 200 \\ 148 & 330 & 0 & 1339 & 330 \\ 245 & 545 & 1480 & 0 & 545 \\ 148 & 330 & 330 & 493 & 0 \end{pmatrix}.$$

Now, consider the assumption that the flows from or to region 1 are considered completely reliable. Thus, these entries of the matrix $\mathbf{Y}$ are automatically fixed. We use the iterative proportional fitting algorithm to generate a full starting value for $\mathbf{Y}$ and then alter the S-PLUS code to ensure that the known flows are not changed for any of the proposals. Again the model was able to find the parameters used to generate the migration flow matrix. Taking the means of the sampled y_{ij} showed that they were all accurate to the nearest integer. The means of the sampled values for γ and τ were 0.992 and 0.004, respectively.

7.6 Northern European migration

In this section, we analyse the same Northern European immigration data used in Raymer (2007). The data represent migration during the 1999–2000 time period and come from the Eurostat NewCronos database. Table 7.3 shows the available reported *immigration* flows (i.e. flows reported by the receiving countries). Overall, we have information on 51 of the 90 total flows. The data provided by the Northern European statistical agencies vary in their migrant definitions and collection methods. (Refer to Chapter 3 for a more detailed review of the data and issues.) Since the reported

Table 7.3 International migration flows based on available *immigration* data for Northern Europe, 1999–2000.

Country of origin	Country of destination									
	Den.	Fin.	Ice.	Nor.	Swed.	Est.	Ire.	Lat.	Lith.	UK
Denmark	—	355	1 446	2 734	2 194					2 025
Finland	448	—	46	1 380	3 647					1 556
Iceland	1 267	54	—	463	384					
Norway	3 188	955	602	—	5 496					3 814
Sweden	2 298	3 229	572	6 044	—					1 539
Estonia	257	784	6	85	262	—				
Ireland	266	41	6	73	199		—			
Latvia	376	44	7	116	169			—		
Lithuania	499	21	33	104	111				—	149
UK	3 965	586	167	2 014	2 447	21 611				—
Total	12 564	6 069	2 885	13 013	14 909					

Source: Table 1 in Raymer (2007).

values of the Scandinavian flows are based on registration methods, they may be considered more reliable than the UK and Ireland estimates, which are based upon surveys (Poulain 1994).

7.6.1   Overview of Raymer's estimation strategy

The goal, as always, is to estimate the full migration flow table based on the available reported values. Again, it is assumed that the margins of the true flow table have already been estimated and are assumed correct. Raymer (2007) provides details of the methods used for the estimation of the margins for this example (see also Chapter 10). A brief overview of his approach is now given. A more detailed explanation of the theory behind the method may be found in Willekens (1999) and in Chapter 6.

1. The reported immigration table (i.e. flows reported by receiving countries) was combined with an emigration table (i.e. flows reported by the sending countries). This is done by favouring, where possible, the immigration flow data since they are assumed to be more reliable than emigration data. This assumption may not hold in all cases, for example, where the emigration data come from a population register and the immigration data come from a survey.

2. A quasi-independent fit on the given margins generates the expected migration flows.

3. The columns of the combined table of reported values are proportionally adjusted to equal the estimated column margins, and where the data are available, the observed-to-expected ratios are calculated.

4. Regression models, based on contiguity, estimate the ratios of observed to expected data for the unreported data, and hence interaction estimates are produced.

5. The known margins of the true migration flow table, and the interaction coefficients calculated and estimated, provide sufficient statistics to fit a saturated log-linear model, which provides the final estimates of the migration flows.

The estimated flows are set out in Raymer (2007: Table 5), who comments that: 'The only flows that appear questionable are those from Lithuania. Here, it is possible that the gross flow of emigration to Northern European countries (i.e. 9221) was overestimated. Otherwise the results appear reasonable.'

One feature of this method is that, in some sense, the structure of the reported values is preserved, i.e. the *cross-multiplicative ratios* or *odds ratios* of the reported values are consistent with those of the predicted values. However, the value of preserving these ratios is unclear; it may be an attractive property where the four reported values which make each ratio come from two different statistical agencies since there may be a cancellation of the errors inherent in the two agencies' estimation techniques. However, since the data are combined from two tables, in many situations the four reported values making each odds ratio may come from three different sources, in which case such cancellation does not exist, and the property seems unlikely to be realistic. A clear weakness of this method is that all of the reported values are considered equally reliable, and it does not make full use of the available information; where two observations of similar magnitude are reported for a migration flow we may have more confidence in the particular reported values, and would perhaps wish to reflect this at some stage of the modelling procedure.

7.6.2   Using the reported immigration data to generate estimates from the Bayesian model

We consider only the reported *immigration* values and use the framework created thus far to generate estimates for the full migration flow table. The conditional posterior distribution for $\mathbf{Y}$ takes a similar form to (7.8), a product of independent normal densities for y_{ij}, together with the marginal constraint. The $\mathbf{Z}$ matrix is incomplete so that the corresponding normal density for y_{ij} depends on whether or not there exists a corresponding reported value z_{ij}. Where we do have a reported value, the corresponding normal distribution is

$$N\left(\frac{\sigma^2(\mu + \alpha_i + \beta_j + \delta_{ij}\gamma) + \tau^2 \log z_{ij}}{\tau^2 + \sigma^2}, \frac{1}{1/\sigma^2 + 1/\tau^2}\right). \qquad (7.19)$$

Where we do not have a reported value, the corresponding normal distribution is determined by the quasi-independent fit and contiguity parameters, i.e.

$$N(\mu + \alpha_i + \beta_j + \delta_{ij}\gamma, \tau^2). \qquad (7.20)$$

For the estimation of the contiguity parameter γ we follow Raymer's approach of classifying the countries, i.e. countries that share a border, Scandinavian countries, and the former Soviet Union countries are contiguous.

In Section 7.4, it was observed that the modelling framework created gave more credence to the observed values when they were consistent with the true margins used for the $\mathbf{Y}$ matrix. Comparing the margins of the reported immigration data (where available) to the true margins shows that the reported margins range from being 25 % higher than the true margins for Sweden, to 22 % lower for Finland. Also, the reported flow of 21 611 from the UK to Ireland is greater than the total margins for both flows from the UK and flows to Ireland. Therefore, because there is some discrepancy between the $\mathbf{Z}$ margins and the true margins, if we set a non-informative prior for σ^2 the resulting estimates are much closer to the quasi-independent fitted values than we would believe to be true. This is natural, since without feeding any prior information into the model, the quasi-independent fitted values should be preferred as they can be made consistent with the margins, with zero variance. The introduction of the contiguity parameter helps to accommodate the observed values to a degree, but its value is underestimated.

Owing to some of the inconsistencies in the data it is difficult to determine how accurate we believe the reported values to be, and some expert insight would be helpful here to set a realistic prior truly reflecting our beliefs in the data quality. In this example, we set a prior on σ^2 centred around 0.01 (reflecting approximately an error in the reported values within ± 20 %, which seems reasonable given the data and the margins) using an IG(64, 0.64) prior. A quasi-independent fit on the known margins was used as the initial $\mathbf{Y}$ matrix and the MCMC simulation was for eight thousand iterations.

Because of the highly improbable starting point (i.e. the quasi-independent fit), the first 1.5 thousand iterations are excluded as *burn-in*, after which the process seems to be close to an equilibrium distribution. The resulting sampled means are shown in Table 7.4. The two largest flows predicted are 10 426 for the UK to Ireland and 7511 from Lithuania to Ireland. Raymer's estimates for these two flows are 15 435 and 3680, respectively. Indeed, the estimate for the UK to Ireland flow is perhaps the most important in the whole table, since it is by far the largest. Thus, for this flow alone, it may seem as if Raymer's estimate is much more feasible. However, considering the remaining estimates for flows originating in the UK, it

Table 7.4 Estimated migration flows from the MCMC simulation based on reported immigration flows.

Country of origin	Country of destination									
	Den.	Fin.	Ice.	Nor.	Swed.	Est.	Ire.	Lat.	Lith.	UK
Denmark	—	426	1 515	4 053	2 056	117	646	193	337	1 794
Finland	486	—	47	1 752	3 466	115	813	175	525	1 497
Iceland	1 195	59	—	459	356	34	111	41	80	123
Norway	2 893	1 227	559	—	3 261	85	255	99	178	2 511
Sweden	1 638	4 151	487	3 603	—	58	146	73	105	1 084
Estonia	271	1 057	6	90	258	—	396	275	500	355
Ireland	280	44	6	79	210	161	—	332	2 980	5 241
Latvia	392	47	7	126	175	325	859	—	2 055	944
Lithuania	524	22	33	114	114	282	7 511	466	—	155
UK	3 933	728	166	2 080	2 023	189	10 426	271	422	—

is clear that the new estimates are much more in tune with the reported values for the Scandinavian countries than Raymer's estimates. For the flows from the UK to Denmark, Norway and Sweden, Raymer's estimates are barely half of the reported values whereas the new estimates are very close to the reported values.

The new estimate of 7511 for the Lithuania to Ireland flow is clearly too high; even Raymer's estimate of 3680 seems likely to be a large overestimate. Comparing the estimated Lithuania to Denmark and Denmark to Lithuania flows shows that, while the new estimates are consistent with the reported values, Raymer's estimates are eight times and three times higher respectively. Overall the estimates produced by the new model are closer to the reported Scandinavian values than Raymer's estimates. A consequence of this is that there are some very unlikely estimates occurring in the flows between other countries. Raymer's estimates are less close to the reported Scandinavian values, but are less unrealistic in the other regions.

In some sense, the estimates that the new model is producing are preferable because the quality of data for the Scandinavian countries is believed to be high. The argument can be made that the overestimate from Lithuania to Ireland actually highlights the inconsistencies with the data and margins provided most clearly, whereas Raymer's method tries to smooth over it. Essentially the new model is saying that, if the Scandinavian reported values are really believed to be accurate, then there is clearly a problem with the estimated Lithuanian margins, which is exactly the same conclusion that Raymer drew from his analysis. Of course, if we did have confidence in the Lithuanian margins, then the model estimates would be telling us that we would have to alter our prior representing the quality of the Scandinavian observations.

Examining the reported values shows that the estimated flows for people migrating to Lithuania from the Scandinavian countries and the UK sum to 619.

The emigration margin for Lithuania is 7180. This leaves 6561 from Estonia, Latvia and Ireland; even allowing for the interaction expected between the former Soviet Union countries, the margin seems to be highly overestimated.

7.6.3   Assumptions about the error of the reported values

One questionable assumption made in the Bayesian model is that the *relative error* of the reported values is the same regardless of their magnitude. In the modelling procedure, for instance, where there is a reported value of 20 and a fitted value of 10, the fitted value is considered to be as deviant from the reported value as if the reported value were 20 thousand and the fitted value 10 thousand. In practice, we are unlikely to believe that this is true. Instead, we would expect the relative error to be much larger for smaller migration flows. There are a couple of possible modifications that we could introduce to reflect this. We could break up σ^2 into a number of different parameters, setting tighter priors where the reported values are large and we believe them to be more accurate and looser priors on the smaller reported values. Alternatively, we could modify the model slightly so that it now becomes

$$\log z_{ij} \sim N\left(\log y_{ij}, \frac{\sigma^2}{y_{ij}}\right), \quad i \neq j. \tag{7.21}$$

The variance is now *inversely proportional* to the magnitude of the reported value. It is relatively straightforward to update the S-PLUS code to accomplish this. However, setting a prior for σ^2 now becomes more complex. For example, if, as before, we believed the error for the reported UK to Ireland migration flow to be within $\pm 20\%$ or come from $N(\log y_{ij}, 0.01)$ distributions, then a prior value for σ^2 would centre on around 200 (since $200/20000 = 0.01$), but now the log of the reported value for the flow from Iceland to Denmark at 1267 would have a variance of $200/1267 = 0.16$, which allows errors as large as $\pm 80\%$, perhaps greater than we would like.

Table 7.5 shows the migration flow estimates under such a model where an $IG(250, 25\,000)$ was used for σ^2. The predicted flow from the UK to Ireland has increased by 2414 from the previous model up to 12\,840, although this is still below Raymer's estimate of 15\,435 and well below the observed value of 21\,165. In addition, the flow between Lithuania and Ireland is now lower than in Table 7.4; however, this comes at a price, as now the other flows to Lithuania appear unrealistically large. In fact the estimates here are now much closer to the quasi-independent fitted values and the mean value of τ^2 is much lower. For example, where the reported flows from Ireland and Latvia to Iceland are 6 and 7, the new estimates are 855 and 420.

The problem with the inversely proportional approach is that, in our case, the largest reported flow appears highly overestimated. Either the UK immigration and Ireland emigration estimates are underestimated or the reported value is somewhere between 25\% and 50\% or higher. Clearly, where we have a model with the reported errors being proportional to the magnitude of the reported value, and the error

Table 7.5 Estimated migration flows from the MCMC simulation based on reported immigration flows with variance inversely proportional to the magnitude of the reported flows.

Country of origin	Country of destination									
	Den.	Fin.	Ice.	Nor.	Swed.	Est.	Ire.	Lat.	Lith.	UK
Denmark	—	1 198	841	2 534	2 109	128	1 536	193	953	1 644
Finland	1 646	—	330	1 406	2 429	88	1 140	134	639	1 065
Iceland	605	286	—	434	407	21	284	33	143	247
Norway	2 180	1 165	403	—	3 083	101	982	139	693	2 322
Sweden	2 037	2 016	397	3 960	—	110	1 013	150	586	1 076
Estonia	377	313	77	267	324	—	514	199	692	445
Ireland	717	662	181	855	807	125	—	189	1 020	4 778
Latvia	460	329	103	420	439	225	769	—	1 478	707
Lithuania	1 389	936	203	1 286	960	385	2 086	555	—	1 419
UK	2 200	855	292	1 191	1 364	182	12 840	334	978	—

of a reported value of 21 611 is around 40 %, the relative error for any reported values under 100 will be huge, and once the variance becomes so large, the quasi-independent fitted model becomes increasingly preferred.

If we did wish to keep the assumption that the error is inversely proportional to the reported magnitude it would be necessary to break up σ^2 into several components. One possibility would be to introduce a separate parameter for the variance of each country's immigration values, since given, the nature of the collection techniques, we probably would consider the reported value of Norway to Denmark of 3188 to have less relative error than the reported value of the UK to Ireland flow. This way we could set a looser prior on the UK to Ireland flow.

7.6.4 Altering the Lithuanian margins

In Section 7.6.2 it was suggested that the estimated margins for Lithuania were too high. In this section, these margins are decreased to see how the estimates generated by the model change. The current populations of Estonia, Latvia and Lithuania are 1.33 million, 2.29 million and 3.60 million, respectively. The immigration margins are 1365, 1924 and 7180, respectively, and the emigration margins are 3206, 4930 and 9221, respectively. Considering the population sizes and the earlier model, it does not seem inconceivable that the emigration margin for Lithuania is overestimated by as much as 4500. The immigration margin is unlikely to be this far out though, so 3500 was subtracted from both margins for Lithuania and the original model set-up was rerun as in Section 7.6.1. The estimates under the new margins are set out in Table 7.6. By comparing with Table 7.4, we can see that most of the 3500 taken away from the Lithuania immigration and emigration

Table 7.6 Estimated migration flows after subtracting 3500 from each of the Lithuanian margins, based on reported immigration.

Country of origin	Country of destination									
	Den.	Fin.	Ice.	Nor.	Swed.	Est.	Ire.	Lat.	Lith.	UK
Denmark	—	439	1 511	3 732	2 122	149	1 009	182	275	1 717
Finland	515	—	48	1 758	3 320	131	1 318	189	243	1 357
Iceland	1 163	56	—	476	361	41	151	46	64	102
Norway	2 775	1 325	579	—	3 314	101	348	138	163	2 321
Sweden	1 741	3 953	477	3 698	—	73	193	94	106	1 011
Estonia	258	1 124	6	91	259	—	601	279	338	250
Ireland	292	44	6	81	209	150	—	342	1 715	6 494
Latvia	397	44	7	125	172	265	3 184	—	438	297
Lithuania	500	22	33	110	116	289	4 113	384	—	154
UK	3 970	751	161	2 284	2 048	166	10 248	271	339	—

totals are reflected in the smaller Lithuania to Ireland and Latvia to Lithuania flows. However, 2200 have now been redistributed to the Latvia to Ireland flow, which has made that flow unrealistically large. The flow from the UK to Ireland remained the same.

Overall, the estimated migration flow table is still far from satisfactory, but we have gained some understanding of how the model is functioning. The UK to Ireland flow is 'stuck' at around 10 248; the 3500 that Ireland has lost to Lithuania has passed to Latvia and not the UK. Despite the reported value of 21 611 for the UK–Ireland flow, the model has no slack for the flow, since the UK–Ireland estimate is being largely fixed by the UK to Scandinavia flows. The model is suggesting that either the UK origin margin is heavily underestimated, or the Ireland immigration margin is overestimated; again this comes with the assumption that the Scandinavian reported values are accurate. Given the reported values, it seems likely that it is actually the UK emigration margin that is being underestimated, and also the UK immigration margins looks to be underestimated, but not to the same extent.

7.6.5   Assessing the model performance

Overall, it is unclear how successful the modelling process has been. The model has not really produced an entirely plausible estimate for the full migration flow table. Instead the model has indicated that there are inconsistencies between the observed data and the margins constructed. Essentially we have seen that the model has tried to fit the Scandinavian estimates as close as possible to the reported values, and this has resulted in other flows in the table becoming unrealistically large.

An important point to reflect on is that the whole modelling framework that we have created is based on the assumption that the true margins are known absolutely. In situations such as this where some of the margins seem questionable it does not matter how sophisticated the modelling process is, the results will always be limited by the quality of the margin estimates. One solution would be to build some uncertainty about the margins into the model instead of making them rigid, somehow incorporating the margin estimation into the modelling process. However, this would also add additional uncertainty into the model and could lead to problems of non-identifiability. For example, in this case it is fairly clear that most of the problems are due to the overestimation of the Lithuanian margins, but in other cases it may not be clear what is causing the inconsistencies.

A natural extension to the modelling process when both immigration and emigration flows are available would be to include two $\mathbf{Z}$ matrices into the model, where $\mathbf{Z}_1$ is the matrix of the reported immigration data and $\mathbf{Z}_2$ is the reported emigration data, and model these as

$$\log z_{1ij} \sim N(\log y_{ij}, \sigma_{1ij}^2), \quad i \neq j,$$
$$\log z_{2ij} \sim N(\log y_{ij}, \sigma_{2ij}^2), \quad i \neq j. \tag{7.22}$$

This approach is appealing as we make full use of the available data, and it allows greater weight to be placed on flows with two reported values in agreement. It is probable that we should set at least $\sigma_{1ij}^2 < \sigma_{2ij}^2$ as the reported immigration data are likely to be more accurate. It is likely that we would use common σ_{ij}^2 parameters for particular groups of flows, e.g. to differentiate between countries that use registration systems to generate the reported values and countries that use surveys. Finally there is the option as in the previous section to take some account of the fact that, for any given country's techniques, we would expect the relative error for smaller flows to be greater than for larger flows. The modelling framework has now become very versatile.

7.7 Conclusion

This chapter has introduced a Bayesian modelling framework to generate place-to-place migration flow estimates. The approach assumes that the margins of an origin–destination migration flow matrix have already been accurately estimated, and starting from a table of reported migration values, which may or may not be complete, an MCMC simulation provides estimates for the individual migration flows. The Bayesian framework created here seems a natural way to combine multiple data sources, with their differing levels of error, and prior information about the structure of the migration process into a single prediction with an associated measure of uncertainty. Overall, the development of the modelling framework has been very much exploratory and the analysis presented here far from being complete. The analysis of the Northern European migration data presented difficulties, but also ideas for further development of the model, which will be pursued in future work.

References

Agresti A. 1990. *Categorical data analysis*. New York: John Wiley & Sons, Inc.

Gamerman D. 1997. *Markov chain Monte Carlo: Stochastic simulation for Bayesian inference*. London: Chapman & Hall.

Poulain M. 1994. *Internal mobility in Europe: the available statistical data*. Working Paper 17. Conference of European Statisticians, Commission of the European Communities, Mondorf-les-Bains, Luxembourg.

Raymer J. 2007. The estimation of international migration flows: a general technique focused on the origin–destination association structure. *Environment and Planning A* 39:985–995.

Willekens FJ. 1999. Modeling approaches to the indirect estimation of migration flows: from entropy to EM. *Mathematical Population Studies* 7(3):239–278.

8

Applying model migration schedules to represent age-specific migration flows

James Raymer[†] **and Andrei Rogers**[‡]

[†] *Division of Social Statistics, School of Social Sciences, University of Southampton*
[‡] *Population Program, Institute of Behavioral Science, University of Colorado, Boulder*

8.1 Introduction

Internal migration and national population redistribution are universal phenomena experienced by all nations. But are their patterns, antecedents and consequences similar worldwide? Are there observable regularities in age patterns exhibited by origin–destination-specific flows of migration? Do such regularities hold all over the world? Why do they? How might demographers take advantage of them? At the International Institute for Applied Systems Analysis in Austria, demographers first began to address these questions in the late 1970s by assembling an international database on contemporary internal migration in the developed nations and successfully fitting these data with a mathematical function, since called the Rogers–Castro multi-exponential model migration schedule (Rogers and Castro 1981b). Rogers and Castro (1981b) put forward three families of multi-exponential

International Migration in Europe: Data, Models and Estimates   Edited by J. Raymer and F. Willekens

model migration schedules: a seven-parameter *standard* model, a nine-parameter *elderly post-retirement migration* model, and an 11-parameter *elderly retirement peak* model. Their analysis of over 500 age profiles of migration found throughout the more developed world made the convincing argument that migration has strong regularities in age patterns, much like fertility and mortality. Several years later, Rogers and Watkins (1987) added a 13-parameter *elderly retirement peak plus post-retirement* model (see also Rogers and Little 1994). Stylised schedules of these four families are set out in Figure 8.1. The most commonly found schedule is the standard schedule, comprising three components: a constant, a negative exponential curve representing the pre-labour force ages, and a double exponential (unimodal) curve representing the labour force ages. Multi-exponential model migration schedules have since been used for describing, smoothing and inferring age-specific migration patterns in a wide variety of contexts (e.g. Bates and Bracken 1982, 1987; Holmberg 1984; Kawabe 1990; Liaw and Nagnur 1985; Potrykowska 1986; Rogers 1988; Rogers and Rajbhandary 1997; Rogers and Raymer 1999b). It seems evident, in consequence, that the multiple-parameter model migration schedule adequately describes the regularities in age profiles exhibited by internal migration flows in the developed world today. But what about migration in the less developed countries? And historical migration patterns? And patterns of *international* migration?

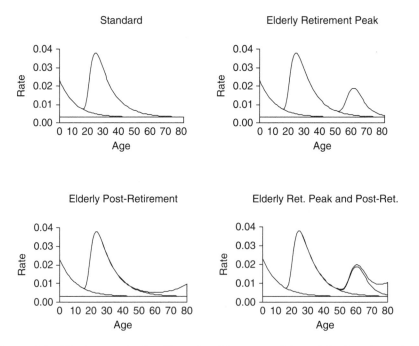

Figure 8.1   The four main families of multi-exponential model migration schedules: standard, elderly post-retirement upslope, elderly retirement peak, and elderly retirement peak and post-retirement upslope.

What commonalities in life course dynamics might explain the widespread observed regularities in migration age profiles? How can such regularities be drawn on to infer age-specific migration patterns in poor countries lacking established data collection systems? How can they be used to improve the quality of collected data? These are the questions that motivate the research described in this chapter. Note that we use the term 'age patterns' to describe age-specific migration data in general, 'age profiles' to describe the shapes of specific age patterns of migration, and 'age compositions' to describe a particular representation of age-specific data, that is, the proportions of all migration from origin i to destination j by age.

Families of model migration schedules are used to illustrate how they can be used to effectively estimate age- and origin–destination-specific flows of migration. Our ideas and models are tested against known age patterns of migration, with the intention of applying them in situations of missing or inadequate patterns of migration (e.g. the findings are used in Chapter 10 to guide the estimation of age-specific international migration flows between countries in Europe). We begin with a review of the life course events that give rise to various patterns, and then we describe how these events are reflected in the Rogers–Castro multi-exponential model migration schedule. Second, we describe the observed 1985–1990 age compositions of interstate migration in the United States West region (US Census Bureau 1993). Third, we estimate these patterns using a family of model migration schedules and compare them with simpler estimation methods that rely on aggregate inmigration and outmigration flows by age. Finally, the applicability of this approach is discussed in a more general migration modelling context. A Bayesian approach to modelling age patterns of migration can be found in Chapter 9.

8.2 Conceptual framework: regularities in the age patterns of migration

8.2.1 Explaining the regularities

Empirical profiles of age-specific migration exhibit remarkably persistent regularities, analogous to those found in other demographic age profiles. Mortality profiles, for example, normally show a moderately high death rate immediately after birth, after which the rates drop to a minimum between ages 10 and 15, then increase slowly until about age 50, and thereafter rise at an increasing pace until the last years of life. Fertility rates generally start to take on nonzero values at about age 15 and attain a maximum somewhere between ages 20 and 30; the curve is unimodal and declines to zero once again at some age close to 50. Similar unimodal profiles may be found in schedules of first marriage, divorce and remarriage. The most prominent regularity in age-specific profiles of migration is the high concentration of migration among young adults; rates of migration also are high among children, starting with a peak during the first year of life, dropping to a low point at about age 16, turning sharply upward to a peak near 20–22, and declining regularly thereafter, except for a possible slight hump at the onset of retirement and possibly an upward slope after that hump.

Underlying these persistent regularities in the age patterns of migration are a collection of different *cause-specific* age patterns (Rogers and Castro 1981a). For example, migrations due to marriage and education are concentrated between the ages of 10 and 30, and are essentially unimodal in age profile. Migrations caused by change of employment and moving closer to the place of work have profiles that are bimodal, with local peaks during infancy and during the early years of labour force participation. The age profiles of housing reasons for migration are similar to those of the aggregate migration profile, exhibiting roughly the same peaks: during infancy, during the early years of labour force participation, and at retirement. Finally, health is apparently an important cause of migration only for the elderly.

The different cause-specific age patterns may be interpreted within a life course framework in which individuals pass through different states of existence (Elder 1985). Starting with birth and then entry into the educational system at the elementary level, the 'passage' may also include entry into military service or university, marriage, multiple entries into and withdrawals from the labour force, perhaps divorce and remarriage, retirement, death of spouse, and moves to enter sanatoria or to rejoin relatives. Life course analysis focuses on the processes of change and ultimately seeks to explain such change.

The formal demography of migration and population redistribution views interregional population transfers as a collection of *independent* individual movements. Yet it is widely recognised that a large fraction of total migration is accounted for by individuals whose moves are dependent on those of others, for example, children migrating with their parents, wives with their husbands, and/or grandparents with their children. Indeed, family migration is such a well established phenomenon that Ryder (1978) has even suggested its use as a criterion for identifying family membership: a family comprises those individuals who would migrate together. Hence, to the extent that migration is undertaken by families as a unit, the age composition of migrants tells us something about family patterns.

To better understand the influences that family and dependence relationships have on migration age compositions, Castro and Rogers (1983a,b) illustrated a number of ways in which the aggregate age profile of migration is sensitive to relative changes in dependence levels and in rates of natural increase and mobility by disaggregating migrants according to age, sex and dependent/independent categories. Viewing the migration process within a framework of dependent and independent movements allows one to observe, for example, that, if the independent component mainly comprises single persons, then the associated dependent migration may be insignificant in terms of its relative share of the total migration. On the other hand, if migration tends to consist primarily of family migration, then the share of dependent children may become a very important component of the aggregate migration age pattern. In short, just as observed population age compositions reflect particular characteristics of past fertility and mortality regimes, so do observed migration age compositions reflect key aspects of a population's age composition. The reverse relationship also holds true (Little and Rogers 2007).

8.2.2 Describing the regularities: interstate migration in the US West, 1985–1990

The patterns of age-specific migration observed between states in the US West region during the 1985–1990 period are examined in this section. The migration data come from the 1990 census (US Census Bureau 1993) and represent place of residence at time of the census by place of residence five years prior to the census (i.e. status or transition migration data). The age composition of migration is defined as

$$p_{ij}(x) = \frac{n_{ij}(x)}{\sum_x n_{ij}(x)},$$

(8.1)

where $p_{ij}(x)$ denotes the proportion of all migration from origin i to destination j in age group x (i.e. 0–4, 5–9, . . . , 80+ years measured at the beginning of the time interval) and $n_{ij}(x)$ denotes the corresponding flow of migration. Origin refers to the place of residence in 1985 and destination refers to the place of residence in 1990. For example, $p_{CA,AZ}(20)$ denotes the proportion of all migrants from California to Arizona who were 20–24 years old in 1985. Note that, in this chapter, we always refer to the migrants' age at the beginning of the time interval. Persons born during the interval are excluded.

The age compositions for 196 flows of migration between internal and external locations for the US West were examined for the purpose of grouping each age composition into a distinct category. We found four distinct age profiles in the patterns: a standard age profile, a standard young age profile, a standard old age profile and an elderly retirement peak age profile. The *standard* age profile of migration contains a single labour force peak centred at the 20–24 year old age group. *Standard young* age profiles refer to age compositions with labour force peaks centred on the 15–19 year old age group, and *standard old* age profiles refer to age compositions centred on the 25–29 year old or 30–34 year old age groups. The *elderly retirement peak* age profile of migration is one that contains both a labour force peak and a peak between ages 50–69 years.

The allocation of the 196 age compositions into the four classifications is set out in Figure 8.2: 60 % (118 flows) of the age compositions exhibited the standard age profile of migration (see e.g. migration from or to Washington); 18 % (35 flows) of the age compositions exhibited the old standard age profile (see e.g. migration from Alaska or Wyoming or to Montana); 17 % (34 flows) of the age compositions exhibited the young standard age profile (see e.g. migration from Idaho or to Utah); and finally, 5 % (nine flows) of the age compositions exhibited the elderly retirement age profile of migration (see e.g. migration from California or to Arizona). Examples of each of the four age profiles are set out in Figure 8.3. They represent the age compositions of migration from California to Utah (i.e. young standard), Colorado (i.e. standard), Montana (i.e. old standard) and Arizona (i.e. retirement peak).

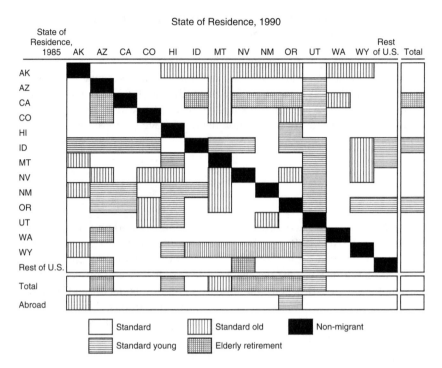

Figure 8.2 The classification of age compositions of interstate migration in the US West region during the 1985–1990 period into standard, standard young, standard old, and elderly retirement age profiles.

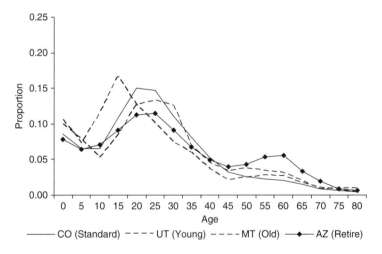

Figure 8.3 Standard, young, old and elderly retirement age profiles of migration: age compositions of migration from California to Colorado, Utah, Montana and Arizona, 1985–1990.

8.3   Fitting multi-exponential model schedules to age patterns of migration

The age-specific counts of interstate migration in the United States, observed during the 1985–1990 period, are set out in Figure 8.4 (along with a corresponding model schedule fit). In total, there were about 7.6 million persons who resided in a different state in 1990 than they did in 1985 (US Census Bureau 1993). Of these migrants, 52 % were between the ages of 15 and 34 years. The age group with the smallest number of migrants was the last age group (80+) with 39 thousand persons. The age group with the largest number of migrants was the 20–24 year old age group with 1.2 million persons. (Note again that the age groups in this chapter are measured at the beginning of the time interval.)

A seven-parameter multi-exponential model migration schedule was fitted to the age-specific counts of interstate migrants (see Figure 8.4). The general model is specified as

$$\hat{n}_{ij}(x) = a_0 + a_1 \exp\left(-\alpha_1 x\right) + a_2 \exp\left\{-\alpha_2 \left(x - \mu_2\right) - \exp\left[-\lambda_2 \left(x - \mu_2\right)\right]\right\}, \quad (8.2)$$

where the a_0, a_1 and a_2 are level parameters and the α_1, α_2, μ_2 and λ_2 parameters are shape parameters. Specifically, the parameter μ positions the unimodal curve on the age axis and the parameters λ and α reflect the steepness of its ascending and descending sides, respectively. If $\alpha > \lambda$, the mode of the function is smaller

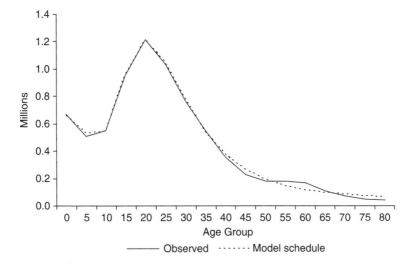

Figure 8.4   Overall age profile of interstate migration totals in the US, 1985–1990: observed and model schedule fit.

than μ. If $\alpha < \lambda$, the mode is larger than μ. The resulting parameter values obtained from fitting the above model to the age-specific counts are

$$\hat{n}_{++}(x) = 48\,099 + 611\,131 \exp(-0.048x)$$
$$+2\,261\,105 \exp\{-0.100\,(x - 17.472) - \exp[-0.175\,(x - 17.472)]\},$$

where $\hat{n}_{++}(x)$ denotes the overall age profile of migration (i.e. aggregated over i and j). We used the nonlinear regression package, TableCurve 2D, to fit the schedule to the observed data (see Appendix for more details [1]). The resulting fit had $R^2 = 0.9965$, which is nearly identical to the observed data. The main differences between the observed data and model schedule are found in the age groups above 45–49 years.

Model migration schedules are flexible and can be fitted to counts, rates or proportions. To illustrate, a seven-parameter migration schedule was also fitted to the age composition (i.e. proportions in each age group) corresponding to the counts set out in Figure 8.4. The resulting parameter values are

$$\hat{p}_{++}(x) = 0.006 + 0.080 \exp(-0.048x)$$
$$+0.297 \exp\{-0.100\,(x - 17.472) - \exp[-0.175\,(x - 17.472)]\}.$$

Notice that the α_1, α_2, μ_2 and λ_2 parameters do not change.

Next, consider the age composition of migration from California to Arizona, which exhibited an elderly retirement peak during the 1985–1990 period (see Figure 8.5). In total, there were 136 thousand persons who migrated between these two states during this time, of whom 41 % were between the ages of 15 and 34 years (measured in 1985) and 14 % were between the ages of 55 and 69 years. Note that, for the overall age composition of interstate migration (see Figure 8.4), only 6 % of all migrants were between the ages of 55 and 69 years.

To estimate an elderly retirement peak, one simply adds an additional unimodal curve to the multi-exponential model. The 11-parameter model migration schedule used for estimating age profiles with retirement peaks is specified as

$$\hat{p}_{ij}(x) = a_0 + a_1 \exp(-\alpha_1 x) + a_2 \exp\{-\alpha_2\,(x - \mu_2) - \exp[-\lambda_2\,(x - \mu_2)]\}$$
$$+a_3 \exp\{-\alpha_3\,(x - \mu_3) - \exp[-\lambda_3\,(x - \mu_3)]\}. \tag{8.3}$$

The estimated parameters of this schedule fitted to the age composition of migration from California to Arizona are

$$\hat{p}_{CA,AZ}(x) = 0.006 + 0.072 \exp(-0.047x)$$
$$+0.229 \exp\{-0.105\,(x - 23.238) - \exp[-0.105\,(x - 23.238)]\}$$
$$+0.000\,004 \exp\{-0.500\,(x - 92.558) - \exp[-0.061\,(x - 92.558)]\}.$$

---

[1] The appendices and/or supplementary material for this chapter are available on the book's website at: http: www.wiley.com/go/raymer

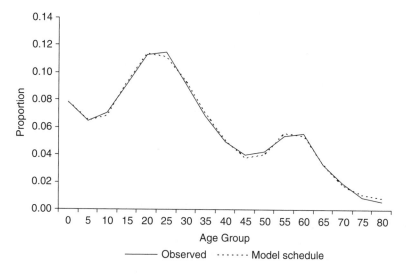

Figure 8.5 Age composition of migration from California to Arizona, 1985–1990: observed and model schedule fit.

As illustrated in Figure 8.5, this model schedule is nearly identical to the observed data ($R^2 = 0.9965$).[2]

8.4 Modelling families of age-specific migration

For the analysis presented in this section, we assume that the levels of aggregate origin–destination-specific migration are known. This allows us to focus on estimating the age compositions of migration and to make use of a simple and straightforward strategy to translate aggregate migration flows, n_{ij}, and age compositions, $p_{ij}(x)$, into age-specific flows, $n_{ij}(x)$:

$$n_{ij}(x) = n_{ij} \times p_{ij}(x), \tag{8.4}$$

which we consider to be our overall modelling objective. This relationship makes the overall estimation process much simpler because it breaks up the modelling of migration into two distinct stages: levels and shapes. Once the aggregate flows are obtained, the age patterns can be readily predicted. In this section, we use various approaches to estimate the 196 age patterns classified in Figure 8.2. For space reasons, however, we focus our presentation of the results on the migration from California to Colorado, Utah, Montana and Arizona (see Figure 8.3). Note that these

---

[2] Note that the software program TableCurve 2D only allows a maximum of 10 parameters to be estimated in its User-Defined Program. To get around this, the constant (a_0) was set to the minimum observed (i.e. 0.006).

four flows also correspond to the four identified families of age-specific migration: standard, standard young, standard old and elderly retirement peak, respectively.

When it comes to the availability of age-specific data, there could be several possible situations. In the worst case scenario, no age-specific data are available. Here, one has no choice but to make a 'best educated guess' based on the characteristics of the location and what the literature on this topic suggests. In other situations, observed age-specific data may be available, but only for aggregate levels of inmigration and outmigration. This is the current situation with international migration flow data in Europe provided by Eurostat (see Chapter 10). Or, some place-to-place age-specific data might be available, but not all. Another situation that could arise is one in which the data were obtained from a relatively small sample of the population, such as in the passenger surveys carried out in Cyprus and the United Kingdom and the household surveys in Portugal and Ireland (see Chapter 3). These data might not exhibit the expected age-specific regularities often observed with migration data. In such situations, it might make sense to smooth the data using model migration schedules. Finally, a situation might arise where one has information about the shapes of the age profiles (i.e. retirement locations), but not necessarily the observed age-specific data for the period of interest. The estimations of the age-specific flows carried out in this section attempt to cover several of these hypothetical situations.

Four families of age-specific profiles have been identified in the observed age compositions of migration. The age compositions were grouped together according to their model family and then fitted with a model migration schedule to obtain an overall (or average) representation of each family. The seven-parameter model migration schedule (Equation (8.2)) was fitted to each set of observed age compositions corresponding to the standard (i.e. 118 age compositions), standard young (i.e. 34 age compositions) and standard old (i.e. 35 age compositions) families of age-specific migration. The 11-parameter model migration schedule was fitted to the set of observed elderly retirement peak age compositions of migration (i.e. nine age compositions). The parameters of these four model migration schedules are set out in Table 8.1. The model schedules captured 89 %, 90 %, 87 % and 94 % of the standard, standard young, standard old and elderly retirement peak age compositions of migration, respectively. Note that these schedules were fitted assuming equal weight for each profile of age-specific migration and represent the average of those profiles. Another strategy would be to aggregate the flows in each of the respective families, and then to fit those age compositions with model migration schedules. This would give more weight to age profiles of larger flows (i.e. from and to California) and potentially improve the overall fit.

Once the four age profiles of migration were estimated, they had to be adjusted (slightly) so that they summed to unity. The resulting model schedules are set out in Figure 8.6. Notice the differences in the shapes. For example, the age composition with a retirement peak curve is associated with a lower labour force peak, because more weight is included in the retirement years. The age composition of a standard young curve's labour force peak is shifted to the left, whereas the standard old

Table 8.1 Model migration schedule parameters of standard, standard young, standard old and elderly retirement peak age compositions of interstate migration in the US West, 1985–1990.

Curve	Parameter	Standard	Standard young	Standard old	Elderly retirement
Constant	a_0	0.0000	0.0000	0.0000	0.0042
Pre-labour force	a_1	0.0973	0.0988	0.1005	0.0746
	α_1	0.0375	0.0468	0.0308	0.0456
Labour force	a_2	0.2763	0.1877	0.2324	0.2102
	α_2	0.1079	0.0627	0.1083	0.0760
	μ_2	17.6860	11.3310	21.2440	17.9213
	λ_2	0.1956	0.3375	0.1700	0.14410
Post-labour force	a_3				0.0000
	α_3				0.5006
	μ_3				92.1693
	λ_3				0.0616
R^2		0.8942	0.8965	0.8672	0.9395
Standard error		0.0169	0.0169	0.0178	0.0088
Number of observations		2 006	578	595	153
Number of age profiles		118	34	35	9

curve's labour force peak is shifted to the right. Finally, the three curves without a retirement peak have basically the same pattern after age 35.

Next, four hypothetical situations with regard to available data are tested. The first situation represents one in which only the observed national age composition of interstate migration, $p_{++}(x)$, is known. The second situation is one in which the observed age compositions of aggregate inmigration, $p_{+j}(x)$, are known. The third situation is one in which the observed age compositions of aggregate outmigration, $p_{i+}(x)$, are known. These first three situations assume that some observed data are available. The fourth situation uses the four model age compositions set out in Figure 8.6, using the classification in Figure 8.2, to represent the 196 origin–destination-specific flows of age-specific migration. Here, the assumption is that the type of migration flow between two states is known or can be estimated (i.e. age-specific migration from i to j belongs to one of the four model migration schedule families). This represents a situation in which a researcher expects the migration

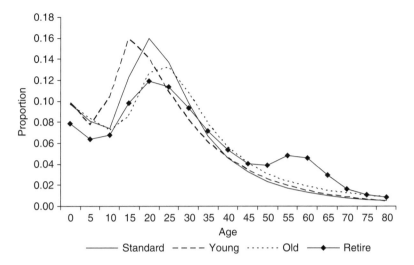

Figure 8.6 Model migration schedules of observed standard, standard young, standard old and elderly retirement peak age compositions of migration in the US West, 1985–1990.

flow between, say, California and Arizona to have a retirement peak because of the existence of many elderly residing in Arizona who came from California. Likewise, one might suspect that migration flows into Utah might exhibit a young standard age profile of migration because of recent enumerations that showed a 'bulge' in the age profile of the population stock data in the 15–19 year old age group not explained by past fertility patterns (i.e. cohort ageing).

The age compositions used in each of the four scenarios discussed above were multiplied by the observed aggregate origin–destination-specific flows, n_{ij}, to obtain the corresponding predicted age-specific migration flows, $\hat{n}_{ij}(x)$. Because the flows were observed, we can evaluate the different scenarios. The overall R^2 goodness-of-fit statistics were 0.984, 0.992, 0.991 and 0.972 for applying the age compositions corresponding to the overall age composition, inmigration age compositions, outmigration age compositions and model schedule age compositions, respectively, to estimate the age-specific flows of migration. The predictions that used the aggregate inmigration and outmigration age compositions for each state gave the best results. The overall fit for the model migration schedule families scenario might have been improved had the model schedules been fitted to four aggregate sets of flows, representing each of the four families, rather than assuming equal weight among many similar age compositions.

The results above represent the overall fit of the five models used to predict all 196 age profiles of migration. Next, the predicted age-specific migration flows from California to Arizona, Colorado, Montana and Utah are selected out and compared with the corresponding observed flows. These R^2 goodness-of-fit statistics are set out in Table 8.2. Interestingly, when the goodness-of-fit values are compared for

Table 8.2 Comparison[a] of different scenarios used to predict age profiles of migration from California to Arizona, Colorado, Montana and Utah, 1985–1990.

Migration flow	Observed aggregate			Model schedule	
	Out	In	Total	Mixture	All flows
CA to AZ	0.9303	0.9316	0.8690	0.9855	0.8717
CA to CO	0.9924	0.9700	0.9430	0.9675	0.9613
CA to MT	0.9397	0.9233	0.8249	0.9535	0.8836
CA to UT	0.7287	0.9836	0.8788	0.9722	0.8546

[a] Values shown are R^2 = goodness-of-fit measure.

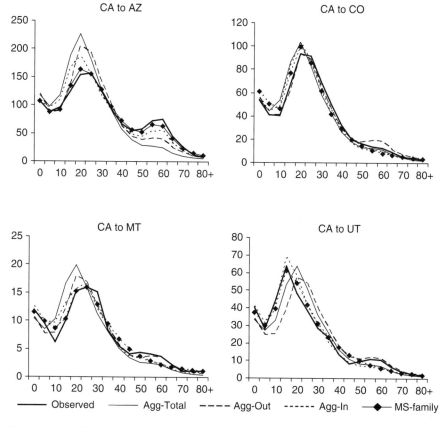

Figure 8.7 Observed and predicted migration flows from California to Arizona, Colorado, Montana and Utah, 1985–1990. Here the *y* axes are in hundreds, and the *x* axes show the age groups. Note that the scales on the *y* axes are different.

the specific flows, the model schedule families scenario produces the best results, except for the California to Colorado flow, where the outmigration model worked best. The message from these results is that, while the overall level produces a better 'average' fit, having more information about the shape of the age profile improves individual fits, at least for the four cases presented here.

The predicted age-specific flows set out in Figure 8.7 illustrate visually the results set out in Table 8.2. For the flow from California to Arizona, we see that the only age profile that is really close to the observed is that from the model schedule families scenario. The other age profiles tend to overpredict the labour force peak and underpredict the observed retirement peak. For the California to Colorado age profiles, all of the flows, while fitting the observed data fairly well, have a labour force peak that is slightly shifted to the left of the observed. Here, the aggregate outmigration curve does a good job of fitting the observed data. The predictions for the California to Montana flow vary substantially from the observed. This is due to it being a fairly unusual age profile in that it has a wide labour force peak. The best model for these data is the model schedule families scenario. Finally, the predictions for the California to Utah age patterns show that only two models come close to capturing its relatively young labour force peak: the model schedule families scenario and the aggregate inmigration model.

8.5   Discussion and conclusion

Model migration schedules (Rogers and Castro 1981b) were applied, in this chapter, to represent specific known patterns of migration. In particular, four model migration schedule families were used to represent 196 flows of migration. This finding simplified the modelling process, by assuming that all one required to model age patterns of migration was some information about the likely shapes of the age profiles. Another useful finding was that the aggregate inmigration and outmigration profiles captured most of the age patterns. Often, origin–destination-specific age patterns of migration are not available (e.g. European migration data from Eurostat only include age patterns of total immigration and total emigration; see Chapter 10). This study suggests that it is quite reasonable to expect that regularities found in observed age patterns of migration can be used to effectively estimate missing or inadequate data.

Four directions of further study are evident. First of all, how stable are the observed age patterns over time? Our preliminary examination of temporal stability suggests that many flows continue to exhibit a surprising degree of constancy in age profiles. Tobler (1995), for example, found that sufficient stability exists for the Rogers and Castro (1981b) model schedule and called it one of the 'laws' of migration. Rogers and Raymer (1999a) found evidence of temporal stability in the interregional migration patterns of the foreign-born in the United States reported by the 1960, 1970, 1980 and 1990 censuses. More recently, Raymer *et al.* (2006) showed strong regularities in the age patterns of Italian interregional migration from 1970 to 2001 with a gradual 'ageing' of the labour force peak. But more definitive findings are needed.

Second, our focus has been directed at internal migration age patterns. What about international patterns? Rogers and Raymer (1999a), for example, found that US immigration age profiles conformed to the Rogers and Castro (1981b) model schedule during the 1955–1990 time period. In this volume, similar evidence for population movements in the European Union is presented in Chapter 10. But, once again, more definitive findings are needed.

Third, what about migration age patterns in the less developed countries? Are there reasons to suspect a lack of conformity? Evidence gathered to date show that the Rogers and Castro (1981b) model migration schedule also describes migration age patterns in the less developed world. Muhidin (2002:272–276) and Rogers *et al.* (2004) present supporting evidence for Indonesia. Partida Bush (2004) does the same for Mexico. It is likely, therefore, that age patterns of migration in less developed countries also exhibit the standard model migration schedule profiles. But further studies are warranted.

Finally, there is the question regarding the feasibility of *indirect estimation of migration*. Demographic techniques for indirectly estimating fertility and mortality age patterns are well developed and widely used in the less developed world. A somewhat dated United Nations manual on the topic is still the best single source for a review of these procedures (United Nations 1983). Unfortunately, it totally ignores migration. Nevertheless, relatively recent efforts to develop such techniques, including the work contained in this volume, are promising and suggest that the development of a formal methodology for such estimation methods is possible with further research (Raymer and Rogers 2007; Rogers and Jordan 2004).

In conclusion, empirical regularities in age pattern characterise observed migration schedules in ways that are no less important than the corresponding well established regularities in observed fertility or mortality schedules. Such regularities in the form of model schedules can be used to assess the *reliability* of empirical migration data and to indicate appropriate strategies for their *correction*, and they may also be used to help resolve problems caused by *missing data*.

Although the basic age profile of migration seems to hold over time, its level and shape may change. Such changes become reflected in temporal variations exhibited by the parameters of the model migration schedule. The impacts of such changes in levels and age profiles may be analysed conveniently by means of population projections. Multiregional and/or multistate projection models assess the numerical consequences, to an observed or hypothetical population, of a particular set of assumptions regarding future patterns of mortality, fertility and interregional transfers. They generally need to keep track of enormous amounts of data. The disaggregations incorporated in such projections are introduced either because forecasts of the specified population subgroups are important in their own right, or because it is believed that simple and regular trends are more likely to be discovered at relatively higher levels of disaggregation.

High levels of disaggregation permit a greater flexibility in the use of the projections by a wide variety of users; they also often lead to a detection of greater consistency in patterns of behaviour among more homogeneous population subgroups. But greater disaggregation requires the estimation of ever greater numbers of data points, both those describing initial population stocks and those defining the future rates of events and flows that are expected to occur. The practical difficulties of obtaining and interpreting such data soon outstrip the benefits of disaggregation. Model schedules offer a means for condensing the amount of information to be specified as assumptions. They also express this condensed information in a language and variables that are more readily understood by the users of the projections, and they provide a convenient way of associating the variables to one another, extrapolating them over time, and relating them to variables describing the economic environment that underlies the projections.

The use of parameterised model schedules in the population projection process allows one to develop an effective description of how the components of demographic change are assumed to vary over time in terms of a relatively few parameters. To the extent that the assumptions correctly anticipate the future, the projection foretells what indeed comes to pass. Also, insofar as the parameters are interpretable by non-demographer users of the projection, they make possible the assessment of the reasonableness of a set of assumptions instead of a set of projected population totals.

As Keyfitz (1982) correctly observed, a trend extrapolation of each age-specific rate in a population projection is an excessive concession to flexibility that can readily produce erratic results. On the other hand, to assume that change in a set of rates occurs uniformly at all ages is to go against experience. Model schedules in general, and model migration schedules in particular, offer a way of introducing flexibility, while at the same time retaining the interdependence between the rates of a particular schedule (Rogers 1986).

Acknowledgements

This work was supported by the National Institute of Child Health and Human Development and the National Science Foundation. The authors would like to thank Frans Willekens for his thoughtful comments and suggestions on earlier drafts of this chapter.

References

Bates J and Bracken I. 1982. Estimation of migration profiles in England and Wales. *Environment and Planning A* 14:889–900.

Bates J and Bracken I. 1987. Migration age profiles for local authority areas in England, 1971–1981. *Environment and Planning A* 19:521–535.

Castro LJ and Rogers A. 1983a. What the age composition of migrants can tell us. *Population Bulletin of the United Nations* 15:63–79.

Castro LJ and Rogers A. 1983b. Patterns of family migration: two methodological approaches. *Environment and Planning A* 15(8.2):237–254.

Elder GH Jr. 1985. Perspectives on the life course. In *Life course dynamics: Trajectories and transitions, 1968–1980*. Elder GH Jr, ed., pp. 23–49. Ithaca, NY: Cornell University Press.

Holmberg I, ed. 1984. *Model migration schedules: The case of Sweden.* Stockholm: Scandinavian Demographic Society.

Kawabe H. 1990. *Migration rates by age group and migration patterns: Application of Rogers' migration schedule model to Japan, The Republic of Korea, and Thailand.* Tokyo: Institute of Developing Economies.

Keyfitz N. 1982. Can knowledge improve forecasts? *Population and Development Review* 8:729–751.

Liaw K-L and Nagnur DN. 1985. Characterization of metropolitan and nonmetropolitan outmigration schedules of the Canadian population system, 1971–1976. *Canadian Studies of Population* 12(8.1):81–102.

Little JS and Rogers A. 2007. What can the age composition of a population tell us about the age composition of its out-migrants? *Population, Space and Place* 13:23–39.

Muhidin S. 2002. *The population of Indonesia: Regional demographic scenarios using a multiregional method and multiple data sources.* Amsterdam: Rozenberg.

Partida Bush V. 2004. Inferring migration flows from birthplace-specific population stocks: the case of Mexico. Presented at the *Colorado Conference on the Estimation of Migration,* Estes Park, Colorado.

Potrykowska A. 1986. Modelling inter-regional migrations in Poland, 1977–81. *Papers of the Regional Science Association* 60:29–40.

Raymer J and Rogers A. 2007. Using age and spatial flow structures in the indirect estimation of migration streams. *Demography* 44(8.2):199–223.

Raymer J, Bonaguidi A and Valentini A. 2006. Describing and projecting the age and spatial structures of interregional migration in Italy. *Population, Space and Place* 12:371–388.

Rogers A. 1986. Parameterized multistate population dynamics and projections. *Journal of the American Statistical Association* 81(393):48–61.

Rogers A. 1988. Age patterns of elderly migration: an international comparison. *Demography* 25(8.4):355–370.

Rogers A and Castro LJ. 1981a. Age patterns of migration: cause-specific profiles. *Advances in multiregional demography.* RR-81-6. Rogers A., ed. Laxenburg: International Institute for Applied Systems Analysis.

Rogers A and Castro LJ. 1981b. *Model migration schedules.* RR-81-30. Laxenburg: International Institute for Applied Systems Analysis.

Rogers A and Jordan L. 2004. Estimating migration flows from birthplace-specific population stocks of infants. *Geographical Analysis* 36(8.1):38–53.

Rogers A and Little JS. 1994. Parameterizing age patterns of demographic rates with the multiexponential model schedule. *Mathematical Population Studies* 4(8.4): 175–194.

Rogers A and Rajbhandary S. 1997. Period and cohort age patterns of US migration, 1948–1993: Are American males migrating less? *Population Research and Policy Review* 16:513–530.

Rogers A and Raymer J. 1999a. Estimating the regional migration patterns of the foreign-born population in the United States: 1950–1990. *Mathematical Population Studies* 7(8.4):181–216.

Rogers A and Raymer J. 1999b. Fitting observed demographic rates with the multiexponential model schedule: an assessment of two estimation programs. *Review of Urban and Regional Development Studies* 11(8.1):1–10.

Rogers A and Watkins JF. 1987. General versus elderly interstate migration and population redistribution in the United States. *Research on Aging* 9(8.4):483–529.

Rogers A, Muhidin S, Jordan L and Lea M. 2004. Indirect estimates of age-specific interregional migration flows in Indonesia based on the mobility propensities of infants. Presented at the *Colorado Conference on the Estimation of Migration*, Estes Park, Colorado.

Ryder NB. 1978. Methods in measuring the family life cycle. In *Proceedings of the IUSSP International Population Conference*, pp. 219–226. Liege: International Union for the Scientific Study of Population.

Tobler W. 1995. Migration: Ravenstein, Thornthwaite, and beyond. *Urban Geography* 16(8.4):327–343.

United Nations. 1983. *Manual X: Indirect techniques for demographic estimation*. New York: Department of International Economic and Social Affairs.

US Census Bureau. 1993. *1990 county to county migration*. Special Project 312. Washington, DC: US Government Printing Office.

9

Models for migration age schedules: a Bayesian perspective with an application to flows between Scotland and England

Peter Congdon

Department of Geography, Queen Mary, University of London

9.1 Introduction

Age schedules of migration rates tend to show regular features such as a peak in the young adult ages and declining migration propensities in old age. For some flows, there may also be elevated migration around retirement age. One way to capture these regularities, whether present in internal or international migration, involves nonlinear regression, whereby the age-specific rates of migration are represented by a sum of exponential or shifted exponential terms; this is known as the multi-exponential migration schedule (Rogers and Castro 1981; see also Chapter 8).

This paper considers a Bayesian modelling approach to migration age schedules. First, a purely parametric model (i.e. the multi-exponential migration schedule)

International Migration in Europe: Data, Models and Estimates   Edited by J. Raymer and F. Willekens
© 2008 John Wiley & Sons, Ltd

is used to estimate age-specific rates of migration. Second, a dynamic general linear model (DGLM) approach is used, a model that avoids a nonlinear regression specification. For related work on mortality graduation, see Neves and Migon (2004). The Bayesian approach used here employs iterative Markov chain Monte Carlo (MCMC) sampling to update information about the density of parameters θ from the observed data, thereby revising existing knowledge summarised in prior densities on the components of θ.

9.2   Parametric vs dynamic general linear model approaches

Let Y_x denote migration flows by single years of age $(x = 1, \ldots, \omega)$ and N_x denote mid-year populations. Poisson sampling is assumed with $Y_x \sim \text{Po}(N_x r_x)$, where r_x are age-specific migration rates. The multi-exponential model for r_x with both labour and retirement peaks can be specified with the following 'purely parametric' model:

$$r_x = a_0 + a_1 \exp(-\alpha_1 x) + a_2 \exp\{-\alpha_2 (x - \mu_2) - \exp[-\lambda_2 (x - \mu_2)]\}$$
$$+ a_3 \exp\{-\alpha_3 (x - \mu_3) - \exp[-\lambda_3 (x - \mu_3)]\}. \tag{9.1}$$

The component with parameters a_1 and α_1 represents child migration associated with the migration of their parents (in young adult ages); the component with parameters a_2, α_2, μ_2 and λ_2 represents young adult migration, which is mainly for labour reasons; and the component with the parameters a_3, α_3, μ_3 and λ_3 represents retirement age migration. One advantage of the parametric model is that one obtains a 'standard' predicted age profile of migration with a labour force peak, even if the data are irregular.

However, if there are irregularities in the observed schedule and/or some degree of overdispersion (excess heterogeneity) relative to the Poisson assumption, then the model fit is likely to be improved by specifying a hierarchical model allowing for uncertainty in the r_x after allowing for the parametric form in (9.1). The conjugate option is to take the r_x as subject to gamma mixing, where a gamma with parameters δ and β has mean δ/β and variance δ/β^2. Substantively, there is often likely to be heterogeneity in migration behaviours at a given age between population subgroups. This leads to the specification

$$Y_x \sim \text{Po}(N_x r_x),$$
$$r_x \sim \text{Ga}(\kappa, \kappa/m_x),$$
$$m_x = a_0 + a_1 \exp(-\alpha_1 x) + a_2 \exp\{-\alpha_2 (x - \mu_2) - \exp[-\lambda_2 (x - \mu_2)]\}$$
$$+ a_3 \exp\{-\alpha_3 (x - \mu_3) - \exp[-\lambda_3 (x - \mu_3)]\}, \tag{9.2}$$

where κ is an additional positive parameter. The variance m_x^2/κ of r_x about the purely parametric model, represented by m_x, declines as κ increases. An additive (e.g. normal) error in a model for $\log(r_x)$ is another possibility.

While the form of the multi-exponential model is based on accumulated substantive evidence, it is a model that presumes a certain regularity that may not be present in all datasets. It may be better to let the data choose the most appropriate nonlinear regression (Stone 1985). In particular, one may avoid parametric regression assumptions by using nonparametric regression, which for migration events Y_x with means $N_x r_x$ has a generic form $\log(r_x) = S(x)$, where $s_x = S(x)$ is an unknown smooth function in age x. For example, one might consider age effects s_x following a low-order random walk (e.g. RW1 or RW2) or an autoregressive model. For example, an RW1 model could take the form

$$\log(r_x) = s_x,$$
$$s_x \sim N(s_{x-1}, 1/\tau_s), \qquad x = 2, \ldots, \omega, \tag{9.3}$$

where τ_s is a precision parameter and s_1 is typically assigned a diffuse prior, such as $s_1 \sim N(0, V_1)$ where V_1 is large. For τ_s one may assume a gamma or lognormal prior with known parameters; for example, Besag *et al.* (1995) used the prior $\tau_s \sim Ga(1, b)$ with b small, e.g. $b = 0.001$. The initial values of the series are assigned fixed effects priors. The RW2 prior penalises deviations from a linear trend $s_x = 2s_{x-1} - s_{x-2}$ with

$$s_x \sim N\left(2s_{x-1} - s_{x-2}, 1/\tau_s\right), \qquad x = 3, \ldots, \omega \tag{9.4}$$

and flat priors on s_1 and s_2. This will tend to produce greater smoothing than an RW1 model and so may be more sensible if the data are irregular due to small sample size.

Random walk priors allow flexibility in the face of possible nonstationarity and also generalise easily to multivariate situations. An alternative is to introduce an autoregression parameter. For instance, a lag 1 autoregressive model could be specified as

$$s_x \sim N\left(\rho s_{x-1}, 1/\tau_s\right), \qquad x = 2, \ldots, \omega, \tag{9.5}$$

where one may assume stationarity *a priori*, or assess it from the posterior probability $\Pr(|\rho| < 1 \,|\, y)$. The latter may be estimated as the proportion of MCMC iterations $t = 1, \ldots, T$ subsequent to convergence for which the condition $|\rho^{(t)}| < 1$ on sampled autocorrelations $\rho^{(t)}$ is satisfied.

Nominally, the models in (9.3) and (9.4) involve $(\omega + 1)$ parameters, as compared to 11 parameters in (9.1). However, the fact that the s_x are drawn from a single hyperdensity means that their effective dimension is typically considerably less than this. A model such as (9.3) may yield a better unpenalised fit (e.g. deviance or error sum of squares) than (9.1) while its effective dimension (or complexity) is not that much greater. Using the approach of Spiegelhalter *et al.* (2002), one may estimate an effective parameter total (d_e) by comparing the deviance at the parameter mean $D(\bar{\theta})$ with the mean deviance $\bar{D}$, where both are obtained from an MCMC run. So $d_e = \bar{D} - D(\bar{\theta})$. A summary measure of fit that penalises for complexity (in a similar way to the Akaike information criterion, AIC) is the deviance information criterion (DIC), with $DIC = \bar{D} + d_e$. Smaller values of DIC indicate a better fit. As a predictive model check (to ensure that the model is reproducing the data satisfactorily), one may sample new migration flows and assess whether the 95 % interval for such predictions includes the actual data; approximately 95 % of the observed data points should lie within the 95 % prediction intervals (Gelfand 1996).

9.3   Pooling strength over different schedules

One advantage of the model structure exemplified in Equation (9.3) is its simple extension to situations where there are several migration schedules, for example, migration totals Y_{xg} from Scotland to England by age x and gender g, or migration schedules Y_{xgc} by age x, gender g and social class c. One may expect modelling assumptions about the underlying rate structure to be exchangeable between schedules specific to socio-demographic groups (for example, male and female young adult migrants from Scotland to England will both be predominantly labour migrants). In particular, one may generalise the models in (9.3) and (9.4) to multivariate random walks in age, for example, multivariate normal effects with covariance $\Sigma = \Omega^{-1}$ expressing the intercorrelation between the migration schedules of, say, males and females. Thus, the male and female labour force peaks and low elderly propensities are likely to be closely similar.

So, for the two schedules by gender, Y_{xg}, one may generalise the model in Equation (9.3) to become

$$
\begin{aligned}
\log(r_{xg}) &= s_{xg}, & g &= 1, 2, \\
s_x &\sim N_2\left(s_{x-1}, \Omega^{-1}\right), & x &= 2, \ldots, \omega,
\end{aligned}
\tag{9.6}
$$

where $s_x = (s_{x1}, s_{x2})$ and the precision matrix Ω may be assigned a Wishart prior. By contrast, there is no clear way that the model in Equation (9.1) can be estimated in such a way as to pool information over the schedules for different groups, except possibly by a constrained model with some parameters equated between models for different schedules.

The above models are for migration flows by single years of age. Migration data, on the other hand, are often available only for aggregated age groups, denoted $[x]$. In the migration from Scotland to England application, data are available for 18 five-year age groups, $[x] = 0, 1, \ldots, 16, 17$ representing 0–4, 5–9, $\ldots$, 80–84, 85+ year olds, respectively. A model may be developed to use aggregated migration data $Y_{[x]}$ but with the likelihood based on a single-year age model. This amounts to a missing data problem and Bayesian methods provide a relatively straightforward approach to imputing the single year of age flows. However, inferences about the parameters in models such as those above can proceed without imputing the missing single years of age flows.

9.4   Case study: Scotland to England migration, 1990–1991

As a case study, data relating to migration from Scotland to England in the year preceding the 1991 UK Census are analysed. Separate schedules for males and females are available. However, an initial analysis comparing the three models above considers only male flows. A migration model with single years of age is used to define the likelihood. Thus, the five-year age groups $[x]$ are centred at a single year of age $5x + 2.5$ and spanning single years of age $5x + 0.5$ to $5x + 4.5$. For example, the five-year age group $[5]$ consists of single years of age $25, 26, \ldots, 29$. The 1990 mid-year male population estimates for Scotland (ages $0, 1, \ldots, 89, 90+$) define the single year of age populations $N_{x+0.5}$.

Next, assume that the five-year migration totals are Poisson with means defined by summed products of single-year age group populations and rates:

$$Y_{[x]} \sim \text{Po}(N_{5x+0.5} r_{5x+0.5} + N_{5x+1.5} r_{5x+1.5} + N_{5x+2.5} r_{5x+2.5}$$
$$+ N_{5x+3.5} r_{5x+3.5} + N_{5x+4.5} r_{5x+4.5}), \tag{9.7}$$

with r_x as defined in Equations (9.1) to (9.3). Notice that this is not an assumption of multinomial sampling in which the five unknown migration flows by single years sum to a known total $Y_{[x]}$. The model form in (9.7) is consistent with the likelihood assumptions for migration data with single years of age as outlined in Section 9.2, since if $Y_{5x+0.5} \sim \text{Po}(N_{5x+0.5} r_{5x+0.5})$, $Y_{5x+1.5} \sim \text{Po}(N_{5x+1.5} r_{5x+1.5})$, etc., then the total $Y_{[x]}$ will be Poisson. Because $N_{x+0.5}$ are input data (i.e. mid-year official population estimates by single years of age) and single-year age rates $r_{x+0.5}$ define the model, predictions of the missing single-year migration data can be sampled. Of course, one could also estimate the parameters assumed to generate the $r_{x+0.5}$ without sampling the missing one-year migration data. A model check is based on sampling new five-year migration totals or 'predictions', namely $Y_{\text{new}[x]}$, obtaining their 95 % intervals, and checking whether the observation $Y_{[x]}$ is included in the 95 % interval of the prediction.

For the Scotland to England migration flows, it is assumed that no retirement element is present. A seven-parameter version of Equation (9.1) is specified as (single outcome model 1 or model S1 for short)

$$r_x = a_0 + a_1 \exp(-\alpha_1 x) + a_2 \exp\{-\alpha_2 (x - \mu_2) - \exp[-\lambda_2 (x - \mu_2)]\} \tag{9.8}$$

for ages $x = 0.5, 1.5, \ldots, 90.5$. Model S1 is estimated using priors on $\theta = \{a_0, a_1, \alpha_1, a_2, \alpha_2, \mu_2, \lambda_2\}$ based on results about typical parameter values presented in Rogers and Raymer (1999) and Rogers *et al.* (2005). The baseline (theoretical minimum) rate a_0 is constrained to have a maximum of 0.005. Furthermore $a_1 \sim \text{Ga}(0.05, 1)$, $a_2 \sim \text{Ga}(0.075, 1)$, $\mu_2 \sim \text{Ga}(20, 1)$, $\lambda_2 \sim \text{Ga}(0.425, 1)$, $\alpha_1 \sim \text{Ga}(0.05, 1)$, $\alpha_2 \sim \text{Ga}(0.15, 1)$ and $a_0 \sim \text{Ga}(0.001, 1) I(0, 0.005)$, where $I(0, 0.005)$ is an indicator function specifying that sampled values are constrained between 0 and 0.005.

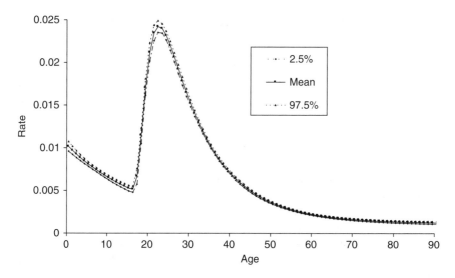

Figure 9.1 Estimated male migration rates from Scotland to England by age: multi-exponential model (model S1).

Table 9.1 Parameter estimates of male migration from Scotland to England: models S1 and S2.

	Model S1			Model S2		
	Mean	2.5%	97.5%	Mean	2.5%	97.5%
a_1	0.009	0.008	0.01	0.009	0.006	0.015
α_1	0.05	0.04	0.06	0.03	0.015	0.11
a_2	0.11	0.10	0.12	0.17	0.07	0.35
μ_2	18.6	18.4	18.8	20.4	17.1	26.2
α_2	0.10	0.09	0.11	0.15	0.09	0.25
λ_2	0.35	0.28	0.42	0.42	0.06	1.90
a_0	0.0010	0.0009	0.0013	0.0003	0.0000	0.0019
κ				32.6	5.4	102.7
DIC	219.0			187.6		
d_e	6.7			15.4		
$D(\bar{\theta})$	205.6			156.8		

A two-chain run of 10 000 iterations shows early convergence, according to Gelman–Rubin diagnostics (Gelman *et al.* 1995). The resulting estimated r_x is highly stylised (Figure 9.1) with precisely defined estimates of rates. The rate plot shows a disjunction at young adult ages, reflecting the much larger migration total for ages 15–19 than for ages 10–14. One might consider some form of moving average to reduce this effect. Table 9.1 contains the estimated parameters and their 95 %

credible intervals. It may be noted that only 13 of the 18 actual flows are included in the 95 % intervals of $Y_{\text{new}[x]}$ so the model has some inconsistencies with the data – in particular, the model rates are too precise to reproduce the actual observations. The DIC is 219 with $d_e = 7$.

Alternatively, consider the gamma mixture form in Equation (9.2) for male migration from Scotland to England (i.e. model S2). The priors are the same as those used in model S1, with the addition of $\kappa \sim \text{Ga}(1, 0.01)$. As illustrated in Figure 9.2, the resulting model schedule is smoother in the late teenage years though less precise (i.e. having wider 95 % intervals) than under model S1. The model dimension increases from 7 to 15.4, but the fit is improved with $D\left(\overline{\theta}\right) = 156.8$ rather than 206. The DIC falls from 219 to 187.6. This model has better predictive accuracy (checks better against the data) with 17 of the 18 actual flows now included in the 95 % intervals of $Y_{\text{new}[x]}$. The exception is the underprediction of migration in the oldest age group.

Models S3 and S4 are nonparametric regressions involving, respectively, first- and second-order random walks in the log scale; see Equations (9.3) and (9.4). A gamma prior on the precision τ_s is assumed with shape parameter 1 and scale parameter 0.1. The posterior mean of the precision parameter under model S3 is 24.5 with 95 % interval (12,42). The DIC for this model improves on that for model S2, namely 176.7 ($d_e = 12.8$, $D\left(\overline{\theta}\right) = 151.1$). As might be expected the resulting migration schedule, set out in Figure 9.3, is less smooth than under models S1 and S2. However, its central features are very similar in form. The plot shows a slight upturn in migration rates among the very old, and now all 18 of the actual flows are included in the 95 % intervals of $Y_{\text{new}[x]}$.

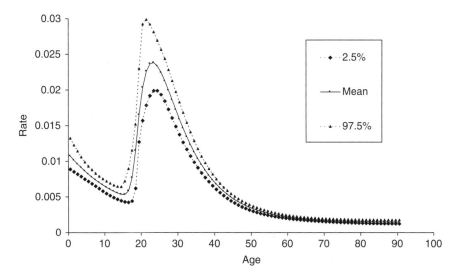

Figure 9.2 Estimated male migration rates from Scotland to England by age: multi-exponential model with gamma mixing (model S2).

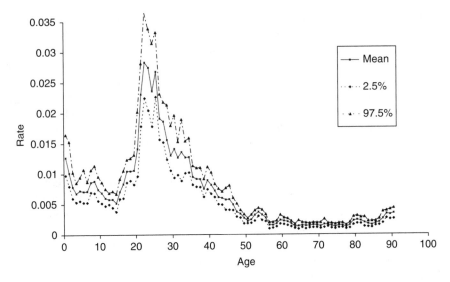

Figure 9.3 Estimated male migration rates from Scotland to England by age: DGLM with RW1 age effects (model S3).

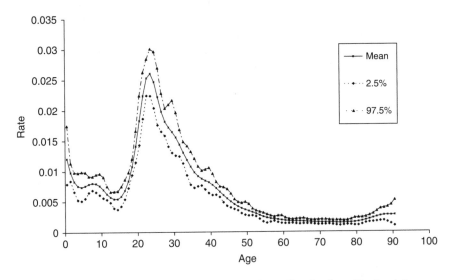

Figure 9.4 Estimated male migration rates from Scotland to England by age: DGLM with RW2 age effects (model S4).

A slightly worse DIC of 181 (with $d_e = 17$) is obtained with an RW2 model (model S4). However, the resulting schedule of rates presented in Figure 9.4 is preferable on substantive grounds in showing a smoothly changing migration propensity over successive ages and with other features not reproduced by the multi-exponential model. These include a relatively flat propensity in young child rates (rather than the decline at these ages as shown in Figure 9.1) and a sharper migration peak in the young adult age groups.

9.5 Multivariate (multiple schedule) model estimates

Migration schedules are often likely to show similar features, which suggests that benefits may be obtained by modelling two or more schedules jointly. For a large collection of schedules one might consider random variation in the parameters of a multi-exponential model. However, a relatively limited multivariate problem is considered in this section, namely joint modelling of male and female migration schedules from Scotland to England. The input data now include mid-year population estimates by single years of age $\{N_{xg}, g = 1, 2\}$ for both males and females for 1990. In the pure parametric model for these data, a distinct parametrisation for males and females is assumed (joint model 1 or model J1 for short):

$$r_{x1} = a_{01} + a_{11} \exp(-\alpha_{11}x) + a_{21} \exp\{-\alpha_{21}(x - \mu_{21})$$
$$- \exp[-\lambda_{21}(x - \mu_{21})]\},$$
$$r_{x2} = a_{02} + a_{12} \exp(-\alpha_{12}x) + a_{22} \exp\{-\alpha_{22}(x - \mu_{22})$$
$$- \exp[-\lambda_{22}(x - \mu_{22})]\}. \tag{9.9}$$

Let θ_1 be the set of male parameters and θ_2 be the female parameters, each of dimension seven. A feature of MCMC estimation by repeated sampling is that one can assess whether particular parameters in the male and female models (say the kth) are effectively equal by monitoring the proportion of iterations where θ_{1k} exceeds θ_{2k}.

If the posterior probability $\Pr(\theta_{1k} > \theta_{2k}|Y)$ is inconclusive (e.g. between 0.1 and 0.9) then there may be grounds for equating the parameters between the two models. One may similarly monitor model outputs of substantive importance, such as the gross migraproduction rate (GMPR) by gender,

$$G_g = \sum_{x=1}^{\omega} r_{xg}, \tag{9.10}$$

and assess, for example, the probability that the male GMPR exceeds the female one, i.e. $\Pr(G_1 > G_2|Y)$. In estimation, the same priors as used in model S1 are assumed, and the second half of a two-chain run of 10 000 iterations is used for inferences. Note that model J1 does not check adequately against the data, with only 27 of the 36 observations lying within the 95 % credible intervals of $Y_{\text{new}[x]}$. Bearing this limitation in mind, the results show that the posterior probabilities $\Pr(\theta_{1k} > \theta_{2k}|Y)$, $k = 1, \ldots, 7$, rule out equating parameters with the clear exception of the decay parameters of the child component, namely $\{\alpha_{11}, \alpha_{22}\}$, where

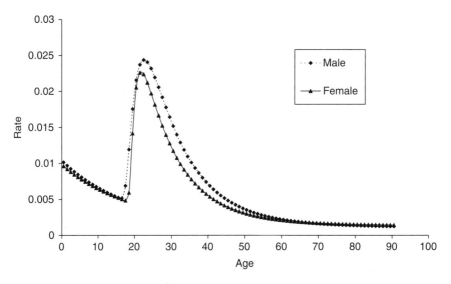

Figure 9.5 Estimated migration rates from Scotland to England by age and gender: purely parametric model (model J1).

$\Pr(\theta_{1k} > \theta_{2k}|Y) = 0.62$, and the posterior means on the two parameters are 0.052 and 0.051, respectively. There are marginal findings (in terms of whether parameters are significantly different) with regard to the baseline constants $\{a_{01}, a_{02}\}$, where $\Pr(\theta_{1k} > \theta_{2k}|Y) = 0.07$, and the young adult exponential decay parameters $\{\alpha_{21}, \alpha_{22}\}$, where $\Pr(\theta_{1k} > \theta_{2k}) = 0.10$. The evidence that the male GMPR exceeds the female one is conclusive with $\Pr(G_1 > G_2 \,|\, Y) = 1$ (see also Figure 9.5). Because of the similarity of the α_1 parameters, the pure parametric model is rerun with $\alpha_{11} = \alpha_{22}$; this leads to a slight fall in DIC, from 432.5 to 430.9, with d_e falling from 13.7 to 12.7.

The next stage is to apply the gamma mixture (Equation (9.2)) to the bivariate migration schedule analysis incorporating the equality constraint on the α_{1g} parameters. However, monitoring $\Pr(\theta_{1k} > \theta_{2k}|Y)$, where θ_1 and θ_2 now include the precision parameters κ_1 and κ_2, suggests that, when the gamma mixing on r_{xg} is introduced, most parameters (including the κ_g) could be equated. The exception are the rate parameters $\{a_{21}, a_{22}\}$ of the young adult migration component, reflecting greater young adult migration intensity for males. So a reduced model is applied (model J2) with

$$r_{xg} \sim \mathrm{Ga}(\kappa, \kappa/m_{xg}),$$
$$m_{x1} = a_0 + a_1 \exp(-\alpha_1 x) + a_{21} \exp\{-\alpha_2(x - \mu_2) - \exp[-\lambda_2(x - \mu_2)]\},$$
$$m_{x2} = a_0 + a_1 \exp(-\alpha_1 x) + a_{22} \exp\{-\alpha_2(x - \mu_2) - \exp[-\lambda_2(x - \mu_2)]\}.$$

$$(9.11)$$

The complexity of this model exceeds that of the pure parametric model with $d_e = 28.1$ but $D\left(\bar{\theta}\right)$ is reduced to 309.1 with a DIC of 365.4. Now, 35 of the 36 data values are contained within the 95 % intervals of predictions from the model, the exception being the oldest male group (i.e. 85+ years).

Finally a multivariate generalisation of the first-order random walk model is fitted (model J3). Thus

$$\log(r_{xg}) = s_{xg}, \qquad g = 1, 2,$$
$$s_x \sim N_2\left(s_{x-1}, \Omega^{-1}\right), \quad x = 2, \ldots, \omega, \tag{9.12}$$

where $s_x = (s_{x1}, s_{x2})$ and the precision matrix $\Omega = \Sigma^{-1}$ is assumed to be Wishart with scale matrix 0.1 $\mathbf{I}$ and two degrees of freedom. In comparison to model J2, the complexity of model J3 ($d_e = 29$) is similar but it has a lower deviance at the parameter mean $\left(D\left(\bar{\theta}\right) = 300\right)$ and deviance information criterion (DIC = 358). All 36 observations are within the 95 % predictive intervals under this model. The 95 % predictive interval for males over 85 is (26,58) with median 41 as compared to the actual flow of 41.

Figure 9.6 confirms the similar shape of the two schedules and the higher GMPR of males, with a posterior mean 0.618 with 95 % interval (0.607,0.629) compared to females, 0.532 95 % (0.522,0.541). The interrelation of schedules is apparent in posterior means of Σ_{11}, Σ_{22} and Σ_{12} of 0.039, 0.045 and 0.031, and a correlation of 0.75 (with 95 % interval (0.42,0.90)) between the male and female migration rate schedules r_{x1} and r_{x2}. Compared to Figure 9.5, the young adult propensity is more peaked and the child rates decline less steeply.

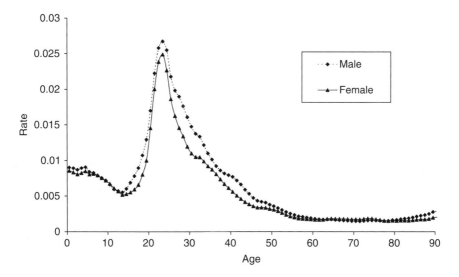

Figure 9.6 Estimated migration rates (posterior means) from Scotland to England by age and gender: multivariate random walk (model J3).

9.6　Discussion and conclusions

This chapter has discussed and implemented model options that go beyond purely parametric forms developed by Rogers and Castro (1981). The MCMC sampling perspective to estimate model parameters has illustrated the additional benefits of a modern Bayesian approach, such as the assessment of hypotheses (e.g. on the relative sizes of GMPRs) that might be difficult in classical estimation. Model choice and comparison for models involving random effects is also difficult under the classical perspective. The analysis of this paper is based on the freeware package WinBUGS, which facilitates Bayesian estimation without the need to specify the full conditional densities or develop application-specific sampling algorithms.[1] The case study exemplifies the benefits of a modelling approach in terms of providing a perspective on the underlying migrant propensities by single years of age, even though the observations are for five-year migration totals.

The analysis has showed the benefits of random effects approaches both in estimating migration schedules with single years of age and in joint estimation of several schedules, for example, in terms of overall fit and in terms of model predictions replicating the actual data. Models that introduce random effects into a fully parametric (multi-exponential) model may be competitive in fit terms with purely nonparametric approaches. However, the latter may be useful for a preliminary smoothing analysis to assess which features (e.g. retirement humps or not) actually seem to be present in the data. The case study has shown that the nonparametric regression approach detects features in the migration data that may not be detected by the multi-exponential model.

References

Besag J, Green P, Higdon D and Mengersen K. 1995. Bayesian computation and stochastic systems. *Statistical Science* 10:3–66.

Gelfand A. 1996. Model determination using sampling based methods. In *Markov chain Monte Carlo in practice*. Gilks W, Richardson S and Spiegelhalter D, eds., Chapter 9. Boca Raton, FL: Chapman and Hall/CRC.

Gelman A, Carlin J, Stern H and Rubin D. 1995. *Bayesian data analysis*. London: Chapman and Hall.

Neves C and Migon H. 2004. Graduação Bayesiana de taxas de mortalidade: uma aplicação na cobertura de sobrevivência e na avaliação da provisão matemática. *Revista Brasileira de Risco e Seguro* 1.

Rogers A and Castro LJ. 1981. *Model migration schedules*. RR-81-30. Laxenburg: International Institute for Applied Systems Analysis.

Rogers A and Raymer J. 1999. Fitting observed demographic rates with the multiexponential model schedule: an assessment of two estimation programs. *Review of Urban and Regional Development Studies* 11:1–10.

---

[1] The Win BUGS code used for this chapter can be found in the Appendix, located at: http://www.wiley.com/go/raymer

Rogers A, Castro L and Lea M. 2005. Model migration schedules: three alternative linear parameter estimation methods. *Mathematical Population Studies* 12:17–38.

Spiegelhalter D, Best N, Carlin B and van der Linde A. 2002. Bayesian measures of model complexity and fit. *Journal of the Royal Statistical Society, Series B* 64:583–639.

Stone C. 1985. Additive regression and other nonparametric models. *Annals of Statistics* 13:689–705.

Part III
Current Estimates

10

Obtaining an overall picture of population movement in the European Union

James Raymer

Division of Social Statistics, School of Social Sciences, University of Southampton

10.1 Introduction

Because of differences in data availability, quality and measurement, no consistent set of migration flow estimates exist between countries in Europe. In fact, only a limited set of information for a small number of countries is known. Given this situation, how does one obtain an overall picture of the population movements, and the related population change, occurring in Europe? The research presented in this chapter addresses this problem by highlighting the key issues and by applying a unifying modelling strategy, based on multiplicative components and log-linear models, to provide estimates of age-specific international migration flows between 25 countries in the European Union (EU) and Iceland, Norway and Switzerland during the 2001–2002 time period. The modelling process starts with the assumption that net migration levels are known, and then works systematically from relatively good, detailed and consistent flow data to situations where the data are inadequate or missing.

The multiplicative component modelling approach used in this chapter to estimate international migration flows is an extension of one used to estimate flows

International Migration in Europe: Data, Models and Estimates   Edited by J. Raymer and F. Willekens
© 2008 John Wiley & Sons, Ltd

(without age) between 10 countries in Northern Europe during the 1999–2000 period (Raymer 2007). Furthermore, this work can be thought of an extension of recent work on describing and estimating age and spatial structures of *internal* migration (Raymer *et al.* 2006; Raymer and Rogers 2007; Rogers *et al.* 2001, 2002a,b, 2003b). Finally, the methodology differs from previous attempts in that it focuses on the underlying structures found in origin by destination by age tables of international migration, which is a categorical data analysis approach to the problem. Poulain (1999), for example, focused on the flows themselves and derived a set of 'correction factors' to adjust the two sets of reported international migration flows by receiving and sending countries (also refer to the double entry matrix described in Chapter 3) to come up with a consistent set of estimates for flows in which data were available.

The organisation of this chapter is as follows. The available data for this study are presented in Section 10.2, along with a discussion of the data types and issues that one has to overcome when trying to obtain an overall picture of European movements. A general background on the estimation of migration and a presentation of the general multiplicative component model for describing and estimating migration flows by origin, destination and age is presented in Section 10.3. The detailed procedure used to obtain the estimates of age-specific flows between countries in the EU and Iceland, Norway and Switzerland during the 2001–2002 period is set out in Section 10.4. The last section presents some possible directions for future research on estimating international migration flows in Europe.

10.2   Migration data

The process of obtaining consistent international migration flow data involves overcoming several major data-related obstacles. Mainly, this involves combining information obtained from independent sources that may (10.1) contain different conceptualisations of migration and (10.2) provide varying levels and qualities of available data (including historical time series). These two issues are briefly reviewed in this section, followed by a description of the available data collected for this study. For a more detailed account of the data issues of international migration and various migration typologies, the reader is referred to Chapters 3, 4 and 6 in this volume, as well as to Champion (1994), Kraly and Gnanasekaran (1987), Poulain (1994), Poulain *et al.* (2006), United Nations (2002) and Willekens (1999).

10.2.1   Data typologies

The United Nations recommends that long-term international migrants be defined as persons who move to a country other than their usual residence for a period of at least a year (United Nations 2002). In reality, countries tend to gather migration data according to their own needs or to be consistent with historical collection methods. Furthermore, there are no real incentives for countries to adjust their data collection methods to provide internationally comparable migration statistics. Therefore, until countries start working together to provide (sets of) harmonised migration statistics, one must have a good sense of the various migration data typologies to produce reliable estimates of migration. This subsection focuses on three main typologies of migration data: event vs status data, place of residence vs nationality data, and the timing criteria used to define migration.

10.2.1.1   Event vs status data

The two main types of migration data are those that capture events and those that capture changes in residential status, providing data on 'migrations' and 'migrants', respectively (Courgeau 1973; Ledent 1980; Rees and Willekens 1986; Willekens 1999). Migration event data typically come from population registers and contain information on the origin, destination and timing of migration. Migration status data, however, typically come from a population census or survey and contain information on place of residence at a particular time, at two points in time (transition data), or at several points in time.

Migrant status data, representing country of residence one year ago or five years ago, are generally preferred for analysing population change. The number of times a person moves during a one-year or five-year period, which is captured by event data, is considered additional information that is unnecessary, particularly if the interest is the condition of a particular population at times t and $t + n$, where n is the length of the time interval. For example, a person who migrates to a country after time t and leaves before time $t + n$ has no direct impact on the change in the population size between t and $t + n$.

10.2.1.2   Place of residence vs nationality data

What is an international migration? Is it a flow of persons moving between two countries (option 1) or a flow of persons of different nationalities moving into a country other than their own (option 2)? Or both? The answer, of course, depends largely on the opinions of the society collecting the data. Most national statistical agencies provide only one of the two options above, i.e. international migrants may represent flows by nationality or by country of origin or destination. This creates a major obstacle for those interested in analysing international migration flows, as methods for harmonising these two data types have yet to be developed.

10.2.1.3   Timing criteria

Ideally, international migrants are persons who have lived (or plan to live) in a country for a period of one year or more. In reality, the timing criteria may vary considerably between countries. For population register data, international migration may refer to persons who have lived in a different country for as long as three months, six months or one year. For census or survey data, the entry date of international migrants is not known, only that they lived outside the country one year or five years prior to the census or survey date. More research is needed to deal with the different timings used to collect or model migration data. At present, only a small body of literature exists on this subject and most of it is focused on internal migration data (Kitsul and Philipov 1981; Liaw 1984; Raymer *et al.* 2007; Rogers *et al.* 2003a; Rogerson 1990).

10.2.2   Missing and inadequate data

Missing data, as used in the context of this study, are either those not collected or those not made available to the public. Inadequate data are data that may be considered unreliable. The solutions to overcoming missing or inadequate data include collecting data (time consuming and expensive), using ancillary data, smoothing the data, or estimating the missing data. Willekens (1999) proposes four steps for estimating migration flows from incomplete data. The first step determines the type of data to predict. The second step specifies a probability model for the 'complete' data, where complete data means the observed and missing data. The third step determines the best model for generating the missing data. And the final step evaluates the results of the estimation process.

10.2.2.1   Missing data

The types of missing data found in European international migration data consist mainly of data not collected or not provided by statistical agencies. The reported flows may have missing groups of persons, for example, illegal migrants who do not participate, for obvious reasons, in a particular survey or register with the local residence authorities. Furthermore, migration data may be available only for the total population, not for more detailed demographic, socio-economic, or spatial characteristics of the population required for a particular study.

10.2.2.2   Inadequate data

International migration data often contain inadequacies due to the collection method or due to non-participation of the migrants themselves (see Chapters 3 and 4). The most common inadequacy found in the reported European international migration data are found in emigration data. Here, it is common that some migrants either fail to notify the population register of their movement or produce statements that are based on intentions, which may or may not be carried out. Surveys, such as the United Kingdom's International Passenger Survey (IPS), are particularly problematic for providing international migration data, as the sample size must be very large in order to provide any level detail for analyses. Without a relatively, large sample size, unexpected irregularities in the data are likely to appear.

10.2.3   The data collected for this study

International migration data were collected for as many European Union (EU) countries as possible for the 2001–2002 period. The available migration data come from published reports produced by the Council of Europe (2002) and Eurostat (2002, 2004). (Note that, at the time of the analysis, the migration data currently available on Eurostat's NewCronos website at http://epp.eurostat.ec.europa.eu/ was not available.) Most of the population and migration data come from a CD-ROM included in the back of a published Council of Europe (2002) report. In addition, data were collected for Iceland, Norway and Switzerland. These countries were included because: (1) they provide relatively good international migration data, which allowed for more observed data to be included in the modelling process; (2) they are members of the European Economic Area (EEA), which basically means that persons from these countries have the same migration rights as EU 15 members (i.e. they have the right to work and reside in any of the EU 15 countries); and (3) they all have strong cultural or proximity connections with other countries in the EU, for example, Iceland and Norway with Sweden and Denmark, and Switzerland with Germany and Austria.

The available *immigration* (I_i), *emigration* (E_i) and *net migration* (NM_i) totals are set out in Table 10.1. Two sets of net migration totals are presented in the Eurostat (2004) report. The first represents the difference between immigration and emigration. The second is indirectly obtained using the demographic accounting equation, i.e.

$$NM_i^{2001-2} = P_i^{2002} - P_i^{2001} + D_i^{2001-2} - B_i^{2001-2}, \tag{10.1}$$

where P_i, D_i and B_i denote population, births and deaths, respectively, and the superscript 2002, for example, refers to the population on 1 January 2002. As can be seen in Table 10.1, the two net migration totals largely correspond with each other, with the exceptions of Belgium, Hungary, Ireland and Spain.

In order to have a set of immigration and emigration totals that correspond with the demographic accounting equation,

$$P_i^{2002} = P_i^{2001} + B_i^{2001-2} - D_i^{2001-2} + I_i^{2001-2} - E_i^{2001-2}, \tag{10.2}$$

emigration totals were estimated based on the simple equation

$$\hat{E}_i^{2001-2} = I_i^{2001-2} - NM_i^{2001-2}, \tag{10.3}$$

where $\hat{E}_i$ denotes the estimated emigrants, I_i denotes the reported immigrants, and NMi denotes the residually obtained net migration total (Equation 10.1). The assumption here is that the reported emigration levels are incorrect. Generally speaking, the difference between the reported and estimated emigration totals is small for most countries. The exceptions are Hungary, Slovenia and Belgium. Note that, the residual net migration for Ireland of 45 876 appeared unrealistic, so the reported immigration (i.e. 46 thousand), emigration (i.e. 20 thousand) and net migration (i.e. 26 thousand) totals were used instead. Malta's reported immigration total of one thousand (i.e. less than the net migration total) and Portugal's reported emigration total of six thousand (i.e. too small relative to the net migration total)

Table 10.1   Available immigration emigration and net migration totals for countries in the European Union, Iceland, Norway and Switzerland: 2001–2002.

Country	Reported			Residual net	Adjusted emigrants
	Immigrants	Emigrants	Net		
Austria	89 928	72 654	17 274	17 265	72 663
Belgium	77 584	52 697	24 887	35 825	41 759
Cyprus	17 485	12 835	4 650	4 650	12 835
Czech Republic	12 918	21 469	−8 551	−8 500	21 418
Denmark	55 984	43 980	12 004	11 996	43 988
Estonia	[a] ...	...	...	142	—
Finland	18 955	13 153	5 802	6 201	12 754
France	...	...	...	64 206	—
Germany	879 217	606 494	272 723	274 850	604 367
Greece	...	...	...	33 900	—
Hungary	21 233	2 591	18 642	9 753	11,480
Iceland	5 002	4 034	968	775	4 227
Ireland[b]	46 200	19 900	26 300	45 876	—
Italy	...	...	...	125 818	—
Latvia	1 443	6 602	−5 159	−5 232	6 675
Lithuania	4 694	7 253	−2 559	−2 512	7 206
Luxembourg	12 135	8 824	3 311	2 750	9 385
Malta[c]	...	...	...	2 226	—
Netherlands	133 404	82 566	50 838	55 985	77 419
Norway	34 264	26 309	7 955	7 966	26 298
Poland	6 625	23 368	−16 743	−16 758	23 383
Portugal[c]	...	...	...	58 740	—
Slovak Republic	2 023	1 011	1 012	951	1 072
Slovenia	7 803	4 811	2 992	4 926	2 877
Spain	414 772	...	...	242 630	172 142
Sweden	60 795	32 141	28 654	28 628	32 167
Switzerland	122 494	82 235	40 259	44 910	77 584
United Kingdom	479 600	307 700	171 900	184 300	295 300

[a] ... Unavailable.
[b] Net migration unrealistic.
[c] Reported immigration or emigration unrealistic.
Source: Eurostat (2004), Tables F-1 and F-3.

also appeared unrealistic and were therefore made 'unavailable'. Refer to Chapters 3 and 4 for more details about the data and their comparability.

Tables containing *international migration flows* in Europe cross-classified by country of origin and country of destination for the 2001–2002 period can be found in a CD-ROM attached to a report published by the Council of Europe (2002).

Country   Aus Bel Cyp CR Den Est Fin Fra Ger Gre Hun Ice Ire Ita Lat Lit Lux Mal Net Nor Pol Por SR Slo Spa Swe Swi UK Oth Out

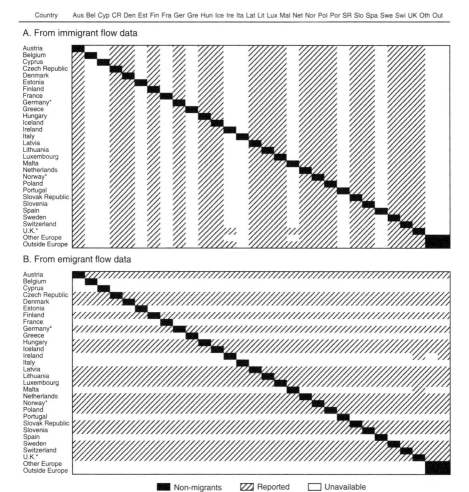

A. From immigrant flow data

B. From emigrant flow data

■ Non-migrants   ▨ Reported   □ Unavailable

Figure 10.1   Available international migration flows between countries in the European Union, Iceland, Norway and Switzerland, 2001–2002.

Both immigration and emigration flows, by previous or next country of residence or by nationality, are reported by the countries that provide data. The available and missing immigration and emigration flows for the complete table are set out in Figure 10.1. Note that, the flows for Germany, Norway and the United Kingdom represent different time periods: the flows for Germany and the United Kingdom represent the 2000–2001 period, and the flows for Norway represent the 2002–2003 period.

   At the time this chapter was written, data on *age-specific international flows* in Europe were unavailable for the period of interest. The age-specific immigration and emigration totals for the 1999–2000 period were used instead. These were

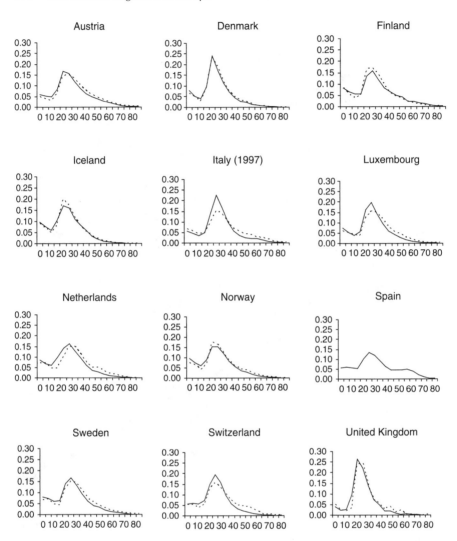

Figure 10.2 Available age compositions of immigration and emigration by country, 1999–2000 (x-axis = proportion; y-axis = age).

found in a Eurostat (2002) report and are presented in Figure 10.2 for the countries that provided data. Note that, age-specific emigration flows were not available for Spain and the flows for Italy represent the 1997–1998 period. The assumption here is that age patterns of migration do not change much over time. Refer to Chapter 8 for a discussion of regularities found in age profiles migration.

10.3 Modelling approach

The models used in this chapter to estimate migration flows are extensions of earlier efforts to estimate *internal* migration using spatial interaction models and log-linear models. There are two types of modelling perspectives relevant to this work: one that relies on population size and geographic distance to estimate migration flows; and another that relies on historical migration flow patterns to estimate migration flows. The latter approach tends to perform better (Snickars and Weibull 1977). However, the general aim should be to combine both historical and covariate information to model migration flows when possible (Rogerson 1984).

Wilson's (1971) introduction of families of spatial interaction models using entropy maximisation techniques, Snickars and Weibull's (1977) minimum information principle, and Bishop, *et al.*'s (1975) text on discrete multivariate analysis were important turning points in the modelling of spatial patterns of migration. They provided the base for many papers on migration estimation (e.g. Alonso 1986; Aufhauser and Fischer 1985; Bennett and Haining 1985; Flowerdew 1991; Flowerdew and Lovett 1988; Plane 1982; Willekens 1977, 1980, 1982, 1983; Willekens and Baydar 1985). Furthermore, linkages and equivalences were made between these various approaches to estimating migration flows (Flowerdew and Aitkin 1982; Plane 1982; Willekens 1980, 1983), which allowed for a generalisation of the estimation process. The link made between spatial interaction models and categorical log-linear models is particularly relevant to this work (see also Chapter 6).

The multiplicative component model is presented next. This model provides a conceptual framework for describing, analysing and estimating migration flow tables (Raymer 2007; Raymer *et al.* 2006; Raymer and Rogers 2007; Rogers *et al.* 2002a). In Section 10.3.2, the conceptual framework is linked with the (statistical) log-linear model for estimating migration flow patterns. In Section 10.3.3, a general plan for estimating a consistent set of age-specific international migration flows between countries in Europe is presented.

10.3.1 Multiplicative component model

The multiplicative component model for an origin (O) by destination (D) by age (A) table of migration flows is specified as

$$n_{ijx} = (T)(O_i)(D_j)(A_x)(OD_{ij})(OA_{ix})(DA_{jx})(ODA_{ijx}), \qquad i \neq j, \qquad (10.4)$$

where n_{ijx} is an observed flow of migration from origin i to destination j for age group x (i.e. 0–4, 5–9, ..., 85+ years). There are eight multiplicative components in total: an overall level, three main effects, three two-way interaction components, and a single three-way interaction component. Note that, for analysis and estimation purposes, the three-way interaction component ODA_{ijx} is generally ignored because (1) the other seven components capture nearly all of the patterns and (2) it has a relatively complex interpretation (Raymer *et al.* 2006).

The components are calculated with reference to the total level in the migration flow tables – see Raymer (2007) for a discussion of the advantages of the 'total reference coding scheme' for migration modelling. The T component represents the total number of all migrants in the system,

$$T = \sum_{ijx} n_{ijx} = n_{+++}. \tag{10.5}$$

The main effect components, O_i, D_j and A_x, represent the proportions of all migrations from each origin, to each destination, and in each age group, respectively, i.e.

$$O_i = \frac{\sum_{jx} n_{ijx}}{\sum_{ijx} n_{ijx}} = \frac{n_{i++}}{n_{+++}}, \tag{10.6}$$

$$D_j = \frac{\sum_{ix} n_{ijx}}{\sum_{ijx} n_{ijx}} = \frac{n_{+j+}}{n_{+++}}, \tag{10.7}$$

$$A_x = \frac{\sum_{ij} n_{ijx}}{\sum_{ijx} n_{ijx}} = \frac{n_{++x}}{n_{+++}}. \tag{10.8}$$

The two-way interaction components represent the ratios of observed migration to expected migration (for the case of no interaction) and are calculated as

$$OD_{ij} = \frac{n_{ij+}}{(T)(O_i)(D_j)}, \tag{10.9}$$

$$OA_{ix} = \frac{n_{i+x}}{(T)(O_i)(A_x)}. \tag{10.10}$$

$$DA_{jx} = \frac{n_{+jx}}{(T)(D_j)(A_x)}. \tag{10.11}$$

These interaction components represent ratios of observed flows or marginal totals to expected ones (i.e. based on the assumption of independence between the variables). The OD_{ij} component captures the association or 'connectedness' between origins and destinations. The OA_{ix} and DA_{jx} components represent the deviations from the overall age profile of migration, n_{i+x}. For estimation purposes, its useful to know that they also represent ratios of the age compositions of emigration and immigration to the overall age composition of migration, i.e.

$$OA_{ix} = \frac{n_{i+x}/n_{i++}}{n_{++x}/n_{+++}}, \tag{10.12}$$

$$DA_{jx} = \frac{n_{+jx}/n_{+j+}}{n_{++x}/n_{+++}}.$$ (10.13)

Finally, although not analysed or estimated in this chapter, the ODA_{ijx} component is calculated as

$$ODA_{ijx} = \frac{n_{ijx}}{(T)(O_i)(D_j)(A_x)(OD_{ij})(OA_{ix})(DA_{jx})}.$$ (10.14)

10.3.2   Log-linear model

The multiplicative component model above (Equation 10.4) can be re-expressed as a saturated log-linear model:

$$\ln n_{ijx} = \lambda + \lambda_i^O + \lambda_j^D + \lambda_x^A + \lambda_{ij}^{OD} + \lambda_{ix}^{OA} + \lambda_{jx}^{DA} + \lambda_{ijx}^{ODA},$$ (10.15)

where the λ denote the parameters or 'effects' of the model. When expressed in this form, migration structures can be modelled using standard statistical techniques for categorical data (see e.g. Agresti 2002). However, because of normalisation restrictions, the total reference coding scheme set out for the multiplicative components in Section 10.3.2 above is not an option in standard statistical packages. Most standard statistical software packages use the cornered-effect coding scheme, where the parameters are interpreted with reference to a single cell in the table. Geometric mean (or effect) coding has also been used for log-linear analyses. Here, the parameters are interpreted with reference to the overall geometric mean of values in the table (see Knoke and Burke 1980; Willekens 1983). The parameter coding scheme used in the multiplicative component model above has two main advantages over the other two coding schemes (Raymer 2007). First, it relies on the information given in the marginal totals. This is important because marginal totals of migration flow tables (e.g. origin by destination by age) are more likely to be more available than the cross-classified counts. Second, the parameters have straightforward interpretations that can be used for describing migration flow tables over time or for estimating sets of two-way interaction effects when the data are incomplete. Unfortunately, this parameter coding scheme is not an option in standard statistical packages, so one is forced to translate between different coding schemes.

Reduced forms of the models set out in Equation (10.9) are considered unsaturated models. For example, the model that only includes the main effects of origin, destination and age is specified as

$$\ln \hat{n}_{ijx} = \lambda + \lambda_i^O + \lambda_j^D + \lambda_x^A.$$ (10.16)

This model assumes independence between each of the categories of origin, destination and age. To control for the diagonal elements in the migration flow table, structural zeros may be inserted into the model (Willekens 1983). The model in Equation (10.16) with structural zeros in the diagonal is a quasi-independent model.

This model provides estimates of age-specific migration flows that assume independence from i to j. Also, the age profile of migration is the same for all flows.

When data are missing or inadequate, offsets can be used to borrow structures from auxiliary data (Rogers *et al.* 2003b). For example, the model

$$\ln \hat{n}_{ijx} = \ln n_{ijx}^* + \lambda + \lambda_i^O + \lambda_j^D + \lambda_x^A + \lambda_{ij}^{OD} \qquad (10.17)$$

provides estimates of migration flows that are consistent with the observed (or estimated) origin–destination-specific totals (i.e. n_{ij+}) but borrow age profiles of migration from the offset, n_{ijx}^*. The offset can be a historical table of migration flows or a constructed table of migration flows. Structural zeros can also be included in the offset to remove cells representing nonmigrants or intranational migrants from the estimation process.

10.3.3   Applying the multiplicative component and log-linear models to estimate international migration flows in Europe

The multiplicative component model is a useful framework for estimating migration flows because it makes a distinction between an overall level, main effects and interaction effects in contingency tables with parameters that can be used to guide the estimation process. This means that one can focus on modelling the underlying structures of migration flows via the multiplicative components. Also, the estimation process can be carried out in a systematic manner working from marginal effects to interaction effects.

The general procedure used in this chapter for estimating international migration flows in Europe is as follows. The first objective is to obtain a consistent set of marginal totals for the origin by destination migration flow table. Here, relatively simple multiple regression models can be used to estimate the missing emigration data (Raymer 2007), following the guidelines discussed in Cadwallader (1992). For international migration, these include: *demographic* variables, such as population size, sex and age; *economic* variables, such as per capita income, cost of living, government expenditure (e.g. education or welfare), taxes, poverty and unemployment; *social* variables, such as education, race, ethnicity, marital status and family ties; and *quality of life* variables, such as climate, natural recreational amenities (e.g. parks, beaches, or mountains), social amenities (e.g. theatres), crime and air pollution. The reason why the focus of the regression models is on estimating emigration flows instead of immigration flows has to do with the fact that emigrants can be associated with a population 'at risk' of migrating (unlike immigrants). Once the missing emigration totals are estimated, the missing immigration flows can then be obtained as a residual, i.e. immigration = net migration + emigration, which forces both to correspond with reported population growth or decline.

The second objective is to obtain the origin-destination–specific flows (without age). The focus here is on estimating the OD_{ij} component (denoted by OD_{ij}^*), i.e. the associations between origins and destinations, which can then be used to produce an offset to be included in a log-linear model. This modelling approach is hierarchical. The reason why the focus is on the OD_{ij} component is because the T, O_i and D_j components have already been obtained (or estimated) in the first objective. The available OD_{ij} components are calculated by dividing the reported flow of migration from i to j by the expected flow, where the expected flow is obtained from the previously estimated marginal totals of the migration flow table (see Equation (10.9)). The unavailable OD_{ij} components may be crudely estimated using the following regression equation (Raymer 2007):

$$OD_{ij} = \alpha + \beta_1 X_{ij} + \varepsilon_{ij}, \tag{10.18}$$

where X_{ij} is a dummy variable representing contiguity (i.e. 1 = neighbours, 0 = non-neighbours) and ε_{ij} is the error term. The assumption here is that migrants are more likely to go to countries nearby or to ones with similar cultural backgrounds, and not to countries that are far away in terms of either distance or culture. More specifically, the associations between neighbouring countries are likely to be very strong (i.e. much greater than unity), whereas those between non-neighbouring countries are likely to be, very weak (i.e. much less than unity).

The set of available and estimated OD_{ij} components can be used together to produce initial estimates of origin-destination–specific migration flows, n_{ij}^*, which can then be inserted as an offset in a log-linear model (without age):

$$n_{ij}^* = (T^*)(O_i^*)(D_j^*)(OD_{ij}^*), \tag{10.19}$$

$$\ln \hat{n}_{ij} = \ln n_{ij}^* + \lambda + \lambda_i^O + \lambda_j^D, \tag{10.20}$$

where T^*, O_i^* and D_j^* are obtained from the marginal totals estimated in the first modelling objective. This forces the initial estimates, which contain the origin–destination interactions, to fit the specified marginal totals of the migration flow table – refer to Raymer *et al.* (2006) for further discussion on the use of offsets and model constraints.

The third and final objective is to estimate the age patterns of international migration. Here, the focus is on estimating the A_x, OA_{ix} and DA_{jx} components (see Equations (10.8), (10.10) and 10.11)). Model migration schedules (see Chapters 8 and 9) may be used both to smooth the available age compositions of migration and to provide an estimate of the overall age composition of migration, represented by A_x^*. The available age compositions of total emigration and total immigration (i.e. n_{+jx}/n_{+j+} and n_{i+x}/n_{i++}, respectively) can then be divided by the estimated A_x to obtain the (available) estimated OA_{ix} and DA_{jx} components. The missing OA_{ix} and DA_{jx} components can simply be set to 1, which means that these flows would have (approximately) the same age profile as the overall age profile included in the A_x^* component. The estimated A_x, OA_{ix} and DA_{jx} components, together with the previously estimated T, O_i, D_j and OD_{ij} components, are then used to produce an offset to be inserted in the log-linear model specified in Equation (10.17). The offset in this case is equal to

$$n^*_{ijx} = (T^*)(O^*_i)(D^*_j)(A^*_x)(OD^*_{ij})(OA^*_{ix})(DA^*_{jx}),$$ (10.21)

where OA^*_{ix} and DA^*_{jx} denote the estimated origin–age and destination–age inter-action components. The assumption in this stage of the modelling process is that most of the age-specific patterns are captured by the marginal structures, which, as demonstrated in Chapter 8, is a fairly reasonable one.

The modelling process just described involves several hierarchical steps begin-ning with the use of available and residually obtained net migration totals. At each stage in the modelling process, there is potential for error, which is then carried forward to the next stage. Some might argue that a reliance on the residually esti-mated net migration totals is particularly problematic as they are likely to include errors associated with the measurements of population stocks, births, or deaths (Siegel and Hamilton 1952; Boulier 1984; Hamilton 1966; Price 1955; Stone 1967). However, the available net migration totals do provide valuable information on the likely impact (i.e. positive or negative) of international migration. Also, for population analysis, it is useful to have migration estimates that correspond with the demographic accounting equation.

This general procedure should be considered a preliminary attempt at overcoming the many obstacles for obtaining an overall picture of movement in the Euro-pean Union. Much work is still needed to refine the method and integrate the key modelling stages. For a start, one could study the relationships between different time intervals used to measure international migration and between migration flows representing nationality and those representing previous or next country of resi-dence. The inconsistencies due to timing differences appear in the marginal totals of the migration flow table, whereas the inconsistencies due to measurement appear in the origin–destination-specific flows.

10.4 Estimation

In this section, the specific procedure used to produce estimates of age-specific international migration between countries in the EU and Iceland, Norway and Switzerland during the 2001–2002 period is described. The main assumptions of this estimation process are as follows: (1) the residual net migration totals reported by Eurostat (2004:Table F-1) are correct; (2) migration by nationality approximates migration by previous or next country of residence; (3) immigration data are better than emigration data; and (4) emigration data are better than no data.

10.4.1 Immigration and emigration totals

The estimated coefficients from the following regression model were used to predict the seven missing emigration totals for Estonia, France, Greece, Italy, Malta,

Portugal and Spain (see Table 10.1) :

$$\ln \hat{n}_{i+} = a_0 + b_1 \ln P_i + b_2 G_i \tag{10.22}$$
$$= 6.460\,1297 + 0.917\,1786(P_i) + 0.000\,0868(G_i),$$

where n_{i+} denotes the adjusted emigration levels, P_i denotes the mid-2001 popula-tions (in millions) collected from the 2001 *World population data sheet* (Population Reference Bureau 2001) and G_i denotes 2001 per capita gross national income in purchasing power parity (GNI PPP) in US dollars collected from the 2003 *World population data sheet* (Population Reference Bureau 2003). The coefficients associ-ated with the two variables were both highly significant. The adjusted R^2 for the fit of the model was 0.771. Finally, the immigration flows for Estonia, France, Greece, Italy, Malta and Portugal were obtained by adding the residual net migration totals set out in Table 10.1 to the estimated emigration levels. The results are presented in Table 10.2.

To obtain estimates of immigrant and emigration totals for Other Europe and Outside Europe, the following procedure was applied. First, the *available* origin–destination-specific 'immigrant flow data' and 'emigrant flow data' identified in Figure 10.1 were rescaled to fit the immigration and emigration totals set out in Table 10.2. From these rescaled data, the estimated numbers from and to Other Europe and Outside Europe were obtained for countries that reported them.

The proportions of migration from and to Other Europe and Outside Europe found in the available data were then used to estimate the corresponding proportions for the countries with missing data.

Four distinct patterns were identified in the *immigrant flow data*:

1. Hungary, Latvia, Lithuania and Slovenia reported an average of 75 % from Other Europe and 12 % from Outside Europe.

2. Denmark, the Netherlands, Norway, Poland, Sweden and Switzerland reported an average of 15 % from Other Europe and 44 % from Outside Europe.

3. Austria, the Czech Republic, Finland and Germany averaged 32 % from Other Europe and 30 % from Outside Europe.

4. Iceland, Luxembourg, the Slovak Republic, the United Kingdom and Malta reported various patterns different from the three clusters above.

Based on geographic proximity, the first cluster of patterns was applied to estimate the numbers from Other Europe and Outside Europe to Estonia, and the second set was applied to estimate the numbers to Belgium, Cyprus, France, Greece, Ireland (Other Europe only), Italy, Malta, Portugal and Spain. The third and fourth sets of patterns were not applied to any country.

Four distinct patterns were also identified in the *emigrant flow data*:

Table 10.2   Estimated immigration and emigration flows for countries in the European Union, Iceland, Norway and Switzerland: 2001–2002.

Country	Immigration		Emigration		
	Flow	%	Flow	%	Net
Austria	89 928	2.16	72 663	1.74	17 265
Belgium	77 584	1.86	41 759	1.00	35 825
Cyprus	17 485	0.42	12 835	0.31	4 650
Czech Republic	12 918	0.31	21 418	0.51	−8 500
Denmark	55 984	1.34	43 988	1.06	11 996
Estonia[a]	**2 154**	0.05	**2 012**	0.05	142
Finland	18 955	0.45	12 754	0.31	6 201
France[a]	**282 625**	6.78	**218 419**	5.24	64 206
Germany	879 217	21.10	604 367	14.51	274 850
Greece[a]	**60 073**	1.44	**26 173**	0.63	33 900
Hungary	21 233	0.51	11 480	0.28	9 753
Iceland	5 002	0.12	4 227	0.10	775
Ireland	46 200	1.11	19 900	0.48	26 300
Italy[a]	**348 010**	8.35	**222 192**	5.33	125 818
Latvia	1 443	0.03	6 675	0.16	−5 232
Lithuania	4 694	0.11	7 206	0.17	−2 512
Luxembourg	12 135	0.29	9 385	0.23	2 750
Malta[a]	**3 089**	0.07	**863**	0.02	2 226
Netherlands	133 404	3.20	77 419	1.86	55 985
Norway	34 264	0.82	26 298	0.63	7 966
Poland	6 625	0.16	23 383	0.56	−16 758
Portugal[a]	**83 326**	2.00	**24 586**	0.59	58 740
Slovak Republic	2 023	0.05	1 072	0.03	951
Slovenia	7 803	0.19	2 877	0.07	4 926
Spain	414 772	9.96	172 142	4.13	242,630
Sweden	60 795	1.46	32 167	0.77	28 628
Switzerland	122 494	2.94	77 584	1.86	44 910
United Kingdom	479 600	11.51	295 300	7.09	184 300
Other Europe[b]	**307 205**	7.37	**656 876**	15.77	**−349 671**
Outside Europe[b]	**574 950**	13.80	**1 437 970**	34.52	**−863 020**
Total	**4 165 992**	**100.00**	**4 165 992**	**100.00**	**0**

[a] Obtained using procedures outlined in the text.
[b] Only include flows from and to the European Union, Iceland, Norway, and Switzerland.

1. Latvia, Lithuania and Slovenia reported an average of 70 % to Other Europe and 13 % to Outside Europe.

2. Denmark, Finland, the Netherlands, Norway, Poland, the Slovak Republic, Sweden and Switzerland reported an average of 7 % percent to Other Europe and 28 % to Outside Europe.

3. Austria, the Czech Republic, Germany and Hungary reported an average of 43 % to Other Europe and 14 % to Outside Europe.

4. Iceland, Luxembourg and the United Kingdom reported various patterns different from the three clusters above.

The first cluster of patterns was applied to estimate the numbers from Estonia to Other Europe and Outside Europe, and the second was applied to estimate the numbers from Belgium, Cyprus, France, Greece, Ireland (Other Europe only), Italy, Malta, Portugal and Spain. The third and fourth sets of patterns were not applied to any country.

Finally, the rescaled and estimated numbers of migration from and to Other Europe and Outside Europe were summed up and then proportionally adjusted (slightly) so that the sum of immigration and emigration for all countries and regions included in the system were equal.

The complete set of estimated immigration and emigration totals for the EU, Iceland, Norway, Switzerland, Other Europe and Outside Europe for the 2001–2002 period are set out in Table 10.2. The total number of persons who migrated is estimated to be 4 165 992. Germany, Outside Europe, the United Kingdom, Spain, Italy and Other Europe received about 72 % of all immigrants, whereas about 72 % of all emigrants came from Outside Europe, Other Europe, Germany and the United Kingdom.

10.4.2   International migration flows

Before the available international migration flow patterns (see Figure 10.1) could be used to estimate the missing flow patterns, several assumptions had to be made. First, the migration flows recorded in the 'unknown' or 'other' categories were assumed to have the same distributions as the country-specific flows. The numbers in these categories were simply proportionally allocated. Second, foreign migrants were assumed to be coming from their country of citizenship. Third, national migrants were assumed to have the same origins and destinations as the foreign migrants. This involved nine countries (i.e. Austria, Denmark, Finland, Hungary, Latvia, Luxembourg, Norway, Slovenia and Sweden) out of about 20 that reported international migration flows. These last two points make the assumption that migration by nationality is a proxy for migration by country of origin.

After the above adjustments, the available international migration flow data from the immigrant and emigrant flow tables (Figure 10.1) were proportionally reallocated to match the total flows presented in Table 10.2. These flows were then combined together to produce one single table of available migration flows. Preference was given to the immigration flow data, except for flows from Iceland, Ireland, Latvia, Luxembourg and Slovenia to the United Kingdom, where the reported emigration flows were used instead. The calculations of the available OD_{ij} components (see Equation (10.9)) were then carried out. The immigration and emigration totals set out in Table 10.2 were used to calculate the expected flows (with the assumption of quasi-independence).

Next, coefficients from the regression model set out in Equation (10.18) were used to estimate the missing OD_{ij} components (see Equations (10.9) and (10.18). The R^2 for this regression was 0.275. The origin–destination interaction between neighbouring countries had a predicted value of 3.5, whereas the interaction between non-neighbouring countries had a predicted value of 0.44. The reference point for both of these ratios is 1. This means that migration between neighbouring countries results in much higher associations (i.e. 35 migrants for every 10 expected) than migration between non-neighbouring countries (i.e. 44 migrants for every 100 expected). The available and estimated OD_{ij} components are set out in Appendix 1.[1] The last step was to use Equations (10.19) and (10.20) to obtain the consistent set of predicted flows. The predicted set of flows from this model and its associated multiplicative components can be found in Appendix 2 and Appendix 3, respectively. In particular, note the differences between the OD_{ij} components used to produce the initial estimates, n_{ij}^*, and the corresponding (smoothed) components resulting from the predictions.

For the most part, the estimated flows maintained the patterns found in the reported flows. For example, consider Iceland's reported and estimated immigration and emigration flows set out in Figure 10.3. The main differences included higher estimated flows associated with Denmark and Germany and lower estimated flows associated with Poland and Outside Europe. Next, consider the United Kingdom's reported and estimated immigration and emigration flows set out in Figure 10.4. Here, the main differences included higher estimated flows associated with Germany and a lower estimated emigration flow associated with Outside Europe. In both cases, migration from and to Germany were predicted to be higher than the reported flows.

Examples of predicted flows for two other countries are set out in Figure 10.5. France, for which no migration data were available, exhibited large predicted flows associated with Belgium (small net loss), Germany (small net loss), Italy (about zero gain), Spain (net loss), the United Kingdom (net loss), and Outside Europe (large net gain). Poland's predicted migration patterns consisted of flows mostly going from and to Germany, with a large net loss resulting in the exchange.

---

[1] The appendices and/or supplementary material for this chapter are available on the book's website at: http://www.wiley.com/go/raymer

Reported

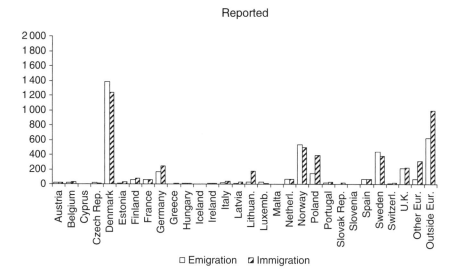

Estimated

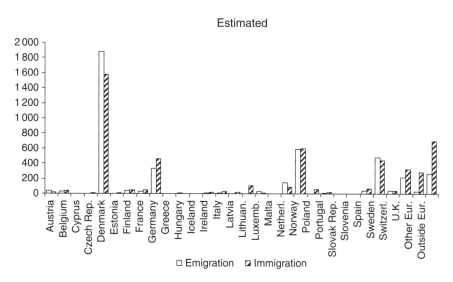

Figure 10.3 Reported and estimated international migration flows from and to Iceland, 2001–2002.

10.4.3   Age-specific flows

Except for the United Kingdom's age profiles of immigration and emigration, all of the reported age profiles appeared reasonable as is (see Figure 10.2). Model migration schedules (see Chapter 8) were used to smooth the United Kingdom's immigration and emigration flows. In particular, there were irregularities found in

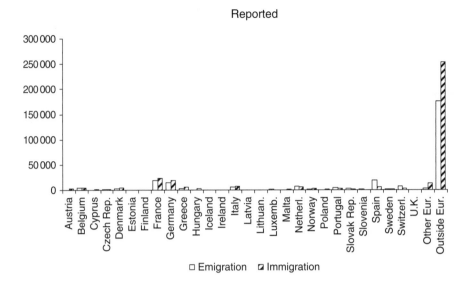

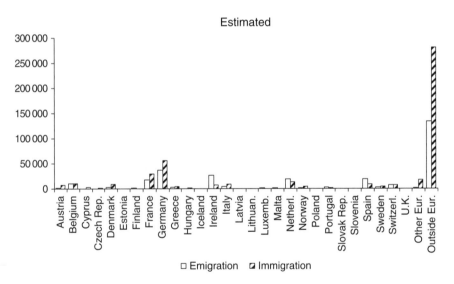

Figure 10.4 Reported and estimated international migration flows from and to the United Kingdom, 2001–2002.

older age groups, including zero values. This irregularity is most likely due to the relatively small sample size of the International Passenger Survey used to obtain the estimates.

The age patterns of international migration were estimated using the available age compositions of immigration and emigration (total flows only), a model migration schedule representing A_x, and the log-linear with offset model set out in

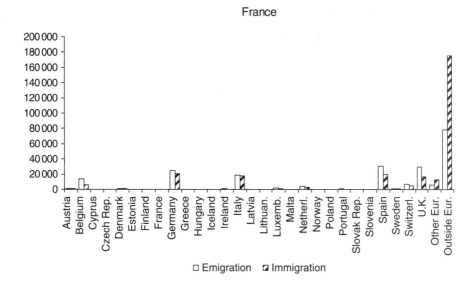

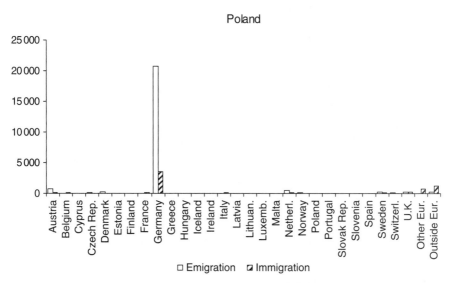

Figure 10.5   Estimated international migration flows from and to France and Poland, 2001–2002. Note that different scales are used in the graphs.

Equation (10.17) – see also Equation (10.21) for specification of the offset. The A_x component was estimated by fitting a seven-parameter multi-exponential model migration schedule (refer to Chapter 8) to the available age-specific compositions of the immigration and emigration totals. The fitted schedule is presented in Figure 10.6, with the following estimated parameter values:

$$A_x = 0.000\,00 + 0.063\,85 \exp(-0.030\,35\,x)$$
$$+0.358\,19 \exp\{-0.110\,02\,(x - 20.624\,79)$$
$$-\exp[-0.204\,98\,(x - 20.624\,79)]\}.$$

All parameter coefficients were highly significant, with the exception of the intercept parameter. The R^2 goodness-of-fit statistic was 0.905, implying that nearly all of the age-specific profiles can be explained by one schedule.

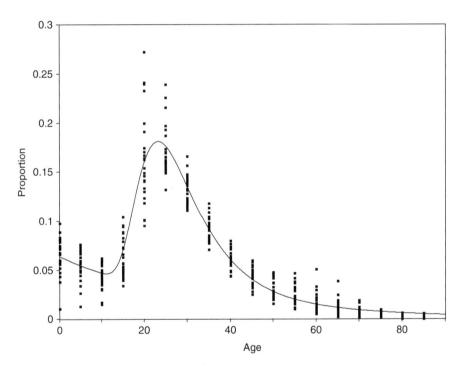

Figure 10.6   Model migration schedule of available immigration and emigration flows in the European Union, Iceland, Norway and Switzerland, 1999–2000.

The next step involved dividing the available age compositions of immigration and emigration totals by the overall age composition, A_x^*, to obtain the available OA_{ix} and DA_{jx} components (see Equations 10.12) and (10.13). The corresponding missing interaction components were given the value of 1. That is, the age patterns for these flows assume the overall (model schedule) shape of age-specific migration.

Examples of predicted age-specific international migration flows are set out in Figure 10.7 for migration from Norway, Portugal, France, the United Kingdom, Other Europe and Outside Europe. The different shapes found in the age profiles reflect the available age-specific data set out in Figure 10.2. For example, flows from and to the United Kingdom exhibit narrow labour force peaks (e.g. United

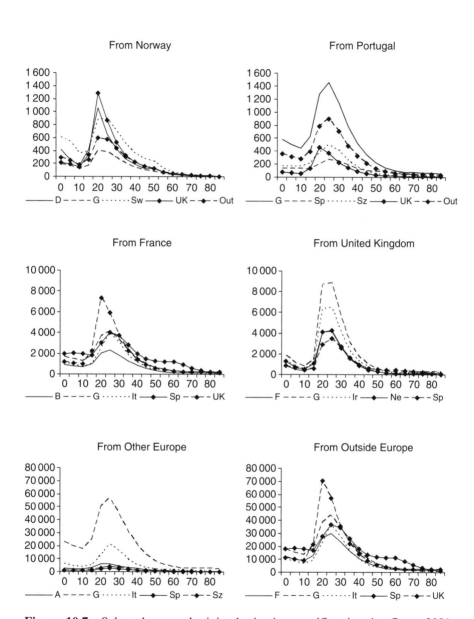

Figure 10.7   Selected age- and origin–destination-specific migration flows, 2001–2002. Note that different scales are used in the graphs. A = Austria, B = Belgium, D = Denmark, F = France, G = Germany, Ir = Ireland, It = Italy, Ne = Netherlands, Sp = Spain, Sw = Sweden, Sz = Switzerland, and Out = Outside Europe.

Kingdom to Germany), whereas immigration flows to Spain exhibit small retirement peaks (e.g. Outside Europe to Spain). Flows such as those from Other Europe to Germany or from France to Belgium exhibit the overall (model schedule) age profile set out in Figure 10.6

10.5   Conclusion

This chapter has presented a starting point for the estimation of international migration in Europe. The many obstacles and details to overcome are enormous. However, with a modelling framework such as the one applied in this chapter, one can obtain an overall picture of the migration patterns. Much work is needed to refine this approach. First, methods to translate between various migration types are needed. In particular, the issues of timing and 'migrant' definitions need to be addressed (see also Chapter 6). Second, the steps in the modelling framework could be integrated together. Finally, work needs to be done to include the various forms of uncertainty in the available data, in the assumptions used and in the resulting predicted flows, so the user has a better sense of the quality of the estimates. In particular, it would be useful to know which estimates are relatively accurate and which ones are not. Here, it appears that a Bayesian framework is ideally suited (see Chapters 7 and 9).

Acknowledgements

Partial support for this work came from a Small Grant from the University of Southampton's School of Social Sciences. The author would like to thank Frans Willekens for his comments on an earlier draft of this chapter and Rene Wetters and Rob van der Erf at the Netherlands Interdisciplinary Demographic Institute for their advice and for providing some of the data used in this chapter.

References

Agresti A. 2002. *Categorical data analysis*, 2nd edn. Hoboken, NJ: John Wiley & Sons, Inc.

Alonso W. 1986. *Systematic and log-linear models: from here to there, then to now, and this to that*. Center for Population Studies, Harvard University, Cambridge, MA.

Aufhauser E and Fischer MM. 1985. Log-linear modeling and spatial analysis. *Environment and Planning A* 17(7):931–951.

Bennett RJ and Haining RP. 1985. Spatial structure and spatial interaction models: modeling approaches to the statistical analysis of geographical data. *Journal of the Royal Statistical Society Series A* 148:1–27.

Bishop YMM, Fienberg SE and Holland PW. 1975. *Discrete multivariate analysis: Theory and practice*. Cambridge, MA: MIT Press.

Boulier BL. 1984. The effects of age mis-statement on estimates of net migration. *Janasamkhya* 2(1):29–37.

Cadwallader M. 1992. *Migration and residential mobility: Macro and micro approaches*. Madison, WI: University of Wisconsin Press.

Champion AG. 1994. International migration and demographic change in the developed world. *Urban Studies* 31(4/5):653–677.

Council of Europe. 2002. *Recent demographic developments in Europe*. Strasbourg: Council of Europe.

Courgeau D. 1973. Migrations et migrants. *Population* 28(1):95–129.

Eurostat. 2002. *European social statistics: Migration*. Luxembourg: Office for Official Publications of the European Communities.

Eurostat. 2004. *Population statistics*. Luxembourg: Office for Official Publications of the European Communities.

Flowerdew R. 1991. Poisson regression modelling of migration. In *Migration models: Macro and micro approaches*. Stillwell J and Congdon P, eds, pp. 92–112. London: Belhaven Press.

Flowerdew R and Aitkin M. 1982. A method for fitting the gravity model based on the Poisson distribution. *Journal of Regional Science* 22:191–202.

Flowerdew R and Lovett A. 1988. Fitting constrained Poisson regression models to interurban migration flows. *Geographical Analysis* 20(4):297–307.

Hamilton CH. 1966. Effect of census errors on the measurement of net migration. *Demography* 3(2):393–415.

Kitsul P and Philipov D. 1981. The one-year/five-year migration problem. In *Advances in multiregional demography*. RR-81-6. Rogers A, ed. Laxenburg, Austria: International Institute for Applied Systems Analysis.

Knoke D and Burke PJ. 1980. *Log-linear models*. Newbury Park, CA: Sage

Kraly EP and Gnanasekaran KS. 1987. Efforts to improve international migration statistics: a historical perspective. *International Migration Review* 21(4):967–995.

Ledent J. 1980. Multistate life tables: movement versus transition perspectives. *Environment and Planning A* 12:533–562.

Liaw K-L. 1984. Interpolation of transition matrices by the variable power method. *Environment and Planning A* 16:917–926.

Plane DA. 1982. An information theoretic approach to the estimation of migration flows. *Journal of Regional Science* 22(4):441–456.

Population Reference Bureau. 2001. *World population data sheet: Demographic data and estimates for the countries and regions of the world*. Population Reference Bureau, Washington, DC.

Population Reference Bureau. 2003. *World population data sheet: Demographic data and estimates for the countries and regions of the world*. Population Reference Bureau, Washington, DC.

Poulain M. 1994. *Internal mobility in Europe: the available statistical data*. Working Paper 17. Conference of European Statisticians, Commission of the European Communities (Eurostat), Mondorf-les-Bains, Luxembourg.

Poulain M. 1999. *International migration within Europe: towards more complete and reliable data?* Working Paper 37. Conference of European Statisticians, Statistical Office of the European Communities (Eurostat), Perugia, Italy.

Poulain M, Perrin N and Singleton A, eds. 2006. *THESIM: Towards Harmonised European Statistics on International Migration*. Louvain-la-Neuve: UCL Presses Universitaires de Louvain.

Price DO. 1955. Examination of two sources of error in the estimation of net internal migration. *Journal of the American Statistical Association* 50:689–700.

Raymer J. 2007. The estimation of international migration flows: a general technique focused on the origin–destination association structure. *Environment and Planning A* 39:985–995.

Raymer J, Abel G and Smith PWF. 2007. Combining census and registration data to estimate detailed elderly migration flows in England and Wales. *Journal of the Royal Statistical Society Series A* 170(4): to appear.

Raymer J, Bonaguidi A and Valentini A. 2006. Describing and projecting the age and spatial structures of interregional migration in Italy. *Population, Space and Place* 12:371–388.

Raymer J and Rogers A. 2007. Using age and spatial flow structures in the indirect estimation of migration streams. *Demography* 44(2):199–223.

Rees P and Willekens FJ. 1986. Data and accounts. In *Migration and settlement: A multiregional comparative study*. Rogers A and Willekens FJ, eds, pp. 19–58. Dordrecht: Reidel.

Rogers A, Willekens FJ and Raymer J. 2001. Modeling interregional migration flows: continuity and change. *Mathematical Population Studies* 9:231–263.

Rogers A, Willekens FJ, Little JS and Raymer J. 2002a. Describing migration spatial structure. *Papers in Regional Science* 81:29–48.

Rogers A, Willekens FJ and Raymer J. 2002b. Capturing the age and spatial structures of migration. *Environment and Planning A* 34:341–359.

Rogers A, Raymer J and Newbold KB. 2003a. Reconciling and translating migration data collected over time intervals of differing widths. *Annals of Regional Science* 37(4):581–601.

Rogers A, Willekens FJ and Raymer J. 2003b. Imposing age and spatial structures on inadequate migration-flow datasets. *The Professional Geographer* 55(1):56–69.

Rogerson PA. 1984. New directions in the modelling of interregional migration. *Economic Geography* 60(2):111–121.

Rogerson PA. 1990. Migration analysis using data with time intervals of differing widths. *Papers of the Regional Science Association* 68:97–106.

Siegel JS and Hamilton CH. 1952. Some considerations in the use of the residual method of estimating net migration. *Journal of the American Statistical Association* 47:475–500.

Snickars F and Weibull JW. 1977. A minimum information principle: theory and practice. *Regional Science and Urban Economics* 7:137–168.

Stone LO. 1967. Evaluating the relative accuracy and significance of net migration estimates. *Demography* 4(1):310–330.

United Nations. 2002. *International migration report 2002*. Population Division, Department of Economic and Social Affairs, United Nations, New York.

Willekens FJ. 1977. *The recovery of detailed migration patterns from aggregate data: An entropy maximizing approach. RM-77-58*. Laxenburg, Austria: International Institute for Applied Systems Analysis.

Willekens FJ. 1980. Entropy, multiproportional adjustment and the analysis of contingency tables. *Systemi Urbani* 2/3:171–201.

Willekens FJ. 1982. Multidimensional population analysis with incomplete data. In *Multidimensional mathematical demography*. Land K and Rogers A, eds, pp. 43–111. New York: Academic Press.

Willekens FJ. 1983. Log-linear modelling of spatial interaction. *Papers of the Regional Science Association* 52:187–205.

Willekens FJ. 1999. Modeling approaches to the indirect estimation of migration flows: from entropy to EM. *Mathematical Population Studies* 7(3):239–278.

Willekens FJ and Baydar N. 1985. Hybrid log-linear models. In *Measuring the unmeasurable*. Nijkamp P, Leitner H and Wrigley N, eds. Dordrecht: Martinus Nijhoff.

Wilson AG. 1971. A family of spatial interaction models, and associated developments. *Environment and Planning* 3:1–32.

11

A simple method for inferring substitution and generation from gross flows: asylum seekers in Europe

Leo van Wissen[†] and Roel Jennissen[‡]

[†] *Population Research Centre Faculty of Spatial Sciences, University of Groningen*
[‡] *Research and Documentation Centre (WODC), Dutch Ministry of Justice, The Hague*

11.1 Introduction

Since the second half of the 1980s, large numbers of asylum seekers have found their way to Europe. Although a European destination is an option for only a small percentage of international refugees, it has been the major destination within the developed world since the 1990s (Jennissen and van Wissen 2003). For the 12 most important destination countries in Europe,[1] the number of asylum applications increased from less than 200 thousand in 1987 to almost 700 thousand in 1992, after which the level decreased sharply to 250 thousand in 1996. By the end of the 1990s, the total inflow stabilised at around 400 thousand applications until 2003.

---

[1] Austria, Belgium, Denmark, Finland, France, Germany, Ireland, the Netherlands, Norway, Sweden, Switzerland and the UK.

International Migration in Europe: Data, Models and Estimates   Edited by J. Raymer and F. Willekens
© 2008 John Wiley & Sons, Ltd

The time trend for individual countries within Europe reflects this overall trend, but not completely, as the total inflow is a function of both an overall trend and relative attractiveness.

In response to the large increase in the number of applications in the early 1990s, governments imposed additional eligibility restrictions on asylum seekers. Initially, the result of these restrictive policies directed asylum seekers to countries with less restrictive policies, which saw their numbers surge, until they also implemented additional measures. When trying to explain the often highly volatile time path of the number of asylum applications in individual European countries, we have to take into account not only the change in the total number of asylum applications in Europe and the asylum policies in each country, but also the change in the number of applications and the implementation of asylum policies in other countries. To some extent, countries are communicating vessels: a decrease in the total inflow in one country due to stricter policies leads, other things being equal, to an increase in other countries.

The extent to which asylum applications in one country is the result of the overall trend in Europe ('generation') or the result of the recent policies in other countries ('substitution') is estimated in this chapter. The generation component applies to all countries similarly. Refugees are 'generated' by war, chaos and famine in various parts of the world. Within Europe they are allocated to different countries through a distribution system that involves airline routes, traffickers, network relations and other mechanisms. These mechanisms are highly sensitive to changes in asylum policies in individual countries. We examine the substitution resulting from the implementation of policies in individual countries, both in the country itself, as well as in other countries with which this substitution takes place. Estimates of asylum substitution between pairs of countries provides insights into the costs and benefits resulting from the implementation of country-specific asylum policies. It also highlights intercountry dependences that exist in the allocation of asylum applicants within Europe.

This chapter is organised as follows. The developments in the number of asylum applications in European countries since 1985 are described and analysed in the next section. In Section 11.3, a method is presented for inferring generation and substitution effects from observed trends. This method is a simple version of a *brand-switching model* used in market research (for an overview, see Böckenholt and Dillon 2000). The results of the estimation are described in Section 11.4. In Section 11.5, we link the results to the most important asylum policies that were implemented in European countries during the period of study. A correspondence between these policies and the estimated substitution effects allows us to test the effectiveness of the method.

11.2   Asylum applications in European countries, 1985–2002

The end of communist regimes in Eastern Europe by the end of the 1980s resulted in a dramatic increase in the number of refugees in Europe as a whole. The war

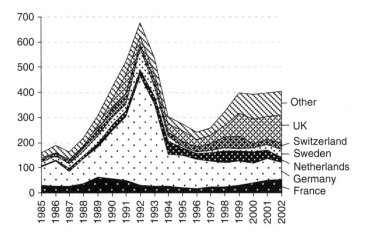

Figure 11.1 Inflow of asylum seekers (in thousands) in selected European countries, 1985–2002. Other = Austria, Belgium, Denmark, Finland, Irish Republic and Norway.
Source: Eurostat (2003), UNHCR (1998, 2003).

in the former Yugoslavia and the political and social turmoil in Romania were the main causes for the peak in the number of asylum seekers in 1992. In that year, 51 % of the total asylum seekers in Europe (i.e. 686 thousand persons) originated from these two countries. In addition to European asylum applicants, many Asians and Africans applied for asylum in Europe. The main countries of origin of asylum seekers from outside Europe in the 1990s were Afghanistan, Iraq and Turkey. The main African countries of origin were Somalia, the Democratic Republic of the Congo and Nigeria (UNHCR 1998, 2003; Eurostat 2003).

Almost all countries saw the number of applications rise sharply in the early 1990s to levels that, from a political point of view, were considered too high to sustain. As a result, nearly all countries imposed stricter asylum policies (see Section 11.5). Some countries placed restrictions earlier than others. This changed the relative attractiveness of alternative destination countries and resulted in unstable allocations of asylum seekers over time (see Figure 11.1). In the first half of the 1990s, Germany received more than half of all applications in Europe. In the second half of the 1990s, this share was reduced to about a quarter, whereas in other countries, their shares increased (notably the United Kingdom, and to a lesser extent France, Sweden, the Netherlands and Switzerland). As demonstrated in Section 11.4, the most important substitution flows during this time came from Germany.

11.3 A method for measuring generation and substitution

The total number of asylum applications in 12 European countries for a period of 18 years represents the information used for this study. The data are presented in

Figure 11.1. In this chapter $Y_i(t)$ denotes the total number of applications in country i ($i = 1, 2, \ldots, 12$) and in year t ($t = 1985, 1986, \ldots, 2002$), representing 218 different totals of country- and time-specific asylum applications. In addition, $Y_+(t)$ is defined as the total number of asylum applications in year t in the 12 countries. The difference in the numbers of asylum seekers in any country i between two years is

$$D_i(t, t+1) = Y_i(t+1) - Y_i(t). \tag{11.1}$$

Note that no links are made between individuals making choices at year t and year $t + 1$, as different persons are represented at each point in time. The method presented in this section decomposes $D_i(t, t+1)$ into a generation component $G_i(t, t+1)$ and substitution components $S_{ij}(t, t+1)$. Here $S_{ij}(t, t+1)$ is considered a pairwise substitution component, i.e. the number of asylum seekers who apply in country j in year $t + 1$ who *would have applied* in country i if they had entered Europe in year t.

Pairwise substitution requires some further explanation. As mentioned above, asylum seekers who enter Europe in year t are not the same as those entering in year $t + 1$. However, for our purposes, a transition table needs to be constructed that links the choices made in year t to the choices made in year $t + 1$. In order to do this, we introduce the term *observational equivalence*, which is borrowed from market research. This study resembles a brand-choice problem, where information is available at different points in time from cross-sectional surveys on the number of customers choosing different brands of the same product. Brand-choice models are designed to estimate the market share of a product, including gains and losses, as a result of stochastic choices made by individuals on a specific purchasing occasion. From these models it may be inferred which customers are loyal to a single product, and which customers switch from product i to product j. When choice-makers have identical attributes over time, they are said to be observationally equivalent. That is, if the properties of the alternative products do not change over time, then the same proportion of people would choose the same product in each time period. This assumption also implies that any change in the relative distribution over the choice alternatives over time is due to changes in the characteristics of the alternatives.

The main attribute that determines a change in attractiveness of each country over time is the set of asylum policies in each country. Therefore, $S_{ij}(t, t+1)$ ($j \neq i$) is the number of refugees who choose country i in year $t + 1$. However, given a set of different circumstances, they would choose country j instead. This switch may be attributed to the relative decrease in attractiveness of country i compared to country j between year t and year $t + 1$. Of course, a relative decrease in attractiveness of country i may be the result of changes in country i, or changes in country j, or both. So $S_{ij}(t, t+1)$ may be viewed as an indicator of the effectiveness of a change in asylum policies imposed in country i and/or j between years t and $t + 1$; and $S_{ii}(t, t+1)$ is the number of equivalent choices in both time periods that result in the identical outcome i. Total substitution from country i to all other countries j is the sum over all other alternatives j:

$$S_{i+}(t, t+1) = \sum_j S_{ij}(t, t+1), \qquad i \neq j. \tag{11.2}$$

Similarly, total substitution to country i out of all other countries is:

$$S_{+i}(t, t+1) = \sum_i S_{ij}(t, t+1), \qquad i \neq j. \tag{11.3}$$

To estimate $S_{ij}(t, t+1)$, consider, for the moment, the assumption that the total numbers of applications in periods t and $t+1$ are the same: $Y_+(t) = Y_+(t+1)$. Later, this assumption is relaxed by including nonzero generation. In principle, $S_{ij}(t, t+1)$ can be estimated in an infinite number of ways, and without additional assumptions there is no unique solution. Therefore, the following assumption is made: the choice outcomes in year $t+1$ must be as close as possible to those in year t. In tabular form we assume that the applications fall as much as possible into the diagonal cells S_{ii}. The solution, therefore, should consist of a table with large diagonal entries and small off-diagonal entries with marginal totals that are consistent with the observed totals $Y_i(t)$ and $Y_i(t+1)$, for all i. This problem can be formulated as an information minimisation problem, in which a given matrix $\mathbf{X}$ with entries x_{ij} and marginal totals r_i and c_j, where $x_{i+} = r_i$ and $x_{+j} = c_j$, is adjusted in such a way that the transformed matrix $\mathbf{Z}$ with entries z_{ij} is as close as possible to $\mathbf{X}$, while conforming to the given marginal totals r_i and c_j, i.e. $z_{i+} = r_i$ and $z_{+j} = c_j$. The marginal totals here are the total number of asylum applications for each country at t and $t+1$: i.e. $z_{i+} = Y_i(t)$ and $z_{+j} = Y_j(t+1)$. This type of technique is used frequently in input–output modelling (Batten 1983), where the matrix consists of inter-industry relations a_{ij} that are scaled to exogenous information on total inputs and outputs per industry. This technique has also been applied in spatial interaction modelling and migration modelling (Willekens 1980, 1983; Scholten and van Wissen 1985). In spatial interaction modelling, the matrix $\mathbf{X}$ is called the historical interaction table, a prior interaction table, or an offset (see e.g. Chapter 6). In our case, the name 'prior substitution table' is most appropriate, since it is an assumed prior distribution of observationally equivalent choices in time periods t and $t+1$. Technically, finding the solution amounts to finding scaling factors α_i and β_j, one for each row and one for each column, such that $z_{ij} = \alpha_i \beta_j x_{ij}$, and such that the scaled matrix $\mathbf{Z}$ fulfils the marginal constraints $z_{i+} = r_i$ and $z_{+j} = c_j$. The scaling factors can be found through the technique of iterative proportional adjustment or fitting (IPF), whereby rows and columns are alternately rescaled to the given marginal totals, until convergence is reached. For more detailed information about this method, see e.g. Willekens (1980, 1983) or Batten (1983).

Consider the simple example of asylum flows to three countries for two consecutive years, set out in Table 11.1. Here, the total number of asylum requests remains the same for the two years, but country 1 experiences an increase in the number of asylum requests from 200 to 400, whereas countries 2 and 3 show a decrease. This

Table 11.1   Marginal totals at times t and $t+1$ for hypothetical example of three countries.

	Time period t	Time period $t+1$
Country 1	200	400
Country 2	300	250
Country 3	500	350
Total	1 000	1 000

Table 11.2   Prior distribution table for hypothetical example of three countries.[a]

Time period t	Time period $t+1$		
	Country 1	Country 2	Country 3
Country 1	1 000	0	0
Country 2	0	1 000	0
Country 3	0	0	1 000

[a] In reality a small positive number (0.5) is entered in the off-diagonal cells (not 0), otherwise scaling of this matrix will not result in positive off-diagonal cells.

example illustrates that the reduced attractiveness of countries 2 and 3 has resulted in a substitution of asylum flows from both of these countries to country 1.

Next, consider another group of asylum seekers identical to those who made the choice for country 1, 2, or 3 at time t. They now enter the system and have to make a similar choice in year $t+1$. The outcome of their choice was identical to the first time period. This assumption is set out in Table 11.2, which may be called a prior distribution table of choice for two points in time. However, we know that this prior distribution was not realised, given the information set out in Table 11.1. To create a new table, which is as close as possible to the prior distribution (Table 11.2), but consistent with the marginal totals of choice at t and $t+1$, as given in Table 11.1, the prior distribution table can be repeatedly scaled to the given marginal totals of time t (i.e. column totals c_i of time t and row totals r_j of time $t+1$) until convergence is reached. Note that, to make this IPF procedure work, we have to replace all '0's with small but positive numbers, otherwise empty cells will always remain empty. Therefore, the '0's in Table 11.2 are replaced with the value 0.5. The minimum information solution obtained by applying the IPF method is set out in Table 11.3. Country 1 increased its number from 200 to 400, by a substitution flow of 50 from country 2 and 150 from country 3. The share of equivalent choices is equal to 800/100 = 0.8. In more complicated examples, involving more countries, and with some countries gaining and some countries losing, the solutions are not always as straightforward as in this example.

Table 11.3 Minimum information solution to hypothetical example of three countries.

Time period t	Time period $t+1$			Total t
	Country 1	Country 2	Country 3	
Country 1	200	0	0	200
Country 2	50	250	0	300
Country 3	150	0	350	500
Total $t+1$	400	250	350	1 000

For most situations, it is unlikely that the total number of asylum requests remains the same over time. Usually, there is a change both in the total number of applications and in the distribution across countries. A structured overview of all the components involved is set out in Table 11.4 for the case of three destination countries. For ease of presentation, the year indices t and $t+1$ are omitted unless required for clarity. As illustrated in Table 11.4, both the 3×3 subtable of substitution and rows and columns of generation are required to estimate the number of asylum applications. The only known information are the marginal totals $Y_i(t)$ and $Y_i(t+1)$.

To completely estimate both substitution and generation, a number of steps are required. In the first step, the total change in the number of asylum seekers for each country is decomposed into a generation component G_i and a substitution component S_i. Increases or decreases in the generation component affect all countries equally. In other words, if the total increase in asylum applications in Europe is 10 %, then the previous year's flows for each country are increased by 10 %. This is true even if the actual number of applications decreases. In this case, a negative substitution component counteracts the increasing effect of the generation component. Thus, the generation component can simply be calculated as

$$G_i(t, t+1) = a(t, t+1)Y_i(t), \qquad (11.4)$$

where $a(t, t+1)$ is the relative change in the total number of applications between year t and $t+1$:

$$a(t, t+1) = D(t, t+1)/Y(t). \qquad (11.5)$$

Note that generation can be positive, zero or negative. If generation is positive, the G_i^+ entries are larger than zero and appear in the in-generation row in the table.

Table 11.4 A generation and substitution table for three destinations.

Destinations t	Destinations $t+1$			Total out-substitution	Out-generation	Total time t
	Country 1	Country 2	Country 3			
Country 1	S_{11}	S_{12}	S_{13}	$Y_1(t) - S_{11} - G_1^-$	G_1^-	$Y_1(t)$
Country 2	S_{21}	S_{22}	S_{23}	$Y_2(t) - S_{22} - G_2^-$	G_2^-	$Y_2(t)$
Country 3	S_{31}	S_{32}	S_{33}	$Y_3(t) - S_{33} - G_3^-$	G_3^-	$Y_3(t)$
Total in-substitution	$Y_1(t+1) - S_{11} - G_1^+$	$Y_2(t+1) - S_{22} - G_2^+$	$Y_3(t+1) - S_{33} - G_3^+$	0	—	$Y(t)$
In-generation	G_1^+	G_2^+	G_3^+	—	G	
Total time $t+1$	$Y_1(t+1)$	$Y_2(t+1)$	$Y_3(t+1)$	$Y(t+1)$		

If generation is negative, the G_i^- entries are larger than zero and appear in the out-generation column. Total *net* generation is given by[2]

$$G_i = G_i^+ - G_i^-. \tag{11.6}$$

The total amount of substitution into or out of each country is determined in the second step. Since $Y_i(t+1)$ is the sum of all column entries in column country i, we know that $S_{2i} + S_{3i} = S_{+i}$ is given by

$$S_{+i} = Y_i(t+1) - G_i^+ - S_{ii}. \tag{11.7}$$

Similarly, we have for S_{i+}, the substitution flows out of country i:

$$S_{i+} = Y_i(t) - G_i^- - S_{ii}. \tag{11.8}$$

Total *net* substitution is

$$S_i = S_{+i} - S_{i+}. \tag{11.9}$$

In the third step, the gross substitution flows S_{ij} and the diagonal entries of equivalent choices are estimated, which represent the 3×3 subtable in the upper left corner of Table 11.4 using the method of information minimisation described above. The starting point for this estimation is the set of marginal totals

$$S_{+i} = S_{+i} + S_{ii} \quad \text{and} \quad S_{i+} = S_{i+} + S_{ii} \quad \text{for} \quad i = 1, 2, 3, \tag{11.10}$$

and the prior substitution table as given in Table 11.2.

11.4 Generation and substitution in twelve European countries

The calculations for obtaining total generation and substitution are almost trivial, but when applied to each country individually, we gain some interesting insights. The decomposition of the change in asylum seekers for each country individually is depicted in Figure 11.2. Since the generation component is assumed to be generic for all countries, there is an overall similarity in the generation curves. It is negative everywhere in 1987, in the period 1993 to 1996, and in 2000. Moreover, the slope of

---

[2] De Beer (see Chapter 13) uses a slightly different method of decomposing the substitution and generation components over time. In his method, demographic substitution is the difference between the actual number of asylum seekers in a country and the number that a country would have had if the total number of asylum seekers in Europe had not changed. Likewise, generation is the difference in the actual number and the number that would result if the proportion of asylum seekers moving to that country had changed. The results of both methods correlate exactly: the size of our substitution effect is exactly a factor $r = (Y_1/Y_2)$ times as large as in de Beer's method. The generation component in our method is $r \times [y_i(t+1)/y_i(t)]$ as large as in de Beer's method. For all years, the correlation between both estimated generation components is 0.9 or higher.

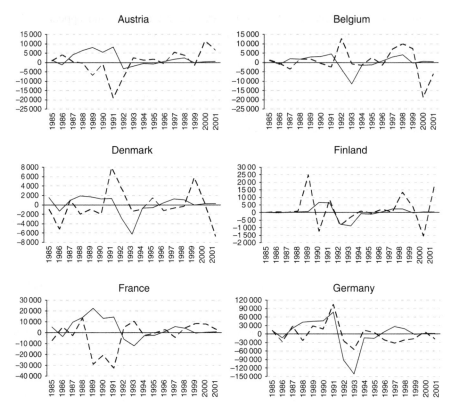

Figure 11.2   Generation (solid line) and substitution (dashed line) in asylum applications for European countries, 1985–2001.

the curve in any particular year is similarly upward or downward for all countries, although it may be very steep in some countries or almost flat in others. In the first period, until 1992, large increases in generation are found in nearly all countries. The exceptions are countries that exhibited an earlier generation peak, followed by a decrease. These include Denmark (peak in 1989), Norway (1988), Sweden (1990) and France (1990). The Irish Republic is also an exception. Asylum seekers did not discover this country as a potential destination until the second half of the 1990s.

Before 1992, generation was the most important component of the increase in asylum seekers. Between 1993 and 1996, the number of asylum applications decreased in Europe and, consequently, negative generation is found everywhere. The negative peak in Germany in 1994 (−140 thousand) is by far the most important example here. In comparison to the substitution flows, the magnitude of generation is usually smaller and less volatile. It appears that countries only serve as a destination for substitution applications for a small period of time. One or two years after receiving refugees through substitution, new substitutions take place from these

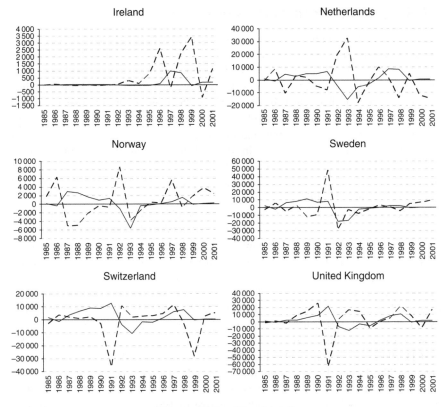

Figure 11.2 (*Continued*)

countries to other destinations. After 1996, positive generation is relatively small compared to earlier values and to substitutions.

The developments in Germany determined, to a large degree, what happened in other countries. Especially important is the change that took place in the period 1991 to 1994. The period 1991–1992 witnessed the largest increase in the number of asylum applications in Europe. Most of them arrived in Germany. Interestingly, this country received more substitution applications (105 thousand) than generation applications (77 thousand) during this time. This means that, in addition to the overall increase, Germany was relatively more attractive for applicants in comparison to other countries. Similarly, a large negative substitution in neighbouring countries is exhibited in this period, particularly in Austria, France, the Netherlands, Switzerland and the United Kingdom. The situation reversed during the next two years. Generation switched from being positive to negative. This was felt most strongly in Germany, which had a negative 140 thousand decrease. Moreover, there was negative substitution amounting to around 60 thousand in 1994, whereas substitution increased substantially in Belgium, France, the Netherlands (more than 30 thousand, or about half of the German substitution flow), Norway, Switzerland and the United Kingdom.

11.5   Substitution and asylum policies

In this section, the relationship between the implementation of asylum policies and resulting substitution flows between pairs of countries is analysed by using the estimated generation–substitution tables for each year in the 1985–2002 period. As an example, the estimates for the transitions between 1993 and 1994 are contained in Table 11.5. The first entry in the table (2685) denotes the number of asylum seekers who applied in Austria in 1993. It also represents the number of applicants, assuming identical characteristics and tastes (i.e. observational equivalence), who would have chosen Austria again in 1994, despite changed conditions. The 29 asylum seekers below represent those who applied in Belgium in 1993 but, under the condition of observational equivalence, would have applied in Austria in 1994, given the changed conditions.

Two examples can be used to demonstrate the effects of policies on other countries. The first example consists of the measures taken in most counties between 1992 and 1994 to counter the substantial increases in applications occurring during these years. All Northern and Western European countries that received considerable numbers of asylum seekers during this time introduced more restrictive asylum procedures in the first half of the 1990s. Austria and Switzerland tightened up their asylum legislation in 1992 (ICMPD 1994; Fassmann 1999). Belgium, Denmark, France, Germany, Norway, Sweden and the United Kingdom introduced more restrictive asylum policies in 1993 or at the beginning of 1994 (ICMPD 1994; OECD 1998; Angenendt 1999a,b). The Netherlands passed stricter asylum policies in 1994. Figure 11.3 presents the estimated substitution flows out of Austria, Germany, Switzerland and the Netherlands. The Austrian and Swiss figures confirm that the stricter policies were already effective in 1992 and resulted in a substantial substitution in that year (i.e. a change in the distribution from 1991 to 1992). The receiving countries involved were foremost Germany and Sweden. In 1993, Germany and most other countries imposed similar rules. The German figures show that these were highly effective by substitution to many other countries. In particular, a large proportion of the 1994 substitution went to the only country that had not yet imposed additional policies, the Netherlands. France and the United Kingdom received a substantial substitution inflow in 1994. Apparently, the stricter policies implemented in these two countries were among the most lenient in Europe. The stricter policies that were imposed in the Netherlands became effective in 1995 and resulted in a reverse flow into Germany, and also into the United Kingdom. These results show that the implementation or lack of implementation of policies is clearly reflected in substitution flows among countries. It shows that countries that are early in implementing policies 'benefit' most, in the sense of reduced numbers of asylum applications, and countries that are late carry the burden of receiving the out-substitution flows from other countries.

The second example focuses on Scandinavian countries. Midway through the 1980s the number of Asian refugees who sought asylum in the Nordic countries started to increase. The Danish government reacted to this increase in October 1986 with a revision to the Aliens Act of 1983 (ICMPD 1994; Jensen 1999; Pedersen

Table 11.5 Estimated generation and substitution for 12 European countries in 1994.

	Austria	Belgium	Denmark	Finland	France	Germany /FRG	Irish Republic	Nether- lands	Norway	Sweden	Switzer- land	UK	Out- substitution	Out- generation	Total 1993
Austria	2685	0	0	0	0	0	0	0	0	0	0	0	0	2059	4744
Belgium	29	14339	0	0	126	0	4	394	0	0	26	202	781	11597	26717
Denmark	55	0	6652	0	236	0	7	741	0	0	49	380	1467	6228	14347
Finland	11	0	0	836	50	0	1	156	0	0	10	80	4	878	2023
France	0	0	0	0	15599	0	0	0	0	0	0	0	0	11965	27564
Germany/FRG	2059	1	0	0	8898	127210	265	27951	0	0	1833	14345	55354	140035	322599
Irish Republic	0	0	0	0	0	0	51	0	0	0	0	0	0	39	90
Netherlands	0	0	0	0	0	0	0	20033	0	0	0	0	0	15366	35399
Norway	145	0	0	0	628	0	19	1973	3379	0	129	1013	3908	5589	12876
Sweden	98	0	0	0	422	0	13	1327	0	18640	87	681	2628	16313	37581
Switzerland	0	0	0	0	0	0	0	0	0	0	14000	0	0	10739	24739
UK	0	0	0	0	0	0	0	0	0	0	0	16129	0	12371	28500
In-substitution	2397	1	0	0	10360	0	309	32543	0	0	2134	16701			
In-generation	0	0	0	0	0	0	0	0	0	0	0	0			
Total 1994	5082	14340	6652	836	25959	127210	360	52576	3379	18640	16134	32830	64445	233181	**537179**

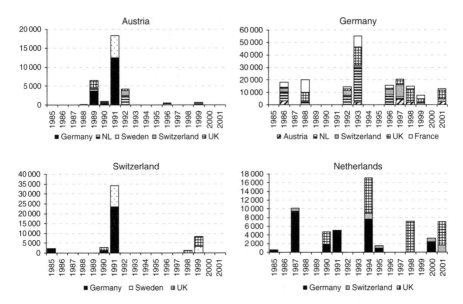

Figure 11.3   Estimated substitution flows from selected countries to other European countries.

1999). Subsequently, the number of refugees who lodged an asylum application in Norway increased considerably in 1987. A likely cause of this increase was the policy measure in Denmark. Subsequently, Norway took policy measures which materialised in 1988 (ICMPD 1994).[3] A similar occurrence happened during the beginning of the 1990s. This time, the refugees were from the former Yugoslavia and entered the Nordic countries on a large scale. Again, Norway was the country that served as alternative destination after a neighbouring country, in this case Sweden, restricted the possibilities for asylum migration. Many Yugoslavian refugees arrived in Sweden in 1992. The Swedish reaction was the introduction of visas for Yugoslav nationals in 1993 (OECD 1998). This led to an increase in the number of Yugoslavian asylum seekers in Norway, which then introduced visas for Bosnians in October 1993 (OECD 1998). The estimated substitution flows for Denmark and Sweden to other Nordic countries and for Norway are shown in Figure 11.4. There are clear signs of correspondence between the timing of the implementation of policies and these substitution flows. Compared to the size of the flows in the 1990s, the numbers of applications were not very large in the 1980s. A large peak was also observed in the 2002 substitution out of Denmark, particularly towards Sweden. In this year, Denmark introduced new immigration laws, which were so strict that the UNHCR questioned the Danish authorities about the new laws being discordant with international laws. Furthermore, Sweden, Belgium and France uttered

---

[3] These policy measures led to substitution to Western Europe (in 1988 to Germany and in 1989 to France, the Netherlands and the UK).

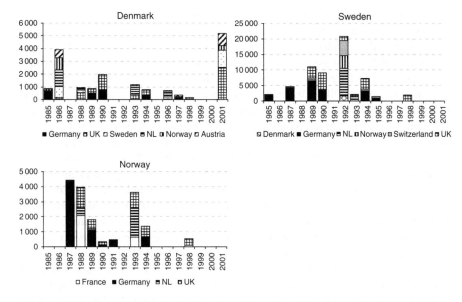

Figure 11.4   Estimated substitution from three Scandinavian countries to selected other European countries.

their 'profound concern' (Osborn 2002). Figure 11.4 shows that, in addition to humanitarian motives, substitution might also have induced Sweden to protest against the introduction of the very strict Danish immigration rules.

11.6   Conclusions

In this chapter, a method was proposed and applied for estimating the generation and substitution of asylum applications in European countries. The main result of this method is that it decomposes overall changes in the number of applications into an 'external' component of generation and an 'internal' component of substitution. In principle, the causes of changes in the overall inflow of refugees into Europe lie elsewhere (Jennissen and van Wissen 2003). The distribution of this inflow over European countries is to a large extent the result of the relative 'attractiveness', or rather 'unattractiveness', since this better reflects the aims of the various asylum policies of European countries. Substitution flows resulting from the implementation of asylum policies are therefore a better indicator of the effectiveness of these policies than overall change in levels, which also includes generation effects. We have demonstrated a clear correspondence between substitution flows and asylum policies. In many cases, opposite generation and substitution effects are found, highlighting effective and ineffective policies.

There are three main assumptions of this method. The first is that asylum seekers are homogeneous, i.e. there are no compositional effects in the change of preferences for alternatives. The only attributes that change are those found in the alternative destinations. Of course, this is a very strong assumption. The composition of asylum seekers does change over time. The large peak in the early 1990s was caused by refugees from the countries of the former Yugoslavia and Romania. In the second half of the 1990s, most asylum applications in Europe originated from Asian countries. Also, due to language and other ties with specific countries, the preferences of asylum seekers are likely to change over time. To account for heterogeneity, one could separate the total inflow into various (homogeneous) subgroups. The problem with doing this is the large amount of input data required, which may not be available. For example, if the country of origin is known, the method presented in this chapter could be used by analysing asylum seekers from each country (or region) of origin separately. If subgroup data are not available, heterogeneity can be introduced by using a multinomial Dirichlet distribution (Böckenholt and Dillon 2000).

The second assumption is that generation and substitution are independent of each other. One might assume that there is an interaction between the attractiveness of a country and its total generation. For instance, if all countries decide to impose the same levels of restriction, then it is likely that the total inflow into Europe will decrease. The alternatives for asylum seekers are to apply in other non-European countries or to remain in neighbouring countries. This assumption would require a higher level of substitution than is presented in this chapter, leading to an inclusion of a nested choice structure, for which Europe is a subset of choices. Given the difficulties of implementing such a structure, we find that the proportionality assumption is a good first approximation.

The third assumption is that choice outcomes in year $t + 1$ are as similar as possible to those in year t. The main consequence of this assumption is that the number of non-changers is as large as possible, and that, for any pair of countries (i, j), there is either a flow from i to j, or from j to i, but not both. One may question the validity of this assumption, but in any case it always holds for the amount of *net* substitution between i and j. So, a less stringent interpretation of this assumption could be based on net substitution between pairs of countries.

In conclusion, the method presented in this chapter provides a good indicator of the effectiveness of asylum policies in terms of (net) substitution flows, which is separated from generation effects. This method is not restricted to the application of asylum flows. In fact, it can be used in any situation where there is repeated cross-sectional information about actors choosing between a fixed set of alternatives, e.g. immigration.

References

Angenendt S. 1999a. Asylum and migration in the EU member states: structures, challenges and policies in comparative perspective. In *Asylum and migration policies in the European Union*. Angenendt S, ed., pp. 6–64. Bonn: Europa Union Verlag.

Angenendt S. 1999b. Germany. In *Asylum and migration policies in the European Union*. Angenendt S, ed., pp. 166–192. Bonn: Europa Union Verlag.

Batten D. 1983. *Spatial analysis of interacting economies*. Dordrecht: Kluwer Nijhoff.

Böckenholt U and Dillon WR. 2000. Inferring latent brand dependencies. *Journal of Marketing Research* 37(1):72–87.

Eurostat. 2003. *NewCronos database.* Luxembourg: Eurostat.

Fassmann F. 1999. Austria. In *Asylum and migration policies in the European Union.* Angenendt S, ed., pp. 65–86. Bonn: Europa Union Verlag.

ICMPD (International Centre for Migration Policy Development). 1994. *The key to Europe: A comparative analysis of entry and asylum policies in Western countries.* Stockholm: Swedish Ministry of Culture.

Jennissen, RPW and van Wissen LJ. 2003. Internationale vluchtelingenstromen op mondiaal en Europees niveau. *Vrede en Veiligheid: Tijdschrift voor Internationale Vraagstukken* 32(1):70–88.

Jensen B. 1999. Thirty years of press debate on 'the foreigners' in Denmark. In *Immigration to Denmark: International and national perspectives.* Coleman DA and Wadensjö E, eds, pp. 191–289. Aarhus: Aarhus University Press.

OECD. 1998. *Trends in international migration.* Paris: Organisation for Economic Co-operation and Development.

Osborn A. 2002. Denmark to get tough on immigration. *The Age*, June 30.

Pedersen S. 1999. Migration to and from Denmark during the period 1960–97. In *Immigration to Denmark: International and national perspectives.* Coleman DA and Wadensjö E, eds, pp. 148–190. Aarhus: Aarhus University Press.

Scholten HJ and van Wissen LJG. 1985. A comparison of the loglinear interaction model with other spatial interaction models. In *Measuring the unmeasurable.* Nijkamp P, Leitner H and Wrigley N, eds, pp. 177–196. The Hague: Martinus Nijhoff.

UNHCR. 1998. *Asylum in Europe: Arrivals, stay and gender from a data perspective.* Geneva: United Nations High Commission for Refugees.

UNHCR. 2003. *Asylum applications lodged in industrialized countries: Levels and trends, 2000–2002.* Geneva: United Nations High Commission for Refugees.

Willekens FJ. 1980. Entropy, multiproportional adjustment and the analysis of contingency tables. *Systemi Urbani* 2:171–201.

Willekens FJ. 1983. Specification and calibration of spatial interaction models: a contingency table perspective and an application to intra-urban migration in Rotterdam. *Tijdschrift voor Economische en Sociale Geografie* 74(4):239–252.

Part IV
Forecasting

12

Bayesian methods in international migration forecasting

Jakub Bijak

Central European Forum for Migration and Population Research, Warsaw

12.1 Introduction

Forecasting international migration is an important, yet difficult, research task, characterised by the highest errors among the forecasts of all the components of demographic change (NRC 2000). Reasons for this include problems associated with inadequate data on international migration (Willekens 1994), identifying and applying potential explanatory variables within a theoretical framework of migration (Kupiszewski 2002) and ignoring forced migration and policy elements (Zolberg 1989). This chapter is devoted to selected methodological aspects of international migration forecasting. The objective is to present an overview of selected methods and to propose an alternative approach based on the paradigm of Bayesian statistics, which allows the formal incorporation of expert judgement. The Bayesian statistical approach does this by combining subjective prior knowledge with statistical information.

Apart from the current introduction, the chapter comprises four sections. In Section 12.2, the issues of uncertainty, subjectivity and expert judgement in migration forecasting are addressed, followed by brief introductory remarks about

International Migration in Europe: Data, Models and Estimates   Edited by J. Raymer and F. Willekens
© 2008 John Wiley & Sons, Ltd

Bayesian statistical inference. The Bayesian philosophy is presented as a formal way to transform prior beliefs by including sample observations, in order to obtain the posterior knowledge – the outcome of the analysis. In Section 12.3, an overview of the existing methods and practices in international migration forecasting is presented, focusing on (a) mathematical models, (b) econometric models, (c) stochastic time series models and (d) Bayesian models. These models are briefly evaluated according to their methodological features, especially related to the issue of uncertainty. In Section 12.4, Bayesian forecasting models are presented for international migration between Poland and Germany, followed by mid-range forecasts for the 2004–2010 period. Finally, in Section 12.5, the main conclusions of the study and an evaluation of potential usefulness of the Bayesian methodology in forecasting international migration are discussed. Throughout this chapter, the terms 'forecast' and 'prediction' are used interchangeably.

12.2   Uncertainty and subjectivity in migration forecasting and in Bayesian statistics

Among the problems in migration forecasting, three issues are of key importance: uncertainty of the processes under study, subjectivity of various components of forecasting models, and the judgemental character of the assumptions made. These issues are briefly discussed in the current section, together with an introduction to Bayesian forecasting.

12.2.1   Uncertainty and subjectivity in migration forecasting

Uncertainty about the values of forecast phenomena is an immanent feature of every prediction. With respect to the sources of uncertainty in population forecasting (and thus also migration forecasting), Keilman (1990:19–20) distinguishes seven types of possible errors, three of which are related to measurement issues (i.e. errors in observed trends, jump-off data, and rounding errors), one to randomness in the parameters of the forecasting model, and three to errors in the forecasts of exogenous variables, possible discontinuity in future trends and to the model specification. Rees and Turton (1998) and Keilman (2001) observed that uncertainty in population forecasting is usually dealt with in a number of ways:

1. Ignored, by constructing single-variant deterministic forecasts.

2. Included, but not quantified in terms of probability, by developing multi-variant scenarios (conventionally: baseline, high and low), which is often done by the national statistical offices, the United Nations (2005) and Eurostat (2005).

3. Accommodated within a stochastic approach, which quantifies uncertainty in terms of probabilities of future events. Keilman (2001) and Wilson and Bell (2004) distinguish three types of stochastic forecasts:

extrapolation of time series (de Beer 1990; Lee and Tuljapurkar 1994; Keilman *et al.* 2001), propagation of historical forecast errors (Keyfitz 1981; Stoto 1983; Alho and Spencer 1985; Alho 1990; NRC 2000), and probabilistic projections based on expert judgement (Lutz *et al.* 1996, 1998, 2004).

In international migration forecasting, all three possibilities are used. Deterministic forecasts are typically based on the judgement of experts, sometimes within the framework of the Delphi method (e.g. Drbohlav 1995). Multi-variant scenarios are often the outcomes of demographic dynamic models, such as the cohort-component or multiregional population projection models. Stochastic forecasts of international migration are usually the outcome of econometric models or time series extrapolation, with a few examples applying the Bayesian approach.

From the probabilistic point of view, the deterministic and scenario approaches are methodologically inconsistent. A deterministic forecast formally has a probability of occurrence equal to zero under any continuous distribution reflecting uncertainty. The scenario approach is criticised for not providing the expected *ex ante* chances that the phenomena under study will be actually observed between the low and high variants (Lutz *et al.* 2004:19). Moreover, the scenario selection (baseline, high or low) often implicitly assumes the presence of a single common underlying factor for all variables (fertility, mortality, migration) and regions under study. The aggregate effects are thus based on the assumption of a perfect correlation between the variables or regions, which is not formally examined and very often not true (NRC 2000:191–192). Unlike in the deterministic and scenario approaches, the stochastic one quantifies uncertainty in terms of probability. Because of this and the methodological consistency of the probabilistic approach to forecasting, many authors argue that the latter will be applied increasingly more often in the future (Lutz and Goldstein 2004:3–4).

As noted by Pittenger (1978), all population forecasts and projections rely heavily on expert judgement. The uncertainty inherent in forecasts requires many subjective elements, including the choice of the forecasting model, its assumptions, and forecasts of the future changes of exogenous variables and other components of population dynamics. This subjectivity can be either explicitly stated in the forecast, or concealed among the assumptions applied. In either case, it is an inherent element of the forecasting model selection and of the assumed scenarios of demographic change (Gjaltema 2001). The incorporation of expert judgement in population forecasting is usually not explicitly addressed by forecasters, with the notable exceptions of Alho and Spencer (1985), Alho (1990) and Lutz *et al.* (1996, 1998, 2004).

International migration is a particularly complex and multidimensional demographic phenomenon, characterised by a large amount of uncertainty, which ideally should be properly addressed and quantified. The existing methods of migration forecasting include various approaches originating in demography, economics, geography, sociology and political science. Improvements in forecasting methodology may require combining expertise from various disciplines. However, the subjective and judgemental elements, inevitable in any forecast, should be explicitly visible in the formulation of the model and its assumptions, as noted by Willekens (1994). This is the basic rationale for selecting Bayesian statistics as a promising framework of forecasting international migration.

12.2.2 Bayesian statistics: introductory notes

In the Bayesian statistical paradigm, sample information is used to transform a researcher's *prior knowledge* with respect to the parameters $\boldsymbol{\theta}$ of the model under study into *posterior knowledge*. The former reflects the subjective opinion (i.e. belief or intuition) on the parameters, without taking observations into account, while the latter is conditional on the sample data vector $\mathbf{x}$ (for multivariate problems, sample data matrix $\mathbf{X}$).

An important issue is the selection of the *prior probability distribution* of the estimated parameters, $p(\boldsymbol{\theta})$, reflecting the knowledge of the researcher, or lack thereof, as in the case of *non-informative* distributions introduced by Jeffreys (1939). The selection of an informative prior is usually supported by expert judgement. In general, Bayes' theorem (Bayes 1763; Laplace 1812) states that the *posterior distribution* $p(\boldsymbol{\theta}|\mathbf{x})$ is proportional to the product of the prior probability $p(\boldsymbol{\theta})$ and the likelihood of the data, $p(\mathbf{x}|\boldsymbol{\theta})$.

The posterior distribution, which is a natural outcome of Bayesian analysis, can be summarised by its point characteristics (e.g. mean, median, or mode) or *credible regions*, analogous to confidence regions in the traditional (Neyman–Pearson) sampling-theory statistics, but without the inconsistencies regarding the interpretation of the latter (Jaynes 1976). Another key concept in Bayesian statistics, distinguishing it from the sampling-theory approach, is subjective probability, independent from the frequency of events under study (Ramsey 1926; De Finetti 1937). A theoretical overview of Bayesian statistics as a complete inference paradigm is given, for example, in Bernardo and Smith (2000), and some introductory information is also provided in Chapter 7 of this volume.

The Bayesian forecasting approach is based on the construction of a *predictive probability distribution* of the vector of future values of the variable under study, $\mathbf{x}^{\mathrm{F}}$, predicted by the assumed forecasting model, conditional on the vector of past (observed) values $\mathbf{x}$, and taking into account the posterior knowledge on the model parameters, $\boldsymbol{\theta}$:

$$p(\mathbf{x}^{\mathrm{F}}|\mathbf{x}) = \int_{\Theta} p(\mathbf{x}^{\mathrm{F}}, \boldsymbol{\theta}|\mathbf{x})\,\mathrm{d}\boldsymbol{\theta} = \int_{\Theta} p(\mathbf{x}^{\mathrm{F}}|\boldsymbol{\theta}, \mathbf{x}) \cdot p(\boldsymbol{\theta}|\mathbf{x})\,\mathrm{d}\boldsymbol{\theta}, \qquad (12.1)$$

where, according to Bayes' theorem, the posterior distribution is

$$p(\boldsymbol{\theta}|\mathbf{x}) = p(\boldsymbol{\theta}) \cdot p(\mathbf{x}|\boldsymbol{\theta})/p(\mathbf{x}).$$

The normalising constant $p(\mathbf{x})$ ensures that the posterior distribution integrates to unity over the space of parameters Θ, and can be obtained by marginalisation:

$$p(\mathbf{x}) = \int_{\Theta} p(\mathbf{x}|\boldsymbol{\theta}) \cdot p(\boldsymbol{\theta}) \, d\boldsymbol{\theta}.$$

The predictive probability distribution of $\mathbf{x}^F$ is calculated according to (12.1) and interpreted as an average from the conditional predictive distribution $p(\mathbf{x}^F|\boldsymbol{\theta}, \mathbf{x})$, weighted by the posterior probabilities of the parameters (Zellner 1971:29).

The Bayesian perspective in prediction-making can thus be summarised as follows. Given the prior beliefs about the parameters $\boldsymbol{\theta}$ of the forecasting model (e.g. regression or autoregression coefficients, variance of the random component, and so on), the sample information $\mathbf{x}$ (i.e. data) on the variable under study (e.g. numbers of migrants or migration rates) is used to update the prior beliefs and to obtain the posterior probability distributions $p(\boldsymbol{\theta}|\mathbf{x})$ according to Bayes' theorem. The posterior distribution is subsequently used to weigh forecasts yielded by models with various possible combinations of the values of their parameters $\boldsymbol{\theta}$ to produce the predictive distributions of the future values of the variable of interest, $\mathbf{x}^F$. Hence, subjectivity in Bayesian forecasting is explicitly included through the selection of the parameters' prior distributions, $p(\boldsymbol{\theta})$. In migration forecasting, expert judgements may include various assumptions on the regression coefficients, indicating theory-based beliefs in the direction of the impact of various migration determinants, as well as *a priori* assumptions on the variance of the random component, which reflect the uncertainty of the phenomenon. In time series models, priors on the autoregressive or moving-average parameters can represent different beliefs on the stochastic nature of the modelled process (e.g. on stationarity).

In some cases, the Bayesian methodology can reduce estimation and prediction errors, if the prior distribution is informative and consistent with the observations. For non-informative priors, the *ex ante* errors in one-dimensional problems are often the same as in the traditional maximum likelihood estimation (Bernardo and Smith 2000:359). This is important in the small-sample studies (e.g. with population disaggregated by sex, age, regions, etc.), where the prior information has relatively more weight in the posterior result than the observations. The extreme estimates obtained from small-sample data are, in this way, corrected towards the prior expectations. The same applies to forecasting models based on a short time series.

In demography, the Bayesian approach has been successfully applied to forecast population (Daponte *et al.* 1997) and the main components of population change: fertility (Tuljapurkar and Boe 1999), mortality (Girosi and King 2004), and migration (Gorbey *et al.* 1999; Brücker and Siliverstovs 2005). The Bayesian methodology may also be used for model selection, by comparing the posterior odds of different models given the data. Some authors suggest that such Bayesian decision criteria favour more straightforward explanations of the phenomena under study, according to the principle of *Ockham's razor* (Jeffreys and Berger 1992). Another possibility is *Bayesian inference pooling*, currently known as *Bayesian model averaging* (Hoeting *et al.* 1999), which allows for combining the features of various predictive models in order to reduce the uncertainty of model specification.

12.3   Overview of forecasting methods for international migration

The current section begins with some general remarks concerning what may be forecasted in the context of international migration. This is followed by a brief overview of selected mathematical, econometric and time series forecasting models, with a focus on the issue of uncertainty. Finally, the few existing examples that apply the Bayesian approach are listed and described. The presented overview has been designed to cover only certain important classes of formal models, primarily related to migration in Europe. A more thorough overview of various forecasting models and methods, covering many more examples than the current section, is presented in Bijak (2006).

12.3.1   General remarks

International migration can be measured using several variables, which may be subject to modelling and forecasting. Besides the size or intensity of annual *flows* of migrants, the variables of interest may concern also migrant *stocks*. In general, stocks are state variables, which reflect the numbers of migrants living in a foreign country as of a certain date, unlike flows, which depict migration processes as such, and are related to the numbers of migrants during a unit time interval. The selected examples of econometric and time series models presented in the current section consider flows and stocks of migrants alike.

Both concepts of measuring migration are interrelated, that is, yearly differences in stocks should correspond to the net of the magnitude of flows into and out of a given country. Migration flows are usually characterised by higher relative variability than stocks, which take into account the balance of inflows and outflows, as well as the cumulated history of migration from the past. On the other hand, predictions of origin–destination-specific rates of migration flows, relative to the population size of a sending country, constitute a crucial input for demographic forecasting models (Rogers 1975). Net migration, although widely used in official statistics, is criticised by demographers for being an artificial measure that obscures the true magnitudes and age profiles of migration, as opposed to the volume and patterns of gross place-to-place migration flows (Rogers 1990).

The modelling and forecasting of migration is determined, in large part, by the availability of appropriate data, which for many European countries is often the main obstacle to making predictions, especially for different categories of migrants (e.g. disaggregated by sex, age, origin, destination, citizenship, country of birth, or type of migration – labour, family reunification or asylum seekers). A general situation concerning availability and comparability of data in the countries of the EU is provided in Poulain *et al.* (2006). Selected problems are also covered in Part I of the current volume, devoted to data issues (Chapters 2–5).

Even if data on the disaggregations of the gross number of migrants are not readily available, there are some model-based approaches, which can facilitate a solution for this problem. For example, the multi-exponential model of Rogers and

Castro (1981) is widely used for modelling the age schedules of migration intensity (see also Chapters 8 and 9). The use of the *iterative proportional fitting* method for disaggregating migration figures by origin and destination regions, as well as by age and sex, under incomplete statistical information, was suggested by van Imhoff *et al.* (1997).

12.3.2 Mathematical models of population flows

According to Kupiszewski (2002), mathematical models of migration emerged predominantly from two different approaches: geographic and demographic. The geographic approach focuses more on the spatial outcomes of migrations, whereas the demographic approach focuses more on the distributions by sex and age, and their impact on the overall demographic dynamics. Both approaches apply mathematical tools to model and forecast migratory flows.

The geographic approach to modelling migration relies heavily on the applications and methodology of Markov chains, as well as various models of spatial interaction. The usage of Markov chains in modelling population movements between regions (states of the chain) evolved from the models with homogeneous transition matrices (Rogers 1966; Brown 1970). The important modifications include models with unobserved heterogeneity, originating from the *mover–stayer* model of Blumen *et al.* (1955), which, however, considered mobility between social groups rather than spatial population flows. Among other methodological variations, there are models with different transition matrices for various subpopulations (Goodman 1961), with heterogeneous transition matrices (Rogers 1966) and non-stationary chains. The latter include semi-Markov processes (Ginsberg 1971), as well as models with the 'cumulative inertia' property, which assumes that, the longer a person does not migrate, the less likely he or she will do so in the future (McGinnis *et al.* 1963).

Spatial interaction models include the 'intervening opportunities' concept of Stouffer (1940), where the number of migrations is proportional to the number of open 'opportunities' for the migrants at the destination, and inversely proportional to the number of similar 'opportunities' located closer to the region of origin. Stewart (1941) and Isard (1960) developed gravity models of migration, in analogy to Newton's laws, with distance as a discounting factor. Other spatial interaction models applied to migration are based on entropy, catastrophe theory and bifurcations (Wilson 1970, 1981).

In the demographic approach, the cohort-component model for demographic projections (originating from Leslie 1945) has been expanded to include migration in population accounting models (Rees and Wilson 1973), multiregional models (Rogers 1975) and multistate models (Keyfitz 1980). The migration component has also been incorporated in the micro-level event-history analysis, with migration as one of the possible demographic events that may happen to an individual (Ginsberg 1979; Courgeau 1985). In the latter case, the analysis can be performed using Monte Carlo micro-simulations.

In addition to the geographic and demographic approaches, there has also been an attempt to model social processes, including migration, using the tools of theoretical physics within the framework of the 'socio-dynamic' approach (Weidlich and Haag 1988). However, the complexity of the proposed model rendered it unexploited in practice.

The main drawback of the mathematical migration models presented above, apart from the event-history analysis, is that they do not explicitly address the issue of uncertainty. Although some of the models apply stochastic tools (e.g. Markov chains), and can therefore be used to assess uncertainty using simulations, this possibility has not been widely explored to date. Nevertheless, the assessments of uncertainty may also be included in a majority of demographic models (cohort-component, multiregional or multistate) by feeding them at input with stochastic forecasts of particular components of demographic change. The latter may involve econometric forecasts and time series models placed in sampling-theory or Bayesian frameworks.

12.3.3   Econometric forecasts of international migration

Another methodology of forecasting migration uses econometric models, sometimes referred to as *explanatory models* (as opposed to pure time series models; see also Chapter 13). Since the 1990s, many such studies have been published, predominantly in Austria and Germany, focusing on forecasting migration in Europe after the anticipated enlargement of the European Union. An overview has recently been presented by Alvarez-Plata *et al.* (2003) and Brücker and Siliverstovs (2005).

In this section, a selection of econometric models is presented, covering different modelling approaches. The notation follows the original studies, with the country subscripts always denoted i and j for the origin and destination countries, respectively, and the normal distributions presented as $N(\mu, \tau)$, with mean μ, and precision parameter τ being the reciprocal of the variance, $\tau = \sigma^{-2}$. Precision notation is used instead of the more common one based on variance (σ^2) to be consistent with the models set out in Section 12.4 and because of the settings of the software used (WinBUGS 1.4, Spiegelhalter *et al.* 2003).

Franzmeyer and Brücker (1997) built a gravity model of net migration between the regions i and j, based on the logarithm of the difference of the GDP *per capita* (PPP-adjusted),[1] $\ln(Y_i/Y_j)$. The model has been calibrated on the basis of the empirical analysis of the elasticity of migration on income differentials in Europe presented by Barro and Sala-i-Martin (1995), where an income gap of 10 % was found to drive between 0.08 % and 0.16 % of the population from less wealthy regions in the system. Owing to this assumption, the model yielded extremely high migration forecasts, with migration from the 10 Central–Eastern European countries to the EU15 in the range of 590–1180 thousand persons a year, depending on the pace of convergence of income levels between the origin and the destination countries.

Fertig and Schmidt (2000) modelled rates of migration to Germany, m, from the four then-candidate countries of the European Union: Poland, the Czech Republic,

---

[1] GDP = gross domestic product; PPP = purchasing power parity.

Hungary and Estonia, with the error term decomposed into country-specific, time-specific and cross-sectional components:

$$m_{i,t} = \mu + \varepsilon_i + \varepsilon_t + \varepsilon_{i,t}. \tag{12.2}$$

Here, i denotes country of origin, $\varepsilon_i \sim N(0, \tau_i)$, $\varepsilon_{i,t} \sim N(0, \tau_{i,t})$, and ε_t is a Gaussian autoregressive process AR(1). This forecast yielded an average population inflow to Germany totalling between 15 and 57 thousand migrants a year in the period 1998–2017.

An example of an econometric forecast of migrant *stocks* is the study of Sinn *et al.* (2001), who predicted the size of foreign population in Germany originating from the largest then-EU candidate countries: Poland, Romania, the Czech Republic, Hungary and the Slovak Republic. The authors used a partial adjustments model,

$$B_t = \lambda[\alpha_0 + \alpha_1 YV_t + \alpha_2 G_t + \alpha_3 EU_t + \alpha_4 FR_t + (1/\lambda - 1 + \alpha_5)B_{t-1}] + \varepsilon_t, \tag{12.3}$$

where the explained variable B_t is the foreign population in Germany in year t. The parameter λ satisfies the condition $B_t = B_{t-1} + \lambda(B_t^* - B_{t-1})$, and B_t^* is a 'long-term equilibrium' of the foreign population stocks under study (Sinn *et al.* 2001:8–9). Among the other variables, B_t^* takes into account the number of migrants in the preceding period, B_{t-1}, according to the hypothesis stressing the role of 'migrant networks' at destination (Taylor 1986), which influence the number of migrants through various self-perpetuation mechanisms. The remaining explanatory variables in Equation (12.3) are: $YV =$ a fraction of GDP *per capita* (PPP-adjusted), $G =$ an output gap, $EU =$ a dummy variable representing EU membership, and $FR =$ a dummy variable representing freedom of movement in the labour force. The model has been partially calibrated on the historical data for German population stocks originating from Greece, Italy, Portugal, Spain and Turkey. The B_t was forecast to increase from the initial 459 thousand to 3.2–4.1 million people by 2015. A similar model has been applied by Brücker and Siliverstovs (2005), with the error term decomposed into country-specific effect and white noise.

Alvarez-Plata *et al.* (2003) prepared a forecast of the post-enlargement migration to the EU15 from 10 countries in Central and Eastern Europe. The dependent variables represent the shares of migrants from country i residing in country j, relative to the population size of the latter (ms), that is, the stocks of migrants expressed in relative terms:

$$ms_{i,j,t} = \alpha + (1-\delta)ms_{i,j,t-1} + \beta_1 \ln(w_{j,t}/w_{i,t}) + \beta_2 \ln(w_{i,t}) + \beta_3 \ln(e_{i,t})$$
$$+ \beta_4 \ln(e_{j,t}) + \beta_5 \ln(P_{i,t}) + \gamma Z_{i,j} + u_{i,j,t}. \tag{12.4}$$

The error term in Equation (12.4) is defined as $u_{i,j,t} = \mu_{i,j} + v_{i,j,t}$, with $\mu_{i,j}$ denoting the origin–destination-specific effects, and $v_{i,j,t}$ being the Gaussian white noise. The other explanatory variables are: $w =$ real income levels, $e =$ employment rates, $P =$ population sizes, and $Z =$ cross-country dummy variables denoting the geographic and cultural proximity of particular countries. Under the assumption of long-term convergence of the economic explanatory variables to their EU15 levels, migration from the 10 Central and Eastern European countries is forecast to decline exponentially from 367 thousand a year, shortly after the end of the restrictions on labour force movements in the enlarged EU, to levels below zero in 2030.

A more general econometric framework for forecasting migration flows has been proposed by Willekens and Baydar (1986), who suggested the use of generalised linear models (GLM). In such models, the effects are specific to the regions of origin and destination and their interactions, as well as to exogenous socio-economic 'environments'. The authors based their models on the observation that overall gross migration flows can be decomposed into the *level effect* (the total number of migrants), the *generation effect* (shares of particular sending areas in all outmigrations), and the *distribution effect* (the distribution of the migrations generated by particular sending areas among all possible destinations). Their model for migration from region i to j in year t($m_{i,j,t}$) can be written as

$$m_{i,j,t} = N_t w_{i,t} p_{i,j,t}, \qquad (12.5)$$

where N_t stands for the total number of migrations in year t, $w_{i,t}$ denotes the probability that a migration originates from region i in the year t, and $p_{i,j,t}$ is the probability that such a migration ends in region j (Willekens and Baydar 1986:207). The authors further suggested decomposing the logits of $w_{i,t}$ and $p_{i,j,t}$, in order to extract region-specific and period-specific effects, as well as their interactions.

Although the econometric models incorporate the analysis of uncertainty, this issue is not given proper attention in many studies devoted to migration forecasting. In some studies there are also problems with the model specification (Kupiszewski 2002). First, if numbers of migrants are forecast, as in Franzmeyer and Brücker (1997), instead of rates denoting the relative migration risk, there is a lack of control on the demographic characteristics of the populations under study (size and age structure), which may lead to extreme results, as shown above. Even when forecasting rates or shares, if the population size is one of the explanatory variables, as in Alvarez-Plata *et al.* (2003), the population movements occur *de facto* outside the model, being another source of bias. Finally, many independent variables used as predictors (GDP, unemployment, etc.) can be equally or more difficult to forecast than the dependent variable of migration.

12.3.4   Stochastic forecasts of migration time series

Another important class of stochastic models used in migration forecasting is based on the analysis and extrapolation of time series. Most frequently this is done by applying various ARIMA models[2] (Box and Jenkins 1976), mainly within the framework of the sampling-theory statistics. For example, de Beer (1997) modelled the total volume of migration from and to the Netherlands using AR(1) autoregressive

---

[2] ARIMA = autoregressive integrated moving average.

models, $x_t = c + \phi x_{t-1} + \varepsilon_t$, where ε_t denotes white noise. The moving average process MA(1), $x_t = c + \varepsilon_t - \theta \varepsilon_{t-1}$, was found to be suitable for net migration. For Finland, Alho (1998) used the ARIMA(0,1,1) models $x_t = c + x_{t-1} + \varepsilon_t - \theta \varepsilon_{t-1}$, for the logarithms of immigration and emigration. Keilman *et al.* (2001) made a probabilistic population forecast for Norway with the ARMA(1,1) model [3] for the log of immigration (i.e. $x_t = c + \phi x_{t-1} + \varepsilon_t - \theta \varepsilon_{t-1}$) and ARIMA(0,1,0), that is, a random walk with drift, for the log of emigration (i.e. $x_t = c + x_{t-1} + \varepsilon_t$). More details on migration forecasting based on the time series approach can be found in Chapter 13.

Multivariate generalisations of time series models can be used to include other explanatory variables, as in the VAR(4) vector autoregression models of Gorbey *et al.* (1999), used for forecasting migration between Australia and New Zealand. Their models, based on quarterly data, have the general form

$$\mathbf{x}_t = \mathbf{C}_0 + (\mathbf{C}_1 L + \mathbf{C}_2 L^2 + \mathbf{C}_3 L^3 + \mathbf{C}_4 L^4)\mathbf{x}_t + \varepsilon_t, \tag{12.6}$$

where $\mathbf{x}_t$ denotes a vector of interdependent variables under study, $\mathbf{C}_k$ are coefficient matrices, L is the lag operator ($L^k(\mathbf{x}_t) = \mathbf{x}_{t-k}$), and ε_t is a multidimensional white noise. The authors tested four alternative models, based on different vectors $\mathbf{x}_t$, which included net migration rates, growth of the real GDP ratio (or the real income ratio) for the two countries, differences in unemployment rates, and country-specific unemployment growth indices. Although the authors remarked that the movements between Australia and New Zealand are largely visa-free and resemble internal migration, similar models can be used for other international population flows.

An example of a partial departure from the sampling-theory statistical paradigm in demographic forecasting is the concept of the *expert-based probabilistic population projections*, developed by Lutz *et al.* (1996, 1998, 2004). The method applies subjective expert judgement in order to set the framework for the stochastic forecasts. In formal terms, with v_t denoting the phenomenon under study (e.g. migration), the forecasting model is $v_t = \bar{v}_t + \varepsilon_t$, where $\bar{v}_t$ is the average trajectory of the process, assumed *a priori* by the experts, and ε_t is a random process, for example AR(p) or MA(q). Lutz *et al.* (2004) applied $\varepsilon_t \sim$ MA(30) assuming that the standard deviation of ε_t, namely $\sigma(\varepsilon_t)$, is equal to a predefined value $\sigma^*(\varepsilon_t)$, also set on the basis of expert judgement. For migration, it was assumed that $\bar{v}_t = \bar{v}$ (i.e. the mean of the process is time-invariant), while $\sigma^*(\varepsilon_t)$ has been selected so that 80 % of the density of the probability distribution of v_t is concentrated between zero and the arbitrarily chosen value v_{max}.

Owing to the explicitly expressed subjectivity, this approach can be seen as a hybrid between the traditional and Bayesian methods. On the one hand, probabilities attached to particular trajectories of variables under study are subjective, but on the other, no observations (data) are used to change the prior beliefs into posterior knowledge, and thus the full Bayesian inferential mechanism is not involved in

---

[3] ARMA = autoregressive moving average.

the 'expert-based' projections. The key advantage of all time series models used in migration forecasting is that they explicitly incorporate the analysis of uncertainty. VAR (vector autoregression) models can also include predictors of migration that do not have to be forecast separately.

12.3.5   Bayesian models and forecasts of population flows

There are relatively few examples of international migration models and forecasts based on the Bayesian framework. Gorbey *et al.* (1999) extended their VAR(4) analysis of migration between Australia and New Zealand to the Bayesian case (see Section 12.3.4 above). They used the *Minnesota priors* for coefficients in the models set out in Equation (12.6). The parameters on the first lags of the corresponding variables were given *a priori* a normal distribution $N(1, \tau_{i,i,1})$ and the remaining parameters $N(0, \tau_{i,j,k})$ for interrelations between the ith variable and the kth lag of the jth variable, i.e. for $k > 1$ or $k = 1$ and $i \neq j$. This reflects a belief that the time series of each variable included in the vector under study is probably generated by a random walk process, and that the processes for particular variables are mutually independent. Further, the authors assumed that

$$s_{i,j,k} = \tau_{i,j,k}^{-0.5} = \gamma g(k) f(i, j)(s_i/s_j),$$

where $\gamma = 0.4$, $g(k) = k^{-1}$, $f(i, i) = 1$, and $f(i, j) < 1$ for $i \neq j$, and where s_i denotes a standard error in the autoregressive model for the ith variable. Note that the s_i values were estimated from the observations and, thus, the priors are data-based and not specified independently from the data, as should be the case in the fully Bayesian approach. In the *ex post* analysis, the authors found that the best Bayesian model (with $\mathbf{x}_t$ comprising net migration rates, the growth of the real GDP ratio for the two countries, and quarterly unemployment growth in Australia) performed slightly worse than the corresponding traditional (sampling-theory) VAR, which is probably due to the disagreement between the priors and the data.

A Bayesian approach to modelling patient-to-hospital flows using gravity models can be found in Congdon (2000, 2001). These models can be generalised to cover different types of spatial movements, such as migration flows. Assuming that the number of patients from region i to the hospital j has a Poisson distribution with a mean $\mu_{i,j}$, the models can be defined as (Congdon 2001)

$$\ln(\mu_{i,j}) = k + \alpha_0 \ln(P_i) + \delta \ln(R_{i,j}) + \xi_1 E_{1,j} + \xi_2 E_{2,j} + \phi S_{i,j}, \tag{12.7a}$$

or

$$\ln(\mu_{i,j}) = k + \alpha_0 \ln(P_i) + \alpha_1 YAN_i + \alpha_2 Aged_i + \delta_i \ln(R_{i,j}) + \xi_1 E_{1,j} +$$
$$\xi_2 E_{2,j} + \phi S_{i,j}, \tag{12.7b}$$

where $\mu_{i,j}$ is the expected number of movements during a year. The main explanatory variables in Equations (12.7a) and (12.7b) are population size of the ith region (P_i) and the 'supply' of medical services ($R_{i,j}$, i.e. the number of beds in the jth hospital, weighted by an average distance, either crow-fly or car time, from the ith region to the jth hospital). The other predictors include an index of demand for the health-care services (YAN_i), the fraction of population aged 65+ years $(Aged_i)$, dummy variables for two selected hospitals $(E_{1,j}$ and $E_{2,j})$, and a dummy indicating whether the jth hospital is located in the ith region $(S_{i,j})$. The prior distributions are normal, diffuse for the constant $(k \sim N(1, 0.0001))$ and more informative for the remaining parameters: $\alpha_i, \beta, \delta \sim N(1, 0.1)$, $\gamma \sim N(2, 0.1)$, ξ_1, ξ_2 and $\phi \sim N(0, 0.1)$. The second model fitted the data better with distance defined in car-time terms.

The study by Brücker and Siliverstovs (2005) also contain some Bayesian analysis. However, it is considered by the authors only as an alternative methodology of estimation. No mention is made of the assumed prior distributions or the *a posteriori* uncertainty assessment. An *ex post* comparison of various estimation methods for a partial-adjustment model showed that the hierarchical Bayes estimator (probably the mean in the appropriate posterior distribution) and the sampling-theory fixed-effects estimator performed best.

In addition to the above-mentioned examples, the Bayesian approach is explored in Chapter 7 to model migration flows and in Chapter 9 to model age schedules of migration intensity. As in the case of the sampling-theory time series models, Bayesian models address uncertainty in an explicit way, through the analysis of predictive distributions of forecast future values of the variables under study, expressed in probabilistic terms. The VAR models allow for including additional explanatory variables, without the necessity to forecast them separately, as the future values of all interdependent variables are predicted together, within a system of simultaneous equations. In the Bayesian approach, subjective expert knowledge on the characteristics and interactions of the processes under study, embodied in the model parameters, can be formally incorporated in statistical inference in the form of the prior distributions.

12.4   Examples of simple Bayesian models for forecasting international migration

The numerical examples presented in this section apply Bayesian methods to forecast the intensity of annual migration flows between Poland and Germany in the period before the expected opening of the German labour market to Polish citizens (i.e. 2004–2010). The calculations are based on the available time series for the period 1985–2003.

12.4.1   Data, specification and estimation of the models

The data on migration flows and population stocks used in the current empirical example predominantly come from Eurostat's NewCronos database. The time series of the economic explanatory variables, GDP and unemployment rates come from the databases of the United Nations Economic Commission for Europe and the World Bank, respectively. The German data prior to 1991 represent West Germany. In Poland before 1990, there was officially no unemployment.[4] For this study the respective rates have been assumed to equal 0.1, an arbitrarily small positive number, to avoid problems with their logarithmic transformations.

As the numbers of migrants reported by the origin and destination countries usually differ, the greater of the two values has been taken as the estimate of the real magnitude of each of the flows, following Kupiszewski (2002:111–112). For the current example, this means that German data for flows in both directions must be relied upon (refer to Chapters 3 and 4 for more detailed discussions of the data). The numbers of registered settlement (long-term) migrants in both directions (from Poland to Germany and vice versa) have been transformed into crude occurrence – exposure emigration rates per 1000 inhabitants of the country of origin at mid-year. Note that the Polish populations for 1988–2002 have been adjusted to account for the underestimation of international emigration by the Polish authorities.

For both flows, three models are considered, presented here for brevity on the example of emigration rates (*MR*) from Poland (P) to Germany (D). Model 1, a *Gaussian autoregressive process* AR(1), is specified as

$$\ln(MR_{P-D}(t)) = c + \alpha \ln(MR_{P-D}(t-1)) + \beta Z(t) + \varepsilon(t), \qquad (12.8)$$

where t denotes year, Z is a dummy variable, equalling one for the years 1988–1990 and zero otherwise (to account for the economic transition during this time), and the error term $\varepsilon(t) \sim N(0,\tau)$. The logarithmic transformation is used to ensure that MR_{P-D} remains positive. For c and α, diffuse (i.e. non-informative) prior distributions $N(0,0.001)$ are assumed. Parameter β is expected to follow a relatively informative prior distribution $N(1,0.1)$, assuming that the system transformation in the years 1988–1990 contributed to the magnification of $\ln(MR_{P-D})$ by one unit on average, and thus of MR_{P-D} itself by a factor of about 2.7. Finally, τ is *a priori* expected to follow a chi-squared distribution with one degree of freedom, reflecting a belief in a low precision. The latter assumption seems to be appropriate given the uncertain nature of the migration.

Model 2, a *vector autoregressive process* VAR(1), is specified as

$$\mathbf{x}_1(t) = \mathbf{c} + \mathbf{A}\mathbf{x}_1(t-1) + \boldsymbol{\beta}Z(t) + \boldsymbol{\varepsilon}(t), \qquad (12.9)$$

where $\mathbf{x}_1(t) = [\ln(MR_{P-D}(t)), \ln(GDP_D(t)/GDP_P(t))]'$, $\mathbf{c} = [c_1, c_2]'$, $\mathbf{A} = [\alpha_{i,j}]_{2\times2}$, $\boldsymbol{\beta} = [\beta, 0]'$, Z is the same as above and $\boldsymbol{\varepsilon}(t) \sim \mathbf{N}(\mathbf{0},\mathbf{T})$. Income differentials (i.e. GDP per capita according to the PPP, expressed in 2003 international dollars) are used to explain international migration (and vice versa, as the VAR is

---

[4] The hidden unemployment rate for Poland was estimated to be about 25% in the late 1980s (Rutkowski 1990). However, this rate has not been considered here, as this category cannot be seen as a push factor of migration. Hidden unemployment occurs when a reduction in excess employment in an economy would not lead to a decline in the output. This is not visibly related to the migration decisions of the individuals, unlike in the case of the real unemployed looking for jobs and better life prospects.

a system of two simultaneous equations). Logarithms have been used for the same reason as in model 1. Analogous to the one-dimensional case, $N(\mathbf{0},\mathbf{T})$ denotes the two-dimensional normal distribution with mean $\mathbf{0} = [0, 0]'$ and precision matrix $\mathbf{T}$ (inverse of the covariance matrix $\mathbf{\Sigma}$). The prior distributions for c_1 and c_2 are both diffuse, following $N(0,0.001)$. The own-variable lag parameters, $\alpha_{1,1}$ and $\alpha_{2,2}$, follow $N(1,1)$, assuming a likely random walk character for each of the variables separately. For the cross-variable lag parameters, $\alpha_{1,2} \sim N(0.5,1)$ reflects the initial hypothesis of a positive impact of the income difference between destination and origin countries on migration and $\alpha_{2,1} \sim N(0,100)$ reflects a prior belief that an inverse relationship does not exist.

The rationale for an informative prior on $\alpha_{1,2}$ stems from the economic theory of migration, which treats income differentials as an important pull factor. For $\alpha_{2,1}$, it is assumed *a priori* that the intensity of annual migration flows does not directly influence the *per capita* income differentials in the following year. The precision matrix $\mathbf{T}$ is assumed to follow a two-dimensional Wishart distribution[5] with two degrees of freedom and the scale matrix is $\mathbf{P} = [p_{i,j}]_{2 \times 2}$, where $p_{1,1} = p_{2,2} = 0.1$ and $p_{1,2} = p_{2,1} = 0.005$, reflecting beliefs in a relatively low precision, although higher for both variables independently, and lower for their interrelations.

Model 3 represents an alternative vector autoregressive process VAR(1) specified as

$$\mathbf{x}_2(t) = \mathbf{c} + \mathbf{A}\mathbf{x}_2(t{-}1) + \boldsymbol{\beta}Z(t) + \boldsymbol{\varepsilon}(t), \tag{12.10}$$

where $\mathbf{x}_2(t) = [\ln(MR_{P-D}(t)), \ln(UR_P(t))]'$ and UR represents unemployment rates. This model reflects the hypothesis that unemployment in the sending country is an important push factor of migration. The prior distributions are assumed to be the same as in model 2.

These examples of migration forecasting models are used to illustrate the Bayesian approach. The rationale behind the selection of autoregressive models is that they address, albeit in an indirect manner, the 'migrant network' hypothesis through the perpetuating effect of the autoregression coefficient α, which can be taken as a proxy of the cumulating features of migratory processes (cf. Sinn *et al.* 2001). The VAR-based models 2 and 3 include economic determinants of migration in a dynamic setting.

In the models (12.8), (12.9) and (12.10), the priors for the parameters $c, \alpha, \beta, \tau, \mathbf{c}, \mathbf{A}, \boldsymbol{\beta}$ and $\mathbf{T}$ are updated using the data vectors $\mathbf{x}$ comprising the time series of the variables used in each model, i.e. the logged migration rates in model 1, the logged migration rates and GDP *per capita* differentials in model 2, and the logged migration rates and unemployment rates in the sending country in model 3.

---

[5] A p-dimensional symmetric and positive definite matrix $\mathbf{X}$ is said to follow the Wishart distribution with a symmetric scale matrix $\mathbf{P}$ and k degrees of freedom if its density $p(\mathbf{X}) = c|\mathbf{P}|^{k/2}|\mathbf{X}|^{(k-p-1)/2}\exp(-\mathrm{Tr}(\mathbf{RX})/2)$ where c is a normalising constant, $|\cdot|$ denotes the determinant of a matrix, and Tr is the matrix trace (the sum of its diagonal elements). Under such parametrisation, the Wishart distribution is a k-dimensional extension of a univariate chi-squared distribution, and as such is often used as a prior for the precision matrix in multivariate normal models (Spiegelhalter *et al.* 2003).

The priors and data are then combined by means of Bayes' theorem, in order to yield the posterior distributions of the parameters θ. The latter are eventually used to produce the forecasts, i.e. the predictive distributions of future values of particular variables, $\mathbf{x}^F$, according to Equation (12.1).

Markov chain Monte Carlo (MCMC) simulations were used in the computations. These were implemented by using the WinBUGS 1.4 software (Spiegelhalter *et al.* 2003). The WinBUGS code for the current models, drawing heavily on the examples presented in Congdon (2003:172–175, 189–191, Programs 5.1 and 5.5), is set out in the Appendix[6] (available online at: http://www.wiley.com/go/raymer). The MCMC algorithm used in WinBUGS (i.e. *Gibbs sampler*) is as follows (Casella and George 1992:168). To estimate the joint posterior distribution $p(\boldsymbol{\theta}|\mathbf{x})$ of the vector of d parameters $\boldsymbol{\theta} = [\theta_1, \dots, \theta_d]'$, $d > 1$, given the data vector $\mathbf{x}$:

1. Initialise with $\boldsymbol{\theta}^{(0)} = [\theta_1^{(0)}, \dots, \theta_d^{(0)}]'$.

2. In the qth cycle of the algorithm, the approximation $\boldsymbol{\theta}^{(q)} = [\theta_1^{(q)}, \dots, \theta_d^{(q)}]'$ is generated in sequential stages from the following d conditional distributions:

$$\theta_1^{(q)} \sim p(\theta_1|\theta_2 = \theta_2^{(q-1)}, \theta_3 = \theta_3^{(q-1)}, \dots, \theta_d = \theta_d^{(q-1)}, \mathbf{x}),$$

$$\theta_2^{(q)} \sim p(\theta_2|\theta_1 = \theta_1^{(q)}, \theta_3 = \theta_3^{(q-1)}, \dots, \theta_d = \theta_d^{(q-1)}, \mathbf{x}),$$

$$\vdots$$

$$\theta_d^{(q)} \sim p(\theta_d|\theta_1 = \theta_1^{(q)}, \theta_2 = \theta_2^{(q)}, \dots, \theta_{d-1} = \theta_{d-1}^{(q)}, \mathbf{x}).$$

In the *burn-in phase*, the second step is repeated S times, until the algorithm converges to a steady state, with respect to the stabilisation of the values of estimated $\boldsymbol{\theta}$. Thereafter, M more iterations of the second step are generated, which ultimately serve as a sample used for an approximation of the required posterior distribution.

12.4.2   Forecasts of international migration between Poland and Germany, 2004–2010

For both migration flows under study, from Poland to Germany and vice versa, the three models defined above were estimated. The posterior distributions of the parameters of each model were calculated on the basis of 100 000 iterations of the MCMC algorithm, obtained after discarding the preceding iterations from the 'burn-in' phase of the procedure. After visual checks of convergence of the simulations following the suggestions of Spiegelhalter *et al.* (2003), the length of the 'burn-in' phase was established as 10 000.

As each of the models required a different set of data, a formal comparison using, for example, the deviance information criterion (DIC), was not possible. Instead, the goodness of fit of the three models has been compared by using the sum

---

[6] Available on the book's website at: http://www.wiley.com/go/raymer

Table 12.1 Summaries of posterior densities of the parameters of three forecasting models: estimated by MCMC.[a]

	Parameter	Estimates for migration from Poland to Germany, $\ln(MR_{\text{P-D}})$					Estimates for migration from Germany to Poland, $\ln(MR_{\text{D-P}})$				
		Mean	St. dev.	0.025	Median	0.975	Mean	St. dev.	0.025	Median	0.975
Model 1	α	0.28	0.23	-0.17	0.28	0.73	0.51	0.24	0.04	0.51	0.98
	β	0.96	0.31	0.35	0.96	1.57	0.53	0.21	0.11	0.53	0.94
	c	0.70	0.25	0.21	0.70	1.19	-0.02	0.08	-0.17	-0.02	0.14
	τ	10.16	3.61	4.35	9.74	18.34	13.63	4.84	5.83	13.07	24.61
Model 2	$\alpha_{1,1}$	0.36	0.14	0.09	0.36	0.63	0.56	0.13	0.31	0.56	0.81
	$\alpha_{1,2}$	-0.53	0.29	-1.09	-0.54	0.04	0.19	0.22	-0.24	0.19	0.65
	$\alpha_{2,1}$	0.10	0.04	0.02	0.10	0.17	-0.13	0.05	-0.23	-0.13	-0.02
	$\alpha_{2,2}$	0.88	0.12	0.64	0.88	1.11	0.82	0.12	0.58	0.81	1.06
	β	0.78	0.19	0.39	0.78	1.16	0.47	0.12	0.23	0.47	0.70
	c_1	1.17	0.29	0.58	1.18	1.74	0.18	0.23	-0.27	0.18	0.65
	c_2	0.02	0.14	-0.24	0.01	0.29	-0.18	0.12	-0.42	-0.18	0.07
	$t_{1,1}$	31.62	11.09	13.79	30.31	56.73	60.64	21.39	26.24	58.13	109.20
	$t_{2,2}$	158.00	54.12	70.32	152.00	280.30	156.40	54.25	68.86	150.10	278.70
	$t_{1,2} = t_{2,1}$	-8.81	17.30	-44.54	-8.23	24.16	-3.21	23.95	-51.09	-3.09	44.01
Model 3	$\alpha_{1,1}$	0.42	0.13	0.17	0.42	0.68	0.33	0.15	0.03	0.32	0.64
	$\alpha_{1,2}$	-0.04	0.03	-0.10	-0.04	0.02	-0.35	0.21	-0.76	-0.35	0.10
	$\alpha_{2,1}$	0.05	0.10	-0.14	0.05	0.25	-0.08	0.09	-0.25	-0.08	0.10
	$\alpha_{2,2}$	0.86	0.10	0.65	0.86	1.06	0.78	0.20	0.40	0.78	1.17
	β	0.72	0.18	0.36	0.73	1.08	0.54	0.09	0.35	0.54	0.72
	c_1	0.62	0.17	0.28	0.62	0.94	0.70	0.44	-0.22	0.70	1.53
	c_2	0.40	0.28	-0.14	0.40	0.98	0.46	0.40	-0.33	0.47	1.22
	$t_{1,1}$	42.11	14.87	18.23	40.36	75.71	75.30	26.54	32.61	72.21	135.60
	$t_{2,2}$	2.15	0.80	0.91	2.04	4.02	47.61	16.33	21.47	45.70	84.64
	$t_{1,2} = t_{2,1}$	5.48	2.95	0.53	5.19	12.04	-3.30	15.60	-34.57	-3.21	27.29

[a] Italic font denotes such posterior estimates, for which the 95 % credible intervals cover zero.

Source: Own calculations in WinBUGS.

Table 12.2   Precision of estimation: summaries of distributions of sums of squares (*SS*) obtained from 100 000 MCMC samples.

	Estimates of *SS* for migration from Poland to Germany, ln(MR_{P-D})					Estimates of *SS* for migration from Germany to Poland, ln(MR_{D-P})				
	Mean	St. dev.	0.025	Median	0.975	Mean	St. dev.	0.025	Median	0.975
Model 1	0.91	0.33	0.60	0.82	1.79	0.43	0.25	0.19	0.36	1.08
Model 2	0.57	0.13	0.43	0.53	0.91	0.24	0.07	0.17	0.23	0.42
Model 3	0.66	0.20	0.44	0.61	1.17	0.18	0.06	0.12	0.16	0.34

Source: Own calculations in WinBUGS.

of squared residuals (*SS*) between the observed and estimated values of ln(*MR*). The results of the estimations are presented in Table 12.1, which contains summaries of posterior distributions of the parameters of all three forecasting models for the two migration flows. The summaries include posterior means, standard deviations, medians and the 0.025 and 0.975 posterior quantiles (delimiting the 95 % credible intervals). The summaries of the empirical distributions of *SS*, obtained from 100 000 iterations of the MCMC algorithm, are presented in Table 12.2. The best fits were obtained by model 2 for the migration flow from Poland to Germany and by model 3 for migration from Germany to Poland, reflecting the roles of income differentials and unemployment, respectively.

With the exception of α, $\alpha_{1,1}$, $\alpha_{2,2}$ and β, the parameter intervals between the 0.025 and 0.975 quantiles are very wide, and, in many cases, cover zero (Table 12.1). For migration from Germany to Poland, the constants and most of the interaction parameters, $\alpha_{1,2}$ and $\alpha_{2,1}$, in models 2 and 3 are not significant. For migration from Poland to Germany, α in model 1, c_2 and $\alpha_{1,2}$ in models 2 and 3, and $\alpha_{1,2}$ in model 3 do not differ significantly from zero. For these reasons, the inferences on the constants and the interactions between ln(*MR*) and the other variables in models 2 and 3 (apart from the dummies) are not unambiguous.

In general, the estimated posterior precision parameters (τ or **T**) were higher than indicated by the prior beliefs, with the exception of the ln($UR_P(t)$) variable in model 3 for flows from Poland to Germany. This means that the data had more weight in constructing the posterior distributions. In future analyses, one could consider alternative prior beliefs about the precision parameters. However, the prior belief of weak precision is reasonable, given the uncertain nature of the migration patterns. Owing to the short time series available for migration flows in most of Europe, the forecasts are very sensitive to the selection of judgemental priors. The data series themselves, without additional prior information, would yield much higher estimates of precision parameters than the ones presented above. Such precise estimates and forecasts, however, would not be appropriate in the context of migration, where high levels of uncertainty (thus, low precision) are expected.

Almost all own-variable autoregression coefficients, with the exception of α in model 1 for migration from Poland to Germany, are likely to be positive and smaller than unity. This indicates a probable long-term stationarity of the logarithmic transformations of the variables under study (*MR*, *GDP* and *UR*). In the case of models 2 and 3 for migration from Poland to Germany, and of model 3 for migration from Germany to Poland, interpretation of the model-specific coefficients $\alpha_{1,2}$ is counter-intuitive. With a decreasing income gap or unemployment in the sending country, one would expect a decreasing $MR_{\text{P-D}}$, thus a positive $\alpha_{1,2}$ rather than a negative one, whereas for the mentioned coefficients most of the probability mass is concentrated below zero. This conclusion, however, needs to be treated with caution because the 95 % credible intervals cover zero in all mentioned cases.

For all models, the dummy-related coefficients (β) are most likely to be positive, equalling between 0.47 and 0.54 for migration from Germany to Poland, and between 0.78 and 0.96 for flows in the opposite direction. This indicates that the system transformation significantly contributed to the increase of migration between the two countries in 1988–1990.

The results from the forecasts are presented in Figures 12.1 and 12.2 for flows from Poland to Germany and in Figures 12.3 and 12.4 for flows from Germany to Poland. Figures 12.1 and 12.3 illustrate the estimated kernel densities of particular predictive distributions, that is, probability distributions of the forecast future values of the respective $\ln(MR)$ forecasts for 2004, 2007 and 2010, yielded by particular models. In addition, Figures 12.2 and 12.4 show the appropriate MR data series for 1985–2003, as well as the out-of-sample forecasts of the (unlogged) migration rates for 2004–2010 for the three models defined in Section 12.4.1. The graphs show the posterior quantiles of rank 0.025, 0.5 and 0.975, the former and the latter delimiting the 95 % predictive intervals of the variables under study.[7] In terms of numbers, the summaries of predictive distributions obtained for 2004, 2007 and 2010 in all three models are presented in Table 12.3, for both $MR_{\text{P-D}}$ and $MR_{\text{D-P}}$.

The median forecasts of migration from Poland to Germany predict a stabilisation of $MR_{\text{P-D}}$ around 2.64 in model 1, or a slight declining tendency in models 2 and 3: from 2.92 to 2.86 throughout the forecast horizon in model 2, and from 2.41 to 2.25 in model 3. The median forecasts yielded by model 1 are thus in between the ones produced by models 2 and 3. All forecasts, especially the one produced by model 1, are characterised by relatively wide uncertainty spans, as indicated by the predictive quantiles shown in Figure 12.2.

The predicted median-forecast rates of emigration from Germany to Poland are expected to be fairly stable throughout the forecast horizon in all models, with values either equal to or just below unity. Again, model 1 produced median forecasts that are in between the ones yielded by models 2 (higher) and 3 (lower) in the whole period 2004–2010. The uncertainty range is large in all forecasting models, especially in model 1 (Figure 12.4). In the presented examples, adding explanatory variables to the model, as well as making different prior assumptions on precision, visibly reduced the uncertainty in the forecasts.

---

[7] Alternatively, the 80 % predictive intervals can be used in stochastic population forecasts to prevent overestimation of uncertainty (see e.g. Alho 1998; Keilman *et al.* 2001; Lutz *et al.* 2004), which in the context of demographic phenomena is already considered to be high – see, for example, the discussion in Lutz *et al.* (2004:37).

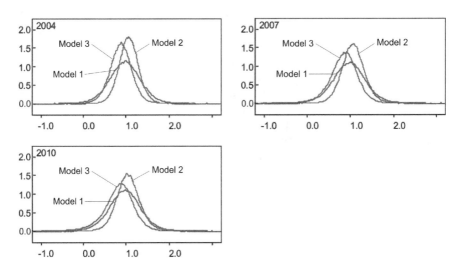

Figure 12.1 Predictive densities of $\ln(MR_{\text{P-D}})$ for 2004, 2007 and 2010: MCMC estimates.

Source: Own calculations in WinBUGS.

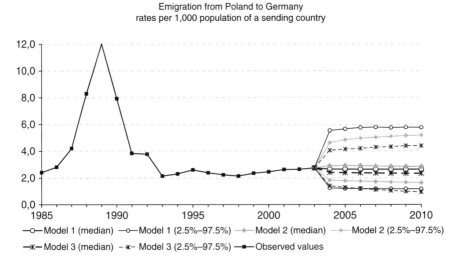

Figure 12.2 Values of $MR_{\text{P-D}}$: observed for 1985–2003 and forecast for 2004–2010 (predictive quantiles).

Source: Data series – Eurostat (as reported by Germany); forecasts – own calculations in WinBUGS.

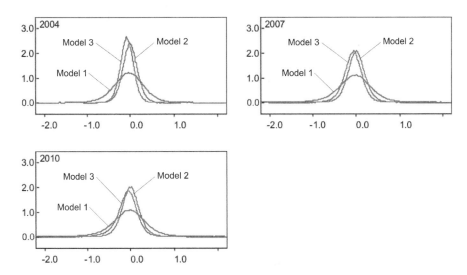

Figure 12.3   Predictive densities of $\ln(MR_{D\text{-}P})$ for 2004, 2007 and 2010: MCMC estimates.
Source: Own calculations in WinBUGS.

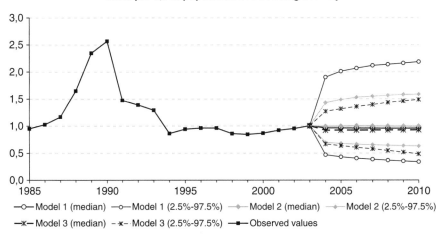

Figure 12.4   Values of $MR_{P\text{-}D}$: observed for 1985–2003 and forecast for 2004–2010 (predictive quantiles).
Source: Data series – Eurostat (as reported by Germany); forecasts – own calculations in WinBUGS.

Table 12.3 Summaries of predictive distributions of the $\ln(MR)$ values forecast for 2004, 2007 and 2010: estimated by MCMC.

	Year	Rates of migration from Poland to Germany, MR_{P-D} (per 1000)					Rates of migration from Germany to Poland, MR_{D-P} (per 1000)				
		Mean	St. dev.	0.025	Median	0.975	Mean	St. dev.	0.025	Median	0.975
Model 1	2004	2.64	1.45	1.26	2.66	5.53	0.96	1.42	0.47	0.96	1.90
	2007	2.64	1.49	1.20	2.64	5.81	0.95	1.55	0.39	0.96	2.12
	2010	2.64	1.52	1.19	2.64	5.75	0.93	1.82	0.34	0.96	2.18
Model 2	2004	2.92	1.26	1.82	2.92	4.62	0.99	1.20	0.70	0.99	1.43
	2007	2.92	1.31	1.73	2.89	5.05	1.00	1.25	0.66	1.00	1.55
	2010	2.86	1.34	1.63	2.86	5.21	0.99	1.27	0.63	0.99	1.58
Model 3	2004	2.41	1.31	1.40	2.41	4.06	0.92	1.17	0.67	0.92	1.27
	2007	2.32	1.40	1.13	2.36	4.31	0.92	1.26	0.58	0.92	1.39
	2010	2.25	1.49	0.95	2.34	4.39	0.91	1.51	0.49	0.93	1.49

Source: Own calculations in WinBUGS.

In general, the 95 % predictive intervals yielded by the various models can be seen as plausible. It is worth noting that, in the case of international migration, any research should not aim to obtain minimal *ex ante* prediction errors, which would be inappropriate, given the uncertain character of the phenomenon. Reasonable migration forecasts should yield predictive intervals of a suitable width (neither too wide, nor too narrow) to properly address the uncertainty associated with the future developments of population flows.

12.5   Conclusions

There are three major advantages of using the Bayesian methodology in the context of international migration forecasting. First, it allows for combining the formality of statistical inference with subjective expert judgement on various parameters of the forecasting models, by means of Bayes' theorem and subjective probability. The approach also offers simple tools for model selection and averaging (Hoeting *et al.* 1999), not exploited in the current study. Second, the quantitative analysis of uncertainty with respect to the future developments of the phenomena under study is inherent in the Bayesian forecasts, which yield whole predictive distributions. The latter can serve as a natural and straightforward way of obtaining projection variants from the selected predictive quantiles. Third, with informative prior distributions, the Bayesian estimates and forecasts are expected to carry more information than the sampling-theory ones, which is especially important in the small-sample cases (e.g. the relatively short time series migration data available for most countries in Europe). This issue is important for *a priori* assumptions on the errors of estimation, expressed in terms of precision or variance of a random component, which are crucial in the context of such uncertain phenomena as migration flows.

Although the major disadvantage of the Bayesian approach is its computational complexity, it can be overcome by using numerical methods e.g. included in the user-friendly and freely available WinBUGS (http://www.mrc-bsu.cam.ac.uk/bugs/). Bayesian statistical inference has the advantage of removing some of the problems associated with interpretation of results, caused by relating probability to the frequency of events, as is usually done in the sampling-theory approach. This is important with respect to the repeatable sample assumption, which is not a natural premise in social studies, e.g. migration research. Further research in Bayesian migration forecasting could address the issues of formal model selection, model averaging, and robustness of the results based on different types of prior information.

Acknowledgements

This chapter was prepared under the project 'Forecasting international migration in Europe: a comparison of traditional methods and the Bayesian approach', financed by the Foundation for Population, Migration and Environment (BMU-PME), Zurich. The author also benefited from the Stipend for Young Scientists of the Foundation for Polish Science. Many thanks go to Professor Jacek Osiewalski for very valuable comments and discussion, and to the editors of this volume for their remarks, which helped to improve the earlier draft.

References

Alho J. 1990. Stochastic methods in population forecasting. *International Journal of Forecasting* 6(4): 521–530.

Alho J. 1998. *A stochastic forecast of the population of Finland*. Reviews 1998/4. Helsinki: Statistics Finland.

Alho J and Spencer B. 1985. Uncertain population forecasting. *Journal of the American Statistical Association* 80(390):306–314.

Alvarez-Plata P, Brücker H and Siliverstovs B. 2003. *Potential migration from Central and Eastern Europe into the EU-15 – An update*. Report for the EC DG of Employment and Social Affairs. Berlin: Deutsches Institut für Wirtschaftsforschung.

Barro RJ and Sala-i-Martin X. 1995. *Economic growth*. New York: McGraw-Hill.

Bayes T. 1763. An essay towards solving a problem in the doctrine of chances. *Philosophical Transactions of the Royal Society* 53:370–418.

Bernardo JM and Smith AFM. 2000. *Bayesian theory*. Chichester: John Wiley & Sons, Ltd.

Bijak J. 2006. *Forecasting international migration: selected theories, models, and methods*. Working Paper 4/2006. Warsaw: Central European Forum for Migration and Population Research.

Blumen I, Kogan M and McCarthy PJ. 1955. *The industrial mobility of labour as a probability process*. Ithaca, NY: Cornell University Press.

Box GEP and Jenkins GM. 1976. *Time series analysis: Forecasting and control*. San Francisco: Holden-Day.

Brown LA. 1970. On the use of Markov chains in movement research. *Economic Geography* 46(Suppl.):393–403.

Brücker H and Siliverstovs B. 2005. *On the estimation and forecasting of international migration: How relevant is heterogeneity across countries*. IZA Discussion Paper 170. Bonn: Institut zur Zukunft der Arbeit.

Casella G and George E. 1992. Explaining the Gibbs sampler. *The American Statistician* 46(3):167–174.

Congdon P. 2000. A Bayesian approach to prediction using the gravity model, with an application to patient flow modelling. *Geographical Analysis* 32(3):205–224.

Congdon P. 2001. The development of gravity models for hospital patient flows under system change: a Bayesian modelling approach. *Health Care Management Science* 4(4):289–304.

Congdon P. 2003. *Applied Bayesian modelling*. Chichester: John Wiley & Sons, Ltd.

Courgeau D. 1985. Interaction between spatial mobility, family and career life cycle: a French survey. *European Sociological Review* 1(2):139–162.

Daponte BO, Kadane JB and Wolfson LJ. 1997. Bayesian demography: projecting the Iraqi Kurdish population, 1977–1990. *Journal of the American Statistical Association* 92(440):1256–1267.

de Beer J. 1990. *Uncertainty of international-migration projections for the 12 EC countries*. Voorburg: Central Bureau of Statistics.

de Beer J. 1997. The effect of uncertainty of migration on national population forecasts: the case of the Netherlands. *Journal of Official Statistics* 13(3):227–243.

De Finetti B. 1937. La prévision: ses lois logiques, ses sources subjectives. *Annales de l'Institute Henri Poincaré* 7:1–68.

Drbohlav D. 1995. Abschätzung der zukünftigen Ost-West-Migration in Europa. Eine Delphi-Studie. In *Die Zukunft der Ost-West Wanderung. Quantitative Entwicklungen, ökonomische Konsequenzen und politische Antworten.* Fassmann H and Münz R, eds, pp. 24–49. Vienna: Österreichische Akademie der Wissenschaften.

Eurostat 2005. *Population projections for the European Union 2004–2050.*

Fertig M and Schmidt CM. 2000. *Aggregate-level migration studies as a tool for forecasting future migration streams.* IZA Discussion Paper 183. Bonn: Institut zur Zukunft der Arbeit.

Franzmeyer F and Brücker H. 1997. *Europäische Union: Osterweiterung und Arbeitskräftemigration.* DIW Wochenbericht 5/97, pp. 89–96. Berlin: Deutsches Institut für Wirtschaftsforschung.

Ginsberg RB. 1971. Semi-Markov processes and mobility. *Journal of Mathematical Sociology* 1(1):233–262.

Ginsberg RB. 1979. Timing and duration effects in residence histories and other longitudinal data: I – stochastic and statistical models. *Regional Science and Urban Economics* 9(4):311–331.

Girosi F and King G. 2005. *Demographic forecasting.* Boston: Center for Basic Research in the Social Sciences, Harvard University. Available at gking.harvard.edu/files/smooth.pdf

Gjaltema TA. 2001. Judgement in population forecasting. Presented at the *European Population Conference*, Helsinki.

Goodman LA. 1961. Statistical methods for the mover–stayer model. *Journal of the American Statistical Association* 56(296):841–868.

Gorbey S, James D and Poot J. 1999. Population forecasting with endogenous migration: an application to trans-Tasman migration. *International Regional Science Review* 22(1):69–101.

Hoeting JA, Madigan D, Raftery AE and Volinsky CT. 1999. Bayesian model averaging: a tutorial. *Statistical Science* 14(4):382–417.

Isard W. 1960. *Methods of regional analysis: an introduction to regional science.* New York: John Wiley & Sons, Inc.

Jaynes ET. 1976. Confidence intervals vs. Bayesian intervals. In *Foundations of probability theory, statistical inference and statistical theories of science 2.* Harper WL and Hooker CA, eds, pp. 175–257. Dordrecht: Reidel.

Jeffreys H. 1939. *Theory of probability.* Oxford: Oxford University Press.

Jeffreys WH and Berger JO. 1992. Ockham's razor and Bayesian analysis. *American Scientist* 80(1):64–72.

Keilman N. 1990. *Uncertainty in national population forecasting: Issues, backgrounds, analyses, recommendations.* Amsterdam: Swets & Zeitlinger.

Keilman N. 2001. Demography: uncertain population forecasts. *Nature* 412(6846):490–491.

Keilman N, Pham DQ and Hetland A. 2001. *Norway's uncertain demographic future.* Oslo: Statistics Norway.

Keyfitz N. 1980. *Multidimensionality in population analysis.* RR-80-33. Laxenburg, Austria: International Institute for Applied Systems Analysis.

Keyfitz N. 1981. The limits of population forecasting. *Population and Development Review* 7(4):579–593.

Kupiszewski M. 2002. *Modelowanie dynamiki przemian ludności w warunkach wzrostu znaczenia migracji międzynarodowych.* Warszawa: Instytut Geografii i Przestrzennego Zagospodarowania PAN.

Laplace PS. 1812. *Théorie analytique des probabilités*. Paris: Veuve Courcier.

Lee RD and Tuljapurkar S. 1994. Stochastic population projections for the United States: beyond high, medium and low. *Journal of the American Statistical Association* 89(419):1175–1189.

Leslie PH. 1945. On the use of matrices in certain population mathematics. *Biometrika* 33(3):183–212.

Lutz W and Goldstein JR. 2004. Introduction: how to deal with uncertainty in population forecasting? *International Statistical Review* 72(1):1–4.

Lutz W, Sanderson WC and Scherbov S. 1996. Probabilistic population projections based on expert opinion. In *The future population of the world. What can we assume today?* Lutz W, ed., pp. 397–428. London: Earthscan.

Lutz W, Sanderson WC and Scherbov S. 1998. Expert based probabilistic population projections. *Population and Development Review* 24(Suppl.):139–155.

Lutz W, Sanderson WC and Scherbov S, eds. 2004. *The end of world population growth in the 21st century: New challenges for human capital formation and sustainable development*. London: Earthscan.

McGinnis R, Myers GC and Pilger J. 1963. Internal migration as a stochastic process. Presented at the *34th session of the International Statistical Institute*, Ottawa.

NRC. 2000. *Beyond six billion*. Washington, DC: National Academies Press.

Pittenger D. 1978. The role of judgement, assumptions, techniques and confidence limits in forecasting population. *Socioeconomic Planning Sciences* 12(5):271–276.

Poulain M, Perrin N and Singleton A, eds. 2006. *THESIM. Towards Harmonised European Statistics on International Migration*. Louvain-la-Neuve: UCL Presses Universitaires de Louvain.

Ramsey FP. 1926. Truth and probability. Published 1931 in *F. P. Ramsey: The foundations of mathematics and other logical essays*. Braithwaite RB, ed., pp. 156–196. London: Kegan, Paul, Trench, Trubner & Co; and New York: Harcourt, Brace and Co.

Rees PH. and Turton I. 1998. Investigation of the effects of input uncertainty on population forecasting. Presented at the *3rd International GeoComputation Conference*, Bristol.

Rees PH and Wilson AG. 1973. Accounts and models for spatial demographic analysis 1: aggregate population. *Environment and Planning A* 5(1):61–90.

Rogers A. 1966. A Markovian policy model of interregional migration. *Papers of the Regional Science Association* 17:205–224.

Rogers A. 1975. *Introduction to multiregional mathematical demography*. New York: John Wiley & Sons, Inc.

Rogers A. 1990. Requiem for the net migrant. *Geographical Analysis* 22(4):283–300.

Rogers A and Castro LJ. 1981. *Model migration schedules*. RR-81-30. Laxenburg, Austria: International Institute for Applied Systems Analysis.

Rutkowski M. 1990. *Labour hoarding and future unemployment in Eastern Europe: the case of Polish industry*. CEP Discussion Paper 6/1990. London: Centre of Economic Performance, London School of Economics.

Sinn HW, Flaig G, Werding M, Munz S, Duell N and Hofmann H. 2001. *EU-Erweiterung und Arbeitskräftemigration*. Wege zu einer schrittweisen Annäherung der Arbeitsmärkte. München: IFO-Institut für Wirtschaftsforschung.

Spiegelhalter DJ, Thomas A, Best NG and Lunn D. 2003. *WinBUGS Version 1.4 Users Manual*. Cambridge: MRC Biostatistics Unit. Available at www.mrc-bsu.cam.ac.uk/bugs

Stewart JQ. 1941. An inverse distance variation for certain social influences. *Science* 93(2404):89–90.

Stoto MA. 1983. The accuracy of population projections. *Journal of the American Statistical Association* 78(381):13–20.

Stouffer SA. 1940. Intervening opportunities: a theory relating mobility and distance. *American Sociological Review* 5(6):845–867.

Taylor JE. 1986. Differential migration, networks, information and risk. In *Research in human capital and dvelopment*, Vol. 4: *Migration, human capital, and development*. Stark O, ed., pp. 141–171. Greenwich, CT: JAI Press.

Tuljapurkar S and Boe C. 1999. Validation, probability-weighted priors and information in stochastic forecasts. *International Journal of Forecasting* 15(3):259–271.

United Nations. 2005. *World population prospects: 2004 revision*. New York: United Nations.

van Imhoff E, van der Gaag N, van Wissen L and Rees PH. 1997. The selection of internal migration models for European regions. *International Journal of Population Geography* 3(2):137–159.

Weidlich W and Haag G, eds. 1988. *Interregional migration: Dynamic theory and comparative analysis*. Berlin: Springer.

Willekens FJ. 1994. Monitoring international migration flows in Europe: towards a statistical data base combining data from different sources. *European Journal of Population* 10(1):1–42.

Willekens F and Baydar N. 1986. Forecasting place-to-place migration with generalized linear models. In *Population structures and models*. Woods R and Rees PH, eds, pp. 203-244. London: Allen and Unwin.

Wilson AG. 1970. *Entropy in urban and regional modelling*. London: Pion.

Wilson AG. 1981. *Catastrophe theory and bifurcation*. London: Croom Helm.

Wilson T and Bell M. 2004. *Australia's uncertain demographic future*. Discussion Paper 2003/04. Brisbane: Queensland Centre for Population Research, University of Queensland.

Zellner A. 1971. *An introduction to Bayesian inference in econometrics*. New York: John Wiley & Sons, Inc.

Zolberg AR. 1989. The next waves: migration theory for a changing world. *International Migration Review* 23(3):403–430.

13

Forecasting international migration: time series projections vs argument-based forecasts

Joop de Beer
Netherlands Interdisciplinary Demographic Institute, The Hague

13.1   Introduction

Since the 1980s, immigration to most European countries has increased substantially. As a consequence, migration has become the main source of population growth in Europe and, therefore, assumptions on the future size of migration are an important input to population projections. Howe and Jackson (2005) argue that 'most official immigration projections . . . are based on little theory and virtually no definable methodology'. The present chapter aims to demonstrate how migration projections may be improved by distinguishing different migration flows. The illustrations come from Dutch data, for which detailed migration data are available over time.

Most national statistical institutes and international organisations, such as the European Union (EU) and the United Nations (UN), use the cohort-component model for making population projections. Typically, these organisations only incorporate assumptions on the future size of *net migration* by age and sex into their

International Migration in Europe: Data, Models and Estimates   Edited by J. Raymer and F. Willekens
© 2008 John Wiley & Sons, Ltd

models. Those projections that incorporate the separate flows of immigration and emigration are preferred because they can be readily associated with predictive variables and can be analysed according to different types of flows. Also, with some time lag, foreign emigration can be linked with foreign immigration. The same is possible for emigration and immigration flows of nationals. Moreover, immigration tends to be positively related with the business cycle and emigration negatively. Thus, there are many advantages in making separate assumptions on future changes in immigration and emigration, which is more difficult with net migration.

The availability of accurate data on immigration and emigration is a problem in many countries (see Chapters 3 and 4). For some countries, information on net migration only may be available. As immigration tends to fluctuate more strongly than emigration, changes in net migration are mainly due to changes in immigration. Hence, assumptions on changes in net migration tend to be mainly based on assumptions on the future direction of immigration. However, in making assumptions on future changes in net migration, one should take into account that, after a period of increase in immigration, there may be an increase in emigration and, thus, a reduction in net migration. The age structure of emigration may also differ from that of immigration. This implies that, if the size of emigration changes in a different way than that of immigration, the age structure of net migration will change.

Projections of immigration and emigration can be based on extrapolations of changes observed in the past. Extrapolations may be based on different time series methods, ranging from simple linear trends to sophisticated stochastic time series models. One special case is to assume that net migration is zero. If net migration fluctuated around zero in the past, this may well be a valid assumption. However, even if total net migration is zero, this usually does not apply to separate age groups. Since emigrants are older than immigrants on average, net migration for young people will be positive and for older people negative. Thus assuming net migration to be zero for all ages may lead to some bias in the population projections.

The use of time series models for projecting immigration, emigration and net migration on the basis of Dutch data is illustrated in Section 13.2. As migration tends to fluctuate rather strongly, different extrapolation methods may produce a wide range of projected outcomes. For making forecasts of future changes in migration, one has to decide to what extent past changes found in immigration and emigration patterns will continue in the future. This requires an identification of the main factors explaining past changes in migration flows. A discussion of how forecasts of migration may be based on explanations or migration theories is presented in Section 13.3.

Different types of immigration and emigration are affected by various factors. For example, while labour migration is primarily affected by the situation in the labour market, marriage migration is affected by the size and composition of the resident migrant population, and asylum migration is affected by asylum policies. For assumptions made on future changes in migration, it is useful to distinguish migrants according to their primary motives, such as labour, family or asylum. This allows one to make argument-based forecasts, rather than simply extrapolating changes observed in the past. As Howe and Jackson (2005) point out, one important benefit of explanatory arguments of forecasts is that these can be objectively evaluated and tested against historical evidence. In many countries, data on different categories of migrants are lacking. In these situations, the population characteristics of age, sex and country of origin may be used as proxies. The main factors affecting the future size of different types of immigration and emigration are discussed in Sections 13.4 and 13.5, respectively. A discussion of how migration categories can be used for making argument-based forecasts is included in Section 13.6.

As time series of net migration tend to exhibit large fluctuations, projections of migration are rather uncertain, even in the short run. The degree of uncertainty of migration forecasts can be assessed on the basis of historic forecast errors or on stochastic time series models. However, different types of immigration and emigration may be assumed to change in different ways in the future. For example, labour migration may be assumed to increase due to the ageing of the labour force, whereas asylum migration may be assumed to decrease due to more strict policies. This raises the question to what extent past developments in migration provide a sufficient basis for assessing the uncertainty of future migration. Therefore, the argument-based approach set out in Section 13.7 may be used for making assumptions about both future changes in migration flows as well as the degrees of uncertainty.

13.2　Extrapolations

In this section, the applications of different extrapolation methods are illustrated with data from the Netherlands. The data represent annual immigration, emigration and net migration totals for the period 1950–2004, which were obtained from the Statistics Netherlands StatLine database (www.cbs.nl). Extrapolations of these data can be made by applying time series models, which include both deterministic and stochastic models. A well known example of a deterministic model is fitting a straight line to a time series of data. Deterministic models are based on the assumption that there is a fixed trend. Random fluctuations do not affect this trend. Stochastic time series models are based on the assumption that the direction of the trend of the time series is subject to random changes. For this, the ARIMA (autoregressive integrated moving average) models introduced by Box and Jenkins (1970) are widely applied.

Linear trends for immigration and emigration in the Netherlands were estimated from 1950 to 2010 (see Figure 13.1). The predicted 2004 values differ substantially from the observed values. Immigration in 2004 dropped sharply below the trend, whereas emigration was well above the trend. Accepting the long-run linear trend implies that the decrease in immigration or the increase in emigration in 2003 and 2004 are assumed to be temporary occurrences.

Stochastic time series models focus on the short run. ARIMA models are identified on the basis of autocorrelation coefficients. These indicate the correlation of a

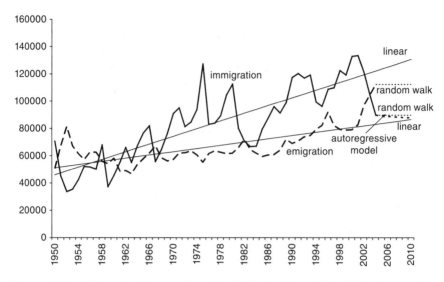

Figure 13.1 Migration from and to the Netherlands, 1950–2010: observations, linear trends and ARIMA models.

time series with the same time series lagged one or two or more years. The auto-correlations for immigration, emigration and net migration for both the time series of observations and the time series of first differences are shown in Table 13.1.

For immigration and emigration, the patterns of the autocorrelation coefficients suggest two models: a first-order autoregressive model, and a random walk model.

Table 13.1 Autocorrelation coefficients for the Netherlands immigration, emigration and net migration data, 1950–2004.

	Immigration		Emigration		
Lag	Levels	First difference	Levels	First difference	Net migration levels
1	0.86	−0.07	0.89	0.13	0.72
2	0.74	−0.15	0.75	−0.17	0.43
3	0.68	−0.25	0.68	−0.22	0.30
4	0.69	0.06	0.66	−0.08	0.32
5	0.69	0.16	0.67	0.02	0.36
6	0.61	−0.15	0.66	−0.05	0.27
7	0.58	−0.11	0.66	0.06	0.29
8	0.58	0.00	0.63	0.09	0.32
9	0.58	0.07	0.53	−0.09	0.26
10	0.56	0.09	0.46	−0.04	0.23

The parameters of the first-order autoregressive model estimated for immigration are

$$IIM_t = 0.86IM_{t-1} + 12\,363 + e_t,$$
$$\quad (0.07) \qquad (6\,258) \tag{13.1}$$

where IM_t = immigration in year t and e_t = random term, which is serially uncorrelated and has an expected value of zero. The numbers between parentheses below the equation are the standard errors. This model implies that, in the long run, the projection tends to a level of 86 445 (i.e. $12\,363/(1 - 0.86)$). The random walk model for immigration is specified as

$$IM_t - IM_{t-1} = c + e_t, \tag{13.2}$$

where the constant term c (usually labelled as 'drift') does not differ significantly from zero. If c is excluded from the model, the projections are equal to the last observed value, i.e. $IM_{t+1} = IM_t$ (because the expected value of e_{t+1} is zero). The autoregressive model and the random walk projections for immigration differ only slightly, as illustrated in Figure 13.1.

For emigration, the estimated autoregressive coefficient for the autoregressive model turned out to equal 0.99. This suggests that a random walk model is more appropriate for projecting emigration flows. Since the constant term does not differ significantly from zero, the following model is used for projecting emigration:

$$EM_t - EM_{t-1} = e_t \tag{13.3}$$

where EM_t is emigration in year t.

Net migration totals can be projected on the basis of outputs from the projections of immigration and emigration or from a time series model applied to the net migration totals themselves. The autocorrelation coefficients in Table 13.1 indicate that a first-order autoregressive model is appropriate for net migration:

$$NM_t = 0.82NM_{t-1} + e_t,$$
$$\quad (0.08) \tag{13.4}$$

where NM_t is net migration in year t. As the constant term does not differ significantly from zero, it is not included in the model. Figure 13.2 shows that the projection based on this model tends to zero in the long run, whereas the projection of net migration based on separate projections of immigration and emigration is equal to around minus 20 thousand. Note that, since immigration and emigration have been modelled as random walk models without drift, the projection of net migration based on these models remains at a constant level.

If the model is estimated for the 1950–2004 period, the random walk model projects a constant level of immigration and emigration, since the constant term in the model does not differ significantly from zero. However, if the model is

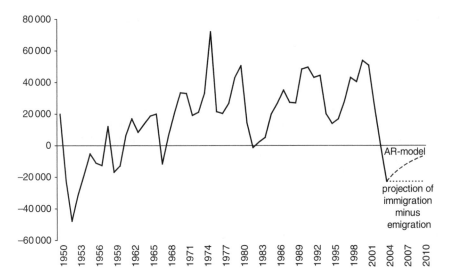

Figure 13.2   Net migration in the Netherlands, 1950–2010: observations and projections.

estimated for the 1980–2004 period, the drift parameter turns out to be positive for the emigration time series (but not for the immigration series). Similar to the linear trend model, the random walk model with drift projects a straight line. The main difference is that the projections from the random walk model start from the last observed value. The projections of the linear trend model and the random walk model with drift are compared in Figure 13.3 for the observation period 1980 to 2004. The trend directions are similar, but the levels are considerably different.

One conclusion that can be made from the extrapolations above is that different methods can lead to very different outcomes. As shown in Table 13.2, the projection of immigration and emigration for the year 2010 ranges from 90 thousand to 133 thousand and from 87 thousand to 125 thousand, respectively. The projection of net migration ranges from −35 thousand to 44 thousand. Clearly, these are considerable differences.

The results of the extrapolations depend on various choices made by the researcher. First, one has to choose between a deterministic or stochastic trend. The deterministic trend emphasises long-run developments. Projections based on this model tend to react slowly to recent changes in the time series. In contrast, projections based on a stochastic model tend to react very quickly, which may result in widely varying projections made in successive years. For example, with a start point of 2001, the random walk model projects emigration in 2010 to be 82 thousand, whereas with a start point of 2004, emigration is projected to be 112 thousand. The deterministic model resulted in 2010 emigration levels that changed much less: from 80 thousand with a start point of 2001 to 87 thousand with a start point of 2004. Second, the choice of the base period makes a difference. For example, on the basis of the 1950–2004 period, it appears that the random walk model does not

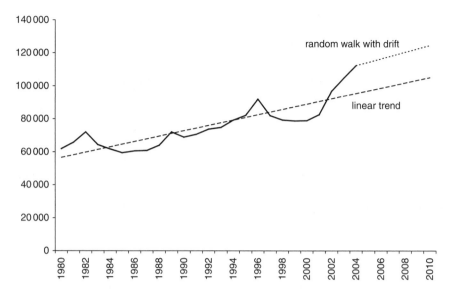

Figure 13.3   Emigration from the Netherlands, 1980–2010: observations and projections.

Table 13.2   The 2010 projections of immigration, emigration and net migration (in thousands) for the Netherlands.

	Immigration	Emigration	Net migration
Base period 1950–2004			
Linear deterministic model	131	87	44
Random walk model	90	112	−23
Base period 1980–2004			
Linear deterministic model	133	105	28
Random walk model	90	125	−35

require a constant term; whereas on the basis of the 1980–2004 period, it appears that a positive constant term is needed. Third, extrapolations of the time series of net migration totals may differ from the difference between separate extrapolations of immigration and emigration. Finally, there is no single extrapolation method that outperforms all other methods under all circumstances. Each one has its pros and cons. One way to decide on a particular model is to examine how the methods performed in the past. However, even this does not lead to a clear solution, as the results tend to vary depending on the choice of the period for which the methods are tested. The logical way to improve projections is to examine the explanations behind the changes in past migration, which can then be used to determine future changes.

13.3   Explanations

Extrapolations are based on the assumption that changes in the past can be projected into the future. However, without knowing the mechanisms affecting past trends, it is difficult to assess to what extent this assumption is valid. Moreover, as discussed in the previous section, different extrapolation methods may lead to widely different projections. Therefore, it is useful to look for explanations of changes in migration by identifying the main factors affecting changes in immigration and emigration. These factors can be assessed on the basis of migration theories. Massey *et al.* (1993) and Howe and Jackson (2005) give overviews of various theoretical frameworks. Most theories focus on push factors creating migration pressure in sending countries (e.g. poverty, unemployment and political turmoil) and pull factors emphasising the importance of the attractiveness of receiving countries which give direction to migration flows. Beyond this, the frameworks focus on aspects such as differentials in wage levels between countries, social networks or the role of policies.

The lack of an overall migration theory makes it difficult to forecast migration. In fact, it is questionable whether one theory is capable of explaining all kinds of changes in migration flows through time. Not only have the levels of migration changed over time, but also the types and mechanisms of migration have changed as well. In the 1960s, there were shortages in the Western European labour market. This created opportunities for large numbers of persons from Southern European countries to migrate in search of jobs. In the late 1960s and early 1970s, the origins of labour migrants shifted to Turkey and the Maghreb area.[1] After the rise of unemployment caused by the economic recession of 1973–1974 and the influx of post-war 'babyboomers' into the labour market, most Western European countries imposed immigration restrictions (Jennissen 2004). As a result, many Southern European migrants returned home. The other labour migrants who stayed brought their families over, which led to an increase of family reunification. While immigration was relatively low in the second half of the 1970s and the first half of the 1980s, immigration rose sharply during the second half of the 1980s. One of the main factors for this was the collapse of communism in Eastern Europe. A large number of ethnic Germans from Poland, the Soviet Union and Romania entered West Germany. Another cause of the rise in immigration was the increase in the number of asylum seekers. In the second half of the 1990s, asylum migration decreased because of the end of the war in Bosnia–Herzegovina and stricter asylum policies (see Chapter 11 for more details). In short, different types of migration were predominant in different periods.

Rather than selecting one theory, one could instead focus on the main types of immigration and emigration, as they tend to be affected by different factors and change in different ways in successive periods. Here, the discussion focuses on labour migration, family-related migration and asylum seekers. Labour migration is primarily affected

---

[1] The Arab states of North-West Africa, especially Morocco, Algeria and Tunisia.

by the situation in the labour market (e.g. wage rates and unemployment rate). Marriage migration is affected by the choice of partners of the resident migrant population and, thus, by networks. Migration of asylum seekers is affected by political turmoil in sending countries and the asylum policies in receiving countries.

Forecasts of migration can be based on explanations by identifying quantitative explanatory models for different types of migration. One problem in estimating quantitative explanatory models is the lack of time series data on different categories of migration. One way of dealing with this problem is to identify specific migration flows distinguished by, for example, country of origin or country of birth, for which data are available and which can be considered to represent a particular type of migration. For example, the immigration flow of EU citizens to the Netherlands comprises mostly labour migrants, whereas the corresponding flows from Turkey and Morocco comprise mostly family migrants. Thus, one may expect the size of immigration of EU citizens to the Netherlands to depend on the situation in the labour market in the Netherlands.

The annual number of EU immigrants to the Netherlands during the 1977–2003 period can be explained by a regression model that includes the number of unemployed persons and a linear trend (de Beer 2004), specified as

$$IMEU_t = -0.020UN_t + 599T_t + 17\,309 + e_t, \quad R^2 = 0.92,$$

$$(0.007) \qquad (36) \qquad (577) \qquad\qquad\qquad (13.5)$$

where $IMEU_t$ is the number of EU immigrants in year t, UN_t is the number of unemployed persons and T_t is a linear trend term. According to this model, a decrease

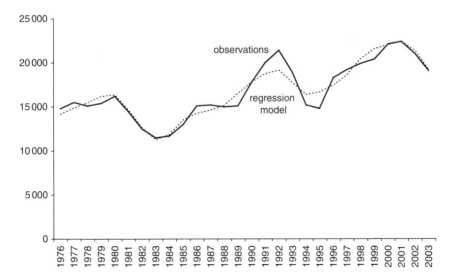

Figure 13.4 Immigration of EU citizens to the Netherlands, 1978–2003: observed and fitted values.

in the number of unemployed persons by 100 thousand leads to an increase in the number of EU immigrants by two thousand. As shown in Figure 13.4, the model is capable of accurately capturing fluctuations in the number of immigrants. This model suggests that the decline in immigration in the last years of the observation period is temporary and should, therefore, not be projected into the future. The estimated model indicates that, apart from short-run fluctuations due to the business cycle, there is a positive long-run trend.

Since time series data for different types of migration are often lacking, expert opinions may be included in the forecasting model to obtain more accurate results. The next three sections discuss how different factors affect the main types of immigration and emigration, and how they may be used to estimate future changes in migration flows.

13.4   Types of immigration

In identifying categories of immigration, it is useful to distinguish between nationals and foreigners. The size of national immigration is related to the size of national emigration in previous years. This relationship depends on the percentage of nationals who return after a stay abroad for, say, at least one year and on the length of their stay abroad. On the basis of Dutch data, it has been estimated that one-half of all nationals who emigrated in 1995 had returned within eight years and that 60 % return in the long run (Nicolaas 2004). Thus, long-run forecasts of national immigration levels are equal to 60 % of the projected national emigration levels.

In the Netherlands, the levels of labour migration, family migration and asylum seekers have changed in different ways (see Figure 13.5). Students, retired persons and other types of immigrants are not included here. Their numbers tend to be considerably smaller than the above categories (i.e. for the Netherlands). Methods for assessing and projecting the size of illegal migration are beyond the scope of this chapter.

13.4.1   Labour migration

In making assumptions about changes in the size of labour migration, one should distinguish short- and long-run developments and skill levels. In the short run, changes in the size of labour migration depend largely on the business cycle. In the previous section, it was shown that the number of EU immigrants to the Netherlands can be explained by the size of unemployment in the Netherlands. Since it is very difficult to project the course of business cycles into the future, this type of immigration cannot be projected with accuracy. However, to the extent that upturns and downturns follow each other, the business cycle does not affect the total flow of immigrants in the long run. For projections, the business cycle can be used to assess recent changes in the size of immigration. For example, if immigration declined in recent years, and this decline was due to an economic downturn, it may be expected that the future level of immigration will be higher than the current level.

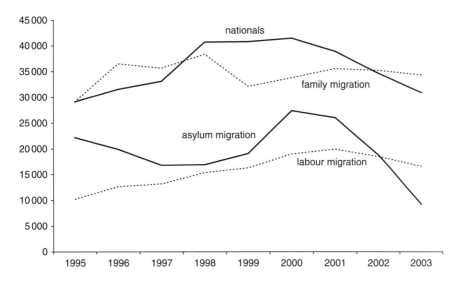

Figure 13.5   Main types of immigration to the Netherlands, 1995–2003.

For long-run forecasts of labour migration, the main question is whether the ageing of the labour force in Europe will lead to shortages in the labour market and whether these shortages will lead to 'replacement migration'. The ageing process, caused by low levels of fertility and mortality, can be partially offset by increases in labour force participation rates or immigration. In 2000, the UN Population Division published a report which contained calculations on the levels of immigration needed to counteract the process of ageing (United Nations 2000). The estimates depend on a number of assumptions, such as the rate of growth in productivity and the rate of growth in gross domestic product (GDP). Johansson and Rauhut (2005) present various calculations on the total number of migrants in the EU in the period 2000–2050 that would be needed to stabilise (1) the size of the population, (2) the number of persons in the working age range and (3) the ratio of the working population to the elderly population. In addition, they assess the effect of different rates of growth on productivity. Their calculations are based on net migration numbers. As shown in Table 13.3, in order to stabilise the number of people in the working age range (15–64 years) in the EU25, the annual size of net migration would need to be around 2.5 million. If the rate of growth of productivity were to be 1 % higher, the annual number of immigrants would be about 100 thousand less. Thus, the effect of the rate of productivity is relatively small. The table also shows that stabilisation of the elderly dependence ratio would require unrealistically high numbers of migrants, e.g. 15 million immigrants *per year* around 2050. This would lead to a doubling of total population size by 2050.

Obviously, these kinds of calculations only give a general sense of the possible sizes of future migration. They do not take into account, for example, changes in the demand for labour, changes in labour force participation, or differences in the

Table 13.3   Average annual net migration and population size in the EU-25, 2000 to 2050.

	Net migration (thousands)			Population size (millions)		
	2000	2025	2050	2000	2025	2050
Constant total	747	1 934	2 706	452	452	452
Constant 15–64 yrs	747	2 677	2 422	452	467	480
Constant 15–64 yrs / 65+ yrs ratio	747	10 412	15 040	452	650	940

Source: Johansson and Rauhut (2005).

qualification structure of labour supply and demand. Moreover, the calculations cannot be directly used for making forecasts of labour immigration, as they refer to total net migration. Only if labour force participation rates of other types of immigrants were the same as those of labour migrants could these calculations be applied to total immigration. Otherwise, total immigration would need to be higher in order to achieve the same effect on the size of the labour force. Furthermore, since the emigration rate of labour migrants tends to be relatively high, the total number of immigrants will have to be considerably higher than the size of net migration. For example, if 50 % of the labour migrants will return after some time, the total number of immigrants will have to be twice the size of net migration in order to have the same effect on the size of the labour force in the longer run. Nevertheless, despite these difficulties in assessing the future size of labour migration, it seems plausible to assume that the ageing of the labour force will cause the structural level of future labour migration to be higher than it used to be in the last decades.

13.4.2   Family-related migration

Four types of family-related migration can be identified. First, a labour migrant may enter a country with family. Second, a labour migrant may bring in family some time after entering a country. Similarly, a refugee may be allowed to bring in family if the asylum request is granted. Third, migrants may marry a partner living abroad. Fourth, nationals may marry a partner from abroad. Generally, family-related migration is only allowed under certain conditions and the rules differ by country.

For migration forecasts, assumptions about the future size of the first two categories can be related to forecasts of labour and asylum migration. In the Netherlands, about one-quarter of total labour migration is accounted for by family members accompanying labour migrants. During the last decade, the number of migrants that arrived because of family reunification was about one-third of the total number of labour and asylum migrants, taking into account some time lag between the arrival of the labour or asylum migrant and corresponding family members.

The number of marriage migrants can be forecast based on assumptions about the choice of partners by migrants. This may differ strongly between the origins of migrants. For example, in the Netherlands, about two-thirds of the Turkish and Moroccan migrants tend to marry a partner from the country of origin. Even a large proportion of their children born in the Netherlands (the so-called 'second generation') tend to marry a partner from their parents' country of origin. For example, over 50 % of second-generation Moroccans and over 60 % of second-generation Turks marry someone from Morocco and Turkey, respectively.

Alders (2005) developed a model for projecting the number of young Moroccans and Turks without a partner residing in the Netherlands on the basis of the current population structure (by age, sex and household position). Assuming that 95 % of all Moroccans and Turks will eventually have a partner, the number of persons who will find a partner can be calculated. On the basis of assumptions about the percentage of these young Moroccans and Turks who will marry a partner from the country of origin and at what age, he calculates the number of marriage migrants to be expected in the next decades. If it is assumed that the rates of marriage with partners from abroad will remain constant, the annual number of marriage migrants would grow for some 20 years (see Figure 13.6). If, however, it is assumed that this rate will decline gradually (a more realistic assumption), the annual number of marriage migrants is expected to decline. As illustrated in Figure 13.6, if it is assumed that the percentage of young (mostly second-generation) migrants who marry a partner from the country of origin will halve, the annual number of marriage migrants will be considerably lower.

In addition to the marriage behaviour of migrants, one should also take into account marriages of nationals to foreigners. These numbers are considerably lower than those of migrants. In the Netherlands, they consist of about 10 % of the total number of marriage migrants. This type of migration is so small that it hardly affects changes in the total size of migration. Therefore, for migration forecasts, one can simply assume constant rates of these migrants over time.

13.4.3 Asylum seekers

For making forecasts of the total number of asylum seekers in each European country, it is useful to distinguish between changes in the total flow of asylum seekers to Europe and changes in the distribution of asylum seekers within Europe (see also Chapter 11). Whereas the total number of asylum seekers to Europe is mainly determined by the situation in the countries of origin, the distribution among European countries is, to an important extent, affected by differences in asylum policies across European countries. In the first half of the 1990s, the total number of asylum seekers entering the EU countries rose sharply from 400 thousand in 1990 to 675 thousand in 1992, and then fell back to 275 thousand in 1995. Since 1996,

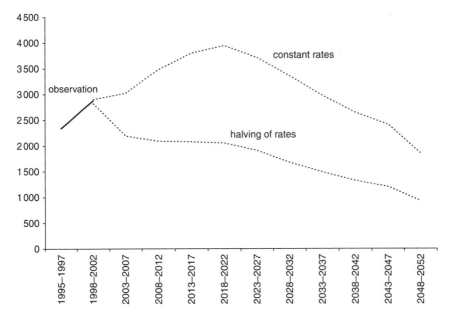

Figure 13.6 Observed and estimated marriage migrants from Turkey to the Netherlands, 1995–1997 to 2048–2052.

the fluctuations have been considerably smaller, increasing from 234 thousand in 1996 to 391 thousand in 2000 and subsequently declining to around 250 thousand in 2004. The average annual change declined from 135 thousand in the years 1991–1995 to 38 thousand in the years 1996–2004.

The effect of changes in the distribution of asylum seekers over the EU countries can be estimated by calculating how much the number of asylum seekers in country i in year t would have changed if the total number of asylum seekers entering the EU had not changed compared with year $t-1$. This is the 'distribution effect'. One alternative method is Chapter 11, in which the 'substitution' effects are estimated between all pairs of countries, rather than total distribution effects for each country separately.

The effect of changes in the total inflow of asylum seekers entering the EU on the number of asylum seekers moving to country i in year t (i.e. 'generation effect') can be estimated by calculating how much the number of asylum seekers in country i in year t would have changed if the fraction of the total inflow of asylum seekers moving to country i had not changed compared with year $t-1$.

In formulae, we can write these dependences as

$$D_{i,t} = A_{t-1}\,\Delta f_{i,t}, \tag{13.6}$$

$$G_{i,t} = f_{i,t-1}\,\Delta A_{t}, \tag{13.7}$$

where $D_{i,t}$ is the distribution effect for country i in year t, A_{t-1} is the total number of asylum seekers moving to the EU in year $t-1$, $f_{i,t}$ is the fraction of the total number of asylum seekers moving to country i, and $\Delta x_t = x_t - x_{t-1}$. It should be noted that these effects added together do not explain the change in the number of asylum seekers between year $t-1$ and t completely. There is also an interaction effect; changes in the total inflow and the distribution can either reinforce each other or offset each other. However, the interaction effects are relatively small; they accounted for only 3 % of the annual changes in the number of asylum seekers in the EU15 countries during the 1991–2004 period.

The average generation and distribution effects for the EU15 countries during the years 1991–2004 are set out in Table 13.4. These estimates differ from those in Chapter 11, as they estimated substitution effects rather than distribution effects. One benefit of their approach is that it provides more detailed estimates, i.e. substitution between all pairs of countries. Their estimates, however, are based on assumptions about unobserved patterns. Table 13.4 shows that in most countries the distribution effects exceeded the generation effects. The exception is Germany. As more than half of the total number of asylum seekers in the early 1990s moved to Germany, the sharp changes in the total inflow in this period strongly affected changes in the number of asylum seekers in Germany. In the Netherlands, Sweden and the UK the distribution effects are considerably higher than the generation effect. Increases in the number of asylum seekers in these countries went together with decreases in other countries. This suggests a substitution

Table 13.4 Average annual change in asylum seekers due to generation and distribution effects in EU-15, 1991–2004.

	Generation	Distribution
Austria	3 023	4 911
Belgium	3 135	3 942
Denmark	1 425	2 419
Finland	375	519
France	6 215	10 295
Germany	34 729	33 242
Greece	589	1 176
Ireland	459	345
Italy	1 391	5 166
Luxembourg	97	129
Netherlands	5 033	15 213
Portugal	109	461
Spain	1 489	2 925
Sweden	5 272	12 853
UK	9 173	20 150
Average EU-15	4 834	7 583

effect, which can be caused by the fact that the asylum procedure in a certain country becomes stricter than in another country. For example, a decrease in the recognition rate in one country may lead asylum seekers to prefer to submit an asylum request in another country. There is a strong negative correlation between the distribution effects of Germany and the UK (-0.75) and Germany and the Netherlands (-0.74). This suggests that there are substitution effects between these countries.

For making assumptions about the future number of asylum seekers, separate assumptions can be made about the total inflow to the EU and the distribution between EU countries. If one assumes that there will be more coordination of asylum procedures in the EU, one would expect that in the short term the distribution effects will change in such a way that the flows of asylum seekers will be distributed more evenly among EU countries according to some criterion, such as the number of asylum seekers per one thousand inhabitants, and that, in the longer run, the distribution effects will become smaller when the distribution has become more even. If future changes in the total flow to the EU do not exceed those in the past, one may expect fluctuations in the number of asylum seekers in separate countries in the long run to be smaller than in the period after 1990. The development in the period since 1990 clearly exhibited the effect of stricter asylum procedures in specific countries. However, as policies in other countries were less strict, the direction of the flow changed. If policies became more strict in all countries, one could expect the total inflow to the EU to become smaller. As discussed above, the ageing of the workforce may lead to an increase of immigration. However, it seems likely that the EU countries will try to direct the immigration flow in order to achieve the goal that those migrants who arrive are qualified to occupy the jobs for which there are vacancies. Hence, some selection procedure seems likely. This means that, even when the total level of migration increases, the number of asylum seekers could still decline.

13.5   Types of emigration

In making forecasts of the total size of emigration, it is useful to distinguish between return migration of foreigners and emigration of nationals. Return (e)migration of foreigners is related to foreign immigration in previous years. Thus projections of foreign emigration can be based on the immigration that occurred in preceding years. Since the patterns of foreigners emigrating to their home country differ between different types of migrants, it is again useful to distinguish between labour migrants, family-related migrants and asylum seekers.

13.5.1   Foreigners

The tendency of foreigners to return to their country of origin differs strongly between categories of migrants. Both the motive for immigration (such as labour, marriage or asylum) and the country of origin (industrialised or developing country) are important determinants. A much larger proportion of labour migrants tend to return to their home country than do marriage migrants or asylum seekers (if granted a residence permit). Immigrants from industrialised countries are more inclined to return than immigrants from developing countries. The return migration rate is higher for males than for females. The return migration rate for immigrants of adult age are higher than those for children and older immigrants. Finally, the return migration rates of Western immigrants are higher than those of non-Western immigrants. In the Netherlands, around 70 % of male immigrants from Western countries, which are mainly labour migrants and students, return to their home country (de Jong and Nicolaas 2005). In contrast, only about 15 % of female immigrants from Morocco, who are mainly marriage migrants, return.

The differences imply that there are strong relationships between the size of immigration and net migration for different types of migration. For labour migration the size of immigration may be more than twice the size of net migration, whereas for marriage migration the difference between the size of immigration and net migration may be considerably smaller. Because of the rather strong relationship between the type of migrants distinguished by immigration motive (labour, asylum or family) and their demographic characteristics, forecasts of emigration can be based on distinctions by age, sex and country of origin – if data on the types of migrants are lacking.

The emigration rate of foreigners decreases with duration of stay. About one-half of all emigrants leave within three years of entry. This implies that the number of emigrants is related to the number of immigrants in preceding years. For example, in the Netherlands the annual number of Western emigrants equals about two-thirds of the number of immigrants three years earlier, whereas the number of non-Western emigrants equals 40 % of the number of immigrants. Hence, if an increase in some immigration category is projected, one would expect the number of emigrants to increase with some time lag. In the long run, one could expect immigration and emigration to move in the same direction. In the short run, however, immigration and emigration may change in opposite directions, as their relationships with the business cycle differ. An economic downturn tends to lead to a decrease of immigration and an increase of emigration.

Return migration of foreigners is not always voluntary. Emigration of asylum migrants depends to an important extent on whether the asylum request is granted. As asylum procedures in one country become more strict, this may have an effect on both immigration and emigration numbers. First, the number of asylum seekers who are not allowed to stay and have to leave the country will increase. Secondly, the number of asylum seekers coming to that country will decline, as they will appeal for a request in another country. The migration from Africa to the Netherlands is illustrated in Figure 13.7. Moroccan migration is excluded because the main motives are marriage migration and family reunion. For other African migrants, the main motive is asylum. In recent years, the emigration of Africans has risen for two reasons. First, the number of immigrants has risen in previous years. Second, the vast majority of asylum seekers are not allowed to stay. Forecasts of emigration

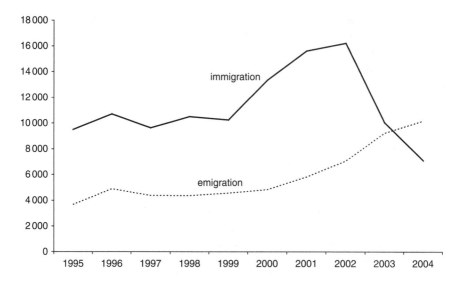

Figure 13.7 Migration of Africans (excluding Moroccans) from and to the Netherlands, 1995–2004.

of Africans can be based on the assumption that emigration will remain high in the short run as a considerable number of asylum migrants have not yet left the country. If it is assumed that the decline of immigration of Africans will be permanent because of stricter asylum procedures, then it can be assumed that emigration will also decline in the longer run.

Finally, the age pattern of emigration of foreigners differs from that of immigration, as illustrated in Figure 13.8. On average, emigrants are three years older than immigrants.

13.5.2   Nationals

Five main categories of emigrating nationals can be identified based on a distinction made between temporary and permanent migration. Of the temporary emigrants, there are two subcategories: students and labour migrants. Students tend to be slightly younger than labour migrants. These emigrants mainly move to other EU countries. Labour migration is inversely related to the business cycle in the home country.

Of the permanent emigrants, that is, those expecting to move for a long, indefinite period, there are three subcategories. The first are nationals marrying a partner from abroad, who choose to move to the country of their partner (those who do, tend to move to other EU countries). This category does not appear to be very large in most Western European countries. Most nationals marrying a foreign partner tend to bring their partner in to their country, particularly if they have found a partner in a non-Western country. The second category are emigrants who want to

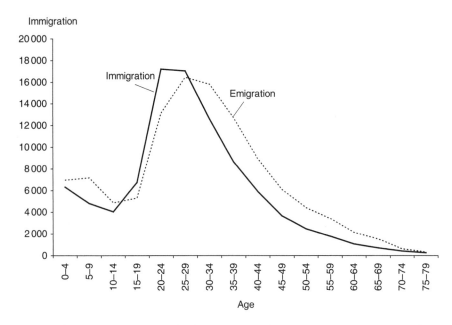

Figure 13.8   Age patterns of foreign migration from and to the Netherlands, 2004.

leave their home country because they are not satisfied with the general situation there. Most of these emigrants move to countries like Canada, Australia and New Zealand. This category represents a relatively small share of the total number of emigrants. A recent NIDI survey shows that only 2 % of the Dutch population aged 15 or over wants to emigrate (Ter Bekke *et al.* 2005). However, only one-tenth of these people have actual plans. This implies that, in 2004, 20 thousand persons had serious plans and 250 thousand persons were thinking about emigrating. On the basis of these results, one would not expect a considerable increase in the annual number of emigrants. The third subcategory represents retired people who move to southern European countries because of the warmer climate and other amenities. France, Spain and Italy are particularly popular countries of destination for these migrants. This category is as yet not very large, but may increase in the future due to the ageing of the population.

Forecasts of number of emigrants can be based on assumptions about the future values of age- and sex-specific emigration rates. Figure 13.9 shows that emigration rates are relatively high for young children (who move together with their parents) and for men between 20 and 35 years of age. Emigration rates of women are considerably lower than those of men. If these rates are held constant over time, the changes in the number of emigrants are determined by changes in the population structure. As elderly people tend to emigrate considerably less than younger persons, ageing may be expected to have a downward effect on the size of emigration in the long run, even though there may be some increase in the number of retirement emigrants.

Figure 13.9   Age-specific emigration rates (per 1000) of persons born in the Netherlands, 2004.

13.6   Assumptions on future changes in immigration and emigration

The identification of the main types of immigration and emigration and their determinants can be the basis for assumptions made on future changes. Even if no
quantitative data on the separate categories of immigration and emigration are available, the distinction of types of migration is useful as a basis for argument-based
forecasts of migration. In specifying assumptions on future changes in migration,
it is important to take into account interdependences between various categories
of immigration and emigration. Therefore the sequence of specifying assumptions
about the separate categories is not random.

One may start with assumptions on future changes in emigration of nationals.
These can be based on assumptions on the level of age- and sex-specific emigration
rates. In the absence of data to estimate these rates, it could be assumed that the
numbers of emigrants decline in the long run due to the ageing of the population,
since emigration rates at older ages tend to be considerably lower than emigration
rates of persons in their 20s or 30s. Subsequently, a forecast of the immigration
of nationals can be based on an assumption about the percentage of emigrating
nationals who return after some time. For example, as mentioned above, 60 % of
emigrating nationals from the Netherlands are expected to return. In formulae, we
can write these dependences as

$$E_{\mathrm{N},t} = e_{\mathrm{N},t}P_{\mathrm{N},t}, \tag{13.8}$$

$$I_{\mathrm{N},t} = r_{\mathrm{N},t}E_{\mathrm{N},t-k}, \tag{13.9}$$

where $E_{N,t}$ is the number of nationals emigrating in year t, $P_{N,t}$ is the number of nationals in the population, $I_{N,t}$ is the number of returning nationals, and $e_{N,t}$ and $r_{N,t}$ are the emigration and return migration rates of nationals. To the extent that more detailed data are available, emigration of nationals may be related to the size of separate age groups (distinguishing different emigration rates) and immigration of nationals may be related to emigration numbers in successive years (distinguishing different immigration rates by duration since emigration).

As for making assumptions about future changes in foreign labour migration, it is useful to start with an analysis of the business cycle effect on the most recent immigration patterns. As labour immigration tends to be positively associated with the business cycle, if there is an economic upturn (downturn) at the moment the projection is made, recent immigration numbers may be higher (lower) than the structural level. Consequently, to the extent that a recent rise (fall) in immigration can be explained by the business cycle, it should not be projected linearly in the long run. Assumptions about the future development of labour migration in the long run can be based on an assessment of future shortages in the labour market caused by ageing, which may lead to an increase in labour migration. As asylum policies in an increasing number of European countries are becoming more restrictive, it can be expected that generation effects will outweigh distribution effects and that the total flows of asylum seekers to Europe will decline. The future levels of family migration are related to the choices of marriage partners in the resident population. The longer they live in a particular country in Europe, the more they will choose a partner already residing in that country. Family migration is also related to the future size of labour and asylum migration. If it is assumed that a large part of labour migrants stay only temporarily, this will have a limiting effect on the size of family-related migration. Moreover, the assumed decline in the number of asylum seekers could have a downward effect on family migration. In formulae, these effect can be expressed as

$$I_t = I_{N,t} + I_{NN,t},\tag{13.10}$$

$$I_{NN,t} = I_{L,t} + I_{A,t} + I_{F,t},\tag{13.11}$$

$$I_{L,t} = p_t V_t,\tag{13.12}$$

$$I_{A,t} = a_t P_t,\tag{13.13}$$

$$I_{F,t} = m_{L,t} I_{L,t-j} + m_{A,t} I_{A,t-j},\tag{13.14}$$

where $I_{NN,t}$ is the number of immigrating non-nationals, $I_{L,t}$ is the number of labour migrants, $I_{A,t}$ is the number of asylum migrants, $I_{F,t}$ is the number of family migrants, V_t is the number of vacancies, and P_t is the population size. It is assumed that the number of labour migrants is related to the number of vacancies, the number of asylum seekers is related to the population size, and the number of family migrants is related to the numbers of labour and family migration j years earlier.

Foreign emigration rates differ between categories of immigrants. As a larger part of labour migrants tend to return to their home country within a particular time period than do family and asylum migrants, the assumption is that the share of labour migrants in total immigration increases and that family and asylum migration decreases. This leads to the expectation that the number of emigrants will decline in the long run. In formulae, we can write

$$E_{\text{NN},t} = e_{\text{L},t} I_{\text{L},t-i} + e_{\text{A},t} I_{\text{A},t-i} + e_{\text{F},t} I_{\text{F},t-i}, \tag{13.15}$$

$$E_t = E_{\text{N},t} + E_{\text{NN},t}. \tag{13.16}$$

Finally, the purpose of this discussion has been to demonstrate how categories of migrants can provide a foundation for argument-based forecasts of migration.

13.7   Uncertainty

There are various reasons why forecasts of migration are uncertain (see also Chapter 12). First, the quality of migration data in many countries of Europe is poor (see Chapters 3–5). If only net migration totals are available, forecasts based on these patterns contain less information about the causes of observed changes. Second, migration patterns tend to exhibit large fluctuations, even in the short run. For example, the size of labour migration tends to change heavily over the course of the business cycle, whereas the number of asylum seekers may change quickly due to changes in policies. Net migration of the EU25 decreased from 1.3 million in 1992 to 0.6 million in 1997 and subsequently rose to 1.9 million in 2003. Migration tends to show much stronger fluctuations than annual numbers of births and deaths. As a result, for the short run, migration is the most uncertain component of population growth. Finally, migration depends on policy changes both in the country of destination and in other countries. Changes in these policies are particularly difficult to forecast.

The degree of uncertainty of migration projections can be addressed with alternative scenarios. Usually, scenarios include a high net migration variant (i.e. combining high immigration with low emigration) and a low net migration variant (i.e. low immigration and high emigration), ignoring the fact that there are usually positive relationships between immigration and emigration patterns. If these relationships are strong, the degree of uncertainty in net migration will be smaller than for immigration and emigration separately. Keep in mind that, in the short run, there may be a negative relationship between immigration and emigration, due to business cycle effects.

In general, there are three ways to assess the degree of uncertainty of migration forecasts. First, one may look at historic forecast errors (e.g. de Beer 1997; Keilman and Pham 2004). Second, stochastic time series models (such as discussed in Section 13.2) produce forecast intervals (e.g. de Beer 1993). The problem with these first two options is that, for many countries in Europe, the available time series on immigration and emigration are short. The third approach uses expert judgements to determine the width of the forecast intervals by including subjective probabilities (see also Chapter 12). This approach incorporates possible explanations for why migration flows could be higher or lower than expected and whether these high and low values can be thought to be permanent or temporary. Assumptions regarding upper and lower boundaries for forecast intervals of immigration and emigration can be based on the same explanations underlying the central or baseline projections. For example, the upper boundary of, say, the 90 % forecast interval of immigration can be based on the assumptions that: (1) the number of labour migrants will equal the number that would be required to stabilise the number of people of working age; (2) the percentage of young foreigners marrying a partner from abroad will not decline; and (3) the recent decline in the number of asylum migrants is only temporary and will rise again to the levels observed several years ago. Accordingly, assumptions can be specified about the lower boundary of immigration and about the upper and lower boundaries of emigration.

13.8   Conclusion

This chapter has discussed the usefulness of distinguishing different types of immigration and emigration, as they are affected by different factors and hence may change in different ways. This allows one to make argument-based forecasts rather than simply extrapolating changes observed in the past. First, one may distinguish migration of nationals and foreigners. Emigration of nationals can be projected on the basis of assumptions on the future values of age- and sex-specific emigration rates. Immigration of nationals can be projected on the basis of an assumption about the percentage of emigrants who will return. Second, for projections of immigration of foreigners, three main categories of migrants can be distinguished: labour, asylum and family migration. For each of these categories, assumptions on future changes can be formulated. The future size of labour migration depends on the effect of ageing on the labour market. The number of asylum seekers depends on the extent to which asylum policies within the European Union will be coordinated. This is particularly important, because a large part of the changes in the number of asylum seekers in individual European countries were due to changes in the distribution of asylum seekers over European countries rather than to changes in the total inflow to the EU. The number of family migrants depends on the tendency of labour and asylum migrants to marry a partner from the country of origin.

Foreign emigration rates differ between the three main categories of immigrants. A relatively high proportion of labour migrants tend to return to the country of origin within a limited number of years. As a relatively large proportion of asylum requests are not granted, many asylum migrants are required to leave. However, those who are allowed to stay tend to remain for a relatively long period. Marriage migrants also tend to stay for a long time.

Finally, because of the poor quality of data and the sharp fluctuations in migration time series, forecasts of net migration tend to be rather uncertain. The degree of uncertainty can be assessed by looking at the size of errors of forecasts made in the past. Another approach is to estimate the width of the forecast interval on the basis of a time series model. Both approaches assume that the uncertainty of future migration can be assessed on the basis of past developments. However, as the sizes of the various categories of migrants tend to change in different ways, one may question the validity of this assumption. Hence it is useful to follow an argument-based approach in which the uncertainty of future migration is assessed by looking for reasons why immigration and emigration could be higher or lower than expected.

References

Alders M. 2005. Prognose van gezinsvormende migratie van Turken en Marokkanen (Forecast of marriage migration of Turks and Moroccans). *Bevolkingstrends* 53(2):46–49.

Box GEP and Jenkins GM. 1970. *Time series analysis: Forecasting and control*. San Francisco: Holden-Day.

de Beer J. 1993. Forecast intervals of net immigration: the case of the Netherlands. *Journal of Forecasting* 12:585–599.

de Beer J. 1997. The effect of uncertainty of migration on national population forecasts: the case of the Netherlands. *Journal of Official Statistics* 13(3):227–243.

de Beer J. 2004. Immigratie uit Europese Unie hangt samen met conjunctuur (Immigration from European Union is related to business cycle). *Bevolkingstrends* 52(2):46–47.

de Jong A and Nicolaas H. 2005. Prognose van emigratie op basis van een retourmigratiemodel (Forecast of emigration based on a return migration model). *Bevolkingstrends* 53(1):24–31.

Howe N and Jackson R. 2005. *Projecting immigration. A survey of the current state of practice and theory*. A report of the CSIS global aging initiative. Washington, DC: Center for Strategic and International Studies.

Jennissen R. 2004. *Macro-economic determinants of international migration in Europe*. Amsterdam: Dutch University Press.

Johansson M and Rauhut D, eds. 2005. *The spatial effects of demographic trends and Migration*. Available at www.espon.lu

Keilman N and Pham DQ. 2004. *Empirical errors and predicted errors in fertility, mortality and migration forecasts in the European Economic Area*. Discussion Paper 386. Oslo: Statistics Norway.

Massey DS, Arango J, Hugo G, Kouaouci A, Pellegrino A and Taylor JE. 1993. Theories of international migration: a review and appraisal. *Population and Development Review* 20:699–751.

Nicolaas H. 2004. Helft Nederlandse emigranten keert terug (Half of Dutch emigrants return). *Bevolkingstrends* 53(2):39–45.

Ter Bekke S, van Dalen H and Henkens K. 2005. Emigratie van Nederlanders geprikkeld door bevolkingsdruk (Emigration of Dutch nationals triggered by population pressure). *Demos* 21:25–28.

United Nations. 2000. *Replacement migration: Is it a solution to declining and ageing populations?* New York: United Nations, Department of Economic and Social Affairs.

Part V
Demographic Consequences

14

International migration component in population dynamics models

Marek Kupiszewski and Dorota Kupiszewska

Central European Forum for Migration and Population Research, Warsaw

14.1   Introduction

The aim of this chapter is to discuss the issues concerning the international migration component in population dynamics models. An emphasis is placed on models that simultaneously project populations of several countries.

Population dynamics models may be used to project, forecast or simulate future population change. In a narrow sense, projections extrapolate trends observed in the past. In a wider sense, projections may incorporate a variety of assumptions regarding future evolutions in the population change components. Forecasts are the projections which forecasters believe are most likely to come true, while simulations include 'What if?' type situations in the modelling, which may be based on a number of assumptions, including highly improbable ones. Although many researchers are aware of the difference between projections and forecasts, the term 'projection' is often used for both.

Throughout this chapter, the following geographical terminology is used to describe population projections. Independently projected populations of single countries are termed national projections. Multiregional projection models simultaneously

International Migration in Europe: Data, Models and Estimates   Edited by J. Raymer and F. Willekens

project subnational (regional) and national populations by incorporating origin–destination-specific internal migration flows. Multinational (or supranational) projections are those covering several countries and prepared using a common methodology. A special case of multinational projections are those in which multi-regional projections are prepared for all the countries, either separately or simultaneously. In the latter case, national populations are linked through international migration flows in a similar way as subnational units are linked in multiregional projections. These models are labelled hierarchical multiregional models.

The structure of this chapter is as follows. The changing perceptions of international migration and its role in population dynamics and models are discussed in Section 14.2. This is followed in Section 14.3 by a general review of multinational projection and forecast models that have been used to estimate future populations in Europe. The complex issues surrounding the incorporation of international migration in population dynamics models are discussed in Section 14.4. In Section 14.5, the hierarchical multiregional model, MULTIPOLES, is described. The chapter ends with a conclusion in Section 14.6.

14.2   The increasing importance of international migration in population dynamics and population modelling

Periodic overviews of forecasting practices in Europe (Keilman 1988; van Imhoff *et al.* 1994; van der Gaag *et al.* 1997a; Kupiszewski and Kupiszewska 2003) show that the most frequently used projection model is a cohort-component model (Leslie 1945), including its multiregional (Rogers 1975) or multistate extensions. Until the late 1970s, the specification and application of these models generally ignored international migration. Early models that accounted for international migration introduced flows between the modelled country and the 'rest of the world' (Rees and Wilson 1977; Rees 1984; Willekens and Drewe 1984). However, this option has not been widely applied, despite substantial inflows of guest workers and their families to European countries. In practice, less than half of countries reviewed by van Imhoff *et al.* (1994) took into account international migration. Booth (2006:549) noted that, in the population forecasts prepared in the 1980s, '. . . migration was treated as an uninteresting constant', whereas now almost all countries incorporate international migration in their models (Kupiszewski and Kupiszewska 2003).

In the second half of the 1980s and early 1990s, the perception of international migration and its role in population development in Europe changed for several reasons. First, with an increasing number of foreigners and second-generation migrants, the widespread conviction that international migration was a temporary process started to fade, both in political circles and in academia. Second, the magnitude of international migration observed in this period, especially from Central Europe and former Yugoslavia to Western Europe, initiated public debates on the role of international migration in demographic, economic and social development.

Third, low levels of fertility in European countries meant that international migration became a leading factor of population change. For example, simulations from the late 1990s showed that the German population could only be sustained if positive net international migration continued (Kupiszewski and Kupiszewska 1997). Moreover, European nations started to face rapid and often region-specific ageing, and some decision-makers perceived international migration as a remedy. Finally, political processes in Europe, especially the enlargement of the European Union (EU) to include Central European countries, generated a strong incentive to consider populations of countries in Europe as the elements of an integrated system.

The increasing importance of international migration can be found by examining official EU population forecasting models over time. The first forecast prepared in the 1980s (Netherlands Economic Institute 1986) assumed that there was no international migration. In the second one, Haverkate and van Haselen (1990, 1992) assumed nonzero net migration in only two countries: Germany and Ireland. It was not until the post-1990 forecasts that international migration was considered a key element in modifying population dynamics (Eurostat 1993, 2006; de Beer and de Jong 1996; de Jong and Visser 1997). Refer also to Rees *et al.* (2001) for a comprehensive list of publications on each of the EU forecasts and to Willekens (2005) for a review of demographic projections at the European level. Rees *et al.* (2001) estimated the effect of ignoring international migration in a 1980 forecast. After 15 years, the *ex post* errors for Luxembourg and Germany were 9.28 % and 7.30 %, respectively. The latter error translates into 4.5 million in absolute numbers.

The appreciation of international migration in population dynamics and in the development of population dynamics models has had two effects. First, it intensified the discussion on the key methodological aspects of incorporating international migration into the models. Wilson and Rees (2005) discuss three issues: (1) strategies for including international migration in population dynamics models; (2) methods for forecasting international migration; and (3) methods for allocating international migrants to subnational units. Second, both researchers and policy-makers began to look at population dynamics as a process occurring within a multinational space, which should be modelled as a system of countries linked together through international migration. In the following two sections, we examine how multinational projections and forecasts are conducted and how they could be improved.

14.3 A review of multinational population projections and forecasts in Europe

Multinational population projections and forecasts may be conducted in a variety of ways. One option is to run a projection or a forecast for large supranational units, such as political or economic groupings of states or continents (see, for example, Lutz *et al.* 1996a,b, 2004). However, such models do not give any detail on the level of individual states, which makes them unsuitable for policy decisions. Another strategy is the one adopted by the United Nations (UN), which prepares the most widely used population projections, published biannually in the *World population prospects* – see United Nations (2005) for the most recent set of projections. The UN uses a cohort-component model and fairly simple assumptions on fertility, mortality and net international migration to project future populations for each country in the world separately (O'Neill *et al.* 2001; United Nations 2005). An analysis of problems concerning the preparation of the UN projections can be found in Keilman (2001).

There are a number of other organisations besides the UN that produce multinational population projections or forecasts at the national level, for example, the US Census Bureau (Johnson 1999) and the World Bank (Vu and Bos 1992; Bos *et al.* 1994; World Bank 2005). Examples for Europe can be found in Eurostat (1991, 1993), Eding *et al.* (1996) and de Beer and van Wissen (1999). More recently, the Central European Forum for Migration and Population Research published the results of a 2002–2052 population and labour force forecast for 27 selected European countries (Bijak *et al.* 2005, 2007, 2008). Unlike the earlier forecasts for Europe, these results were obtained by simultaneously modelling populations of all the countries in the system by linking them together by international migration flows.

The above projections and forecasts do not examine the regional dimension of population processes and, therefore, have a limited attraction for decision-makers, spatial planners and geographers, who are more interested in population processes at smaller spatial units. Notable efforts to produce regional projections for the Member States of the EU, European Free Trade Area and the Candidate Countries have been made by the European Commission and Eurostat. There have been five rounds of population projections for the entire EU at a regional level (i.e. NUTS 2[1] level). The first two covered 12 European Community (EC) Member States in 1980 and 1985 and spanned 30 years to 2010 and 2015, respectively (Netherlands Economic Institute 1986, 1990). The main advantage of the EC projections over those conducted by individual countries was that they were based on a uniform methodology, the same starting year and a unified set of assumptions.

The Netherlands Economic Institute's DEMETER model used in the EC projections was a major step forward at the time of its construction. The main shortcoming of this model was that international migration was either neglected (in DEMETER 2010[2]) or taken into account only for Germany and Ireland (in DEMETER 2015). The scenario-setting capability of the model was limited to mortality and fertility only, which was unfortunate, given the importance of migration in European population dynamics. This shortcoming was rectified in the most recent model (DEMETER 2020), which took into account net international migration for all countries in the 1990 base-year forecast (Netherlands Economic Institute 1994).

In 1990, Harry Cruijsen was made responsible for Eurostat population projections. He initiated large-scale methodological research aimed at improving the techniques for use in the 1995 projection model. Since then, Eurostat contractors,

---

[1] NUTS – Nomenclature of territorial units for statistics, a three-level hierarchical division of the EU. NUTS divides each Member State into NUTS 1 regions, which are in turn subdivided into NUTS 2 regions and so on.
[2] DEMETER models were labelled with the final year of forecast; for example, DEMETER 2010 covered the period 1980–2010.

Statistics Netherlands and the Netherlands Interdisciplinary Demographic Institute (NIDI) have been busy conducting research on forecasting mortality (Tabeau *et al.* 1997), fertility (de Jong 1997), internal migration (van der Gaag *et al.* 1997b) and international migration (de Jong and Visser 1997). The results have been included in a population forecast (EUROPOP 1995[3]). An external review of Eurostat's population projections suggested some further improvements, mostly expanding the research agenda proposed by Cruijsen (see Rees *et al.* 2001). The most recent round of Eurostat forecasts was produced jointly by Eurostat and NIDI for the 2004 base year (i.e. EUROPOP 2004) by using the LIPRO model (van Imhoff and Keilman 1991). For this set of projections, no major methodological changes were made in the treatment of international migration (Eurostat 2006).

The major difference between Eurostat's projections and those prepared by other organisations is the level of spatial detail. Eurostat is the only international organisation which produces projections of both national and regional populations. However, these projections were prepared separately for each country, with net migration used instead of origin–destination-specific flows.

The first to create a model capable of simultaneously handling a system of both internal and international migration flows in multinational projections was Philip Rees and his colleagues (Rees *et al.* 1992; Rees 1996). They constructed a model called ECPOP, for the European Community Member States, which was a female-dominant hierarchical multiregional model with migration being handled on three levels: internal migration, international migration between countries within the system, and international migration to and from countries outside of the system. The design of ECPOP made it possible to set separate scenarios for all categories of migration, as well as for mortality and fertility. It was used for projecting regional populations at the NUTS 1 level. In many ways, Rees' model was a major improvement in population projection practice. From the methodological point of view, it was an implementation of the state-of-the-art population projection theory. Finally, Rees' model inspired us to develop the MULTIPOLES model (presented in Section 14.5), which was initially used to study the population development in Central and Eastern European countries.

14.4   The international migration component in national and multinational population dynamics models

There are three main issues concerning the inclusion of international migration in multiregional population dynamics model. The first is methodological and involves the manner in which international migration is taken into account, including the selection of the migration variables to be used. The second problem concerns

---

[3] EUROPOP projections are labelled with the base year of the projection.

the regional allocation of international migrants. The third issue relates to the availability, comparability and quality of migration data (see Chapters 3 and 4). This issue should be anticipated at the model development stage. Often, population forecasts are much more restricted by the availability of international migration data than is the case for fertility, mortality and internal migration.

14.4.1   The incorporation of international migration into population dynamics models

The way of handling international migration is interconnected with the geographical coverage of the model. For example, for a single-country model, the international migration flows represent flows to and from all other countries of the world. Alternatively, a model might include a group of countries, with international migration flows specified within the group and between each country in the group and the rest of the world. International migration flows are measured in terms of either absolute numbers or rates. Occurrence–exposure migration rates appear in the movement approach, whereas survival probabilities appear in the transition approach (Rees and Willekens 1981, 1986; Rees 1989).

The various ways of incorporating international migration in population dynamics models are presented in Table 14.1, along with references either to the formulation of the model or to its application. The simplest option is to set international flows to zero (Rogers 1975; Willekens and Rogers 1978; Netherlands Economic Institute 1986; Scherbov *et al.* 1986; Scherbov and Grechucha 1988). This option is appropriate only if the international migration component is insignificant compared to other components of population dynamics. The second possibility is to treat the population of the world as a closed system, divided into smaller units, which, from the computational point of view, are equivalent to regions in multiregional forecasts and projections, and to use any multiregional population dynamics model (Rees 1986) with migration defined in terms of emigration rates. Such a solution may be an option for forecasts and projections of continental populations or other large supranational units. Lutz *et al.* (1996a) prepared a projection of the world population using the DIAL model, which was originally designed as a multiregional model for projecting regional and national populations. Theoretically, one can imagine a closed system model using emigration rates, involving a single country (divided into regions) and the rest of the world. In practice, such models are not used due to the difficulty with estimating population and emigration rates from the rest of the world.

Population accounting models (Stone 1971; Rees and Wilson 1977; Rees 1986) take into account the absolute number of immigrants and emigrants from the rest of the world, without including the population of the rest of the world in the modelled system. Some models use net international migration or net international migration rates (Rogers 1971; Espenshade *et al.* 1982; Liaw 1979; Espenshade 1987). Rogers (1990) criticised the concept of relying on net migration in population projection models. He argued that the use of net migration rates over time is inappropriate

Table 14.1 Methods of incorporating international migration in population dynamics models.

Methods of incorporating international migration	Examples of model formulation or application	
	Single-nation projection	Multinational projection
Ignoring international migration	Rogers (1975), Willekens and Rogers (1978), Scherbov *et al.* (1986), Scherbov and Grechucha (1988, DIAL)	Netherlands Economic Institute (1986)
Emigration rates in a closed population system	Rees (1986)	Lutz *et al.* (1996a, DIAL)
Inflow and outflow (absolute numbers)	Stone (1971), Rees and Wilson (1977), Rees (1986)	
Net migration	Rogers (1971), Liaw (1979), Rees (1986)	Netherlands Economic Institute (1994), Haverkate and van Haselen (1990, 1992), Eurostat (2006, EUROPOP 2004), United Nations (2005), Bos *et al.* (1994)
Emigration rates and immigration numbers	Willekens and Drewe (1984), Willekens (1995, MUDEA), van Imhoff and Keilman (1991, LIPRO), Rees (1986)	
Hierarchical models (emigration rates for migration between modelled countries and net migration for migration between modelled countries and the rest of the world)		Rees *et al.* (1992), Rees (1996, ECPOP), Kupiszewski and Kupiszewska (1997, 1999), Kupiszewska and Kupiszewski (2005, MULTIPOLES)

because they do not account for changes in the size of populations at risk of moving. Also, the age patterns of immigrants and emigrants are hidden.

Rees' (1984, 1986) MOVE model was the most universal. Its input could be defined through a combination of emigration numbers or occurrence–exposure emigration rates and immigration numbers or immigration admission ratios (defined as the number of immigrants divided by the population in the receiving country). The main drawback of immigration admission ratio, as noted by Willekens (1995), is that the number of immigrants is related to the population in the destination area, which differs from the population at risk of migration. Willekens recommends the use of the emigration rates and immigration numbers whenever possible. This solution was proposed for a multiregional, single-country system by Willekens and Drewe in the MUDEA model (Willekens and Drewe 1984; Willekens 1995).

In other population projections, international migration has been either ignored, as in the DEMETER 2010 model, or included with net migration, as in all other EU population forecasts. The specification of hierarchical ECPOP and MULTIPOLES models (Rees *et al.* 1992; Rees 1996; Kupiszewski and Kupiszewska 1997, 1999; Kupiszewska and Kupiszewski 2005) were an improvement, but not perfect. International migration flows within the system were correctly included in the model by using origin–destination-specific emigration rates. Migration from and to the rest of the world was included by using net migration totals for each country in the system. A further improvement could be considered by estimating emigration rates for flows from each of the modelled countries to the rest of the world and absolute numbers in the opposite direction, though data availability might be a problem.

14.4.2   Subnational allocation of international migrants

Until the 1990s, little research had taken place on the regional distribution of the flows of international migrants in Europe, despite an ample literature on population and ethnic group distributions. Rees *et al.* (2001) recommended to the European Commission that a careful consideration of this issue might improve the quality of population projections. In general, there were two lines of enquiry. One looked at the relationships between international migration and some specific characteristics of the regional population, such as size, foreign population stocks or past flows of international migrants. The second looked at the relationships between internal and international migration (Korcelli 1994; Stillwell *et al.* 1999).

In the European Economic Area national forecasts at the end of the 1980s and in the early 1990s, Belgium, the Netherlands, Italy and England specified the regional distribution of international migrants (van Imhoff *et al.* 1994). The allocation of net migrants to subnational units was also considered in the EC forecasts (Netherlands Economic Institute 1994; de Jong and Visser 1997). In the DEMETER model, it was assumed that net migration was proportional to the share of the foreign population in the regions (Netherlands Economic Institute 1994). In practical terms, large urban agglomerations were the primary destinations for international migrants. Later, van der Gaag and van Wissen (2001) noted that the 10 largest German planning regions (*Regierungsbezirke*) accounted for 60 % of immigrants and only 40 % of total population. A higher share of immigration to the urban regions compared to their share in total population was also visible in other European countries, such as Austria, Finland, Greece, the Netherlands and Spain (van der Gaag and van Wissen 1999, 2001).

In the EUROPOP 1995 projections, a relatively complex set of assumptions was adopted. In the low scenario, the distribution of incoming international migrants was in line with the initial distribution of migrants' stocks. In the base and high scenarios, the distributions evolved from the one proportional to the initial distribution of migrants to the one proportional to the distribution of the entire population (de Jong and Visser 1997).

The regional distribution of international migrants could be linked to the numbers or shares of foreigners, the overall size of population in the destination region or to non-demographic variables, such as the level of unemployment or the level of urbanisation. In research conducted for Austria, Finland, Greece, Italy, the Netherlands and Spain, van der Gaag and van Wissen (1999, 2001) showed that larger stocks of foreigners meant larger, but less than proportional, numbers of immigrants. The same observation was found for the relationships between the flows of immigrants and total populations. A more detailed investigation for Germany proved that the distribution of foreign immigrants was proportional to the size of the foreign population, but such a relationship was not found for national immigrants. As expected, high unemployment decreases the attractiveness of the region for immigrants.

In another study comparing the predictive capability of various alternative models for Sweden, van der Gaag and van Wissen (2002) concluded that the best estimates of the distribution of foreign immigrants could be obtained by using historical distributions, or by using historical distributions combined with information on foreign population stocks, as a predictor. More complicated models that specified nonlinear relationships between flows and stocks of foreigners or internal migration indicators usually only improved the results marginally. The authors noted that, if data on regional shares of foreigners are not available, then the distribution of total population may be used as a proxy to predict immigration flows. However, the results are not as good as using data on historical shares of flows.

14.4.3   Issues concerning international migration data

The various ways of incorporating the international migration component into population dynamics models were discussed in Section 14.4.1. In this section, we examine the problems associated with assembling country-level data on international migration, but not the corresponding age patterns, which are needed as model inputs. Age-specific migration data are often not available and have to be estimated, for example, by using the Rogers–Castro model (Rogers and Castro 1981; see also Chapters 8 and 9).

There is an ample literature describing the problems associated with European international migration data (refer to Chapters 3 and 4). They were first raised more than 30 years ago and the efforts to solve them have been summarised by

Kelly (1987) and Herm (2006). Research on the comparability and usability of international migration flow data, primarily in the European Union, has been carried out by Poulain and his colleagues (Poulain *et al.* 1991; Poulain 1993, 1996) and more recently within the THESIM project (Nowok and Kupiszewska 2005; Nowok *et al.* 2006). As discussed in Chapter 3, data on international migration are not comparable across countries. This is a result of differences in migration definitions and in the administrative procedures used to collect migration data. Another problem is the lack of data on migration flows disaggregated by country of previous or next residence. For some countries, even data on total immigration or emigration flows are not even available.

One of the key problems involves constructing a consistent matrix of flows from a double entry matrix (see Chapter 3). Three problems may be distinguished: (i) estimation of flows when there are two values in one cell – one reported by the receiving country and the other by the sending country; (ii) estimation of the missing values in the matrix; and (iii) harmonisation of the data in the matrix so that they correspond to a unified definition. Various solutions have been proposed to tackle the first problem. One solution is to prefer the receiving country data over the sending country data (e.g. see Chapter 10) with the argument being that the receiving country's registration system covers more migrants than the sending countries' registration system. That is, migrants tend to avoid deregistration in the country of origin, trying to keep links and access to social security benefits and health insurance for as long as they can. Simultaneously, they tend to register in the destination country to gain access to these benefits. On the basis of these arguments, data from the sending countries are not considered if there is a choice. Another approach is to take the larger of the two values in each double entry matrix cell (Kupiszewska and Kupiszewski 2005). The argument here is that many migrants escape registration, so those who registered were most likely migrants. Poulain (1993) proposed an algorithm for the estimation of international migration flows within the EU using data from both sending and receiving countries. His work was based on the assumption that all migration flows to (or from) a given country were subject to the same distortion, which arose from differences in legal, economic or social systems. Finally, for tackling the problem of missing data, the methodologies used for the estimation of internal migration from inadequate data, recently reviewed by Rogers *et al.* (2003), have been applied in Chapter 10 to estimate a full set of age-specific flows between 25 countries in the EU.

All the above methods have two fundamental deficiencies. First, the definition of migrant and migration in the final matrix of flows (model output) is not specified. No guarantee is made that the same definition applies in all estimated cells. When data from sending or receiving countries were used directly, different cells were calculated according to different definitions. When statistical methods were used, the calculated flows did not have a meaningful interpretation as concerns migration definition. Both approaches are fundamentally flawed, as they do not correctly estimate the stocks of population after the migration, i.e. the definition of a migrant would be either incompatible with the definition of population stocks or unknown. Therefore, at this stage, we have to accept the fact that in all multinational population projections there are inherent errors due to the lack of harmonisation of definitions in the data on migration flows and stocks of population.

In the future, improvements in the modelling of multinational populations will require better estimates of international migration flows that are both numerically consistent and harmonised in terms of definitions. The methodology to do this should, first of all, aim at the recalculation of the available (inconsistent) migration data to a common definition. To a large extent, this is *terra incognita* for migration research. A double entry matrix of corrected migration flows should then be assembled. In this matrix, the differences in the numbers recorded by sending and receiving countries could be attributed to two factors: underregistration and modelling errors. Only after obtaining such a corrected double entry matrix may one attempt to estimate a single entry migration matrix.

14.5   MULTIPOLES: a model with a multilevel treatment of international migration

14.5.1   General description and comparison with earlier models

As noted earlier, in the mid-1990s, it became clear that population projections required the inclusion of international migration in order to be more accurate. Traditional population dynamics models can only handle this in the case of a projection for a single country, e.g. by setting scenarios for emigration rates and immigration numbers. Problems occur when several national populations were projected simultaneously. Using traditional models separately for each country, it is difficult, or even impossible, to ensure the consistency of international migration flows between the modelled countries. In particular, it is difficult to ensure that the sum of all immigrations is equal to the sum of all emigrations.

The best solution to the problem is a system approach originally designed for modelling internal migration in multiregional models and proposed for modelling international migration by Rees (1996). Following Rees' ECPOP model, we have implemented a systematic approach to the treatment of international migration in the MULTIPOLES model. In this approach, the populations of individual countries are not treated as isolated entities but as the subpopulations of a larger system, with interactions (international migration flows) between the elements of the system (national populations). In essence, the principles of multiregional population projections are applied to a hierarchical system.

MULTIPOLES is a multilevel generalisation of the traditional multiregional model (Rogers 1975). The populations in this model are disaggregated by sex and 18 five-year age groups. There are several features that distinguish the MULTIPOLES model from traditional multiregional models and, in particular, from Willekens and Drewe's (1984) MUDEA model, which was an earlier extension of a multiregional model that included international migration. In MULTIPOLES, the population hierarchy has been extended from a two-level one (national and regional populations) to a three-level hierarchy composed of supranational, national and regional populations. The migration component has been extended from two levels, i.e. internal and external (international) migration, to three levels: level 1, origin–destination-specific internal migration; level 2, origin–destination-specific international migration within the system; and level 3, net international migration from outside the system (i.e. between each of the modelled countries and the rest of the world). Finally, this model represents a movement approach, which differs from the transition approach utilised in ECPOP – see Rees and Willekens (1981) for a discussion of both approaches. A movement approach is better suited for the European countries, where most migration data come from population registers.

The main benefit of applying MULTIPOLES is the consistency between immigration and emigration flows within the system and related consistency in the projected population stocks. The main drawback is the large datasets required to run the model. Internal consistency would also be ensured if one used a multiregional model that would refer to the system of all regions of the group of countries. The advantage of the three-level hierarchical modelling approach applied in the MULTIPOLES model is that the migration scenarios are easier and more transparent to incorporate.

It is worth noting that migration from the rest of the world is treated differently in MULTIPOLES than in MUDEA (Willekens and Drewe 1984). MUDEA used emigration rates and absolute numbers of immigrants, whereas the MULTIPOLES model uses net migration. Both approaches have their advantages, but the one implemented in MUDEA might be considered better from the methodological point of view. Its suitability might be justified by the fact that immigration is often subject to the state's control (Rees 1996). In MULTIPOLES, net migration between the countries of the system and the rest of the world was used for practical reasons of better data availability and the ease of setting scenarios for net migration.

14.5.2   An outline of MULTIPOLES structure, data requirements and applications

In the MULTIPOLES model, age-, sex- and region-specific fertility, mortality and interregional outmigration and emigration rates for the benchmark year are used to project regional populations forward. The rates may be kept constant over time or adjusted according to specific scenarios. The calculations are based on the assumption that, for each projection cohort, except the first one (representing births added to the population), the population at the end of the five-year projection interval may be expressed as

$$\mathbf{P}_{a+5}(t+5) = \mathbf{P}_a(t) - \mathbf{M}_a(t)\frac{\mathbf{P}_a(t) + \mathbf{P}_{a+5}(t+5)}{2} + \mathbf{N}_a(t), \qquad (14.1)$$

where $\mathbf{P}_a(t)$ is a vector of regional populations at age a at time t, $\mathbf{M}_a(t)$ is a matrix constructed from mortality rates, outmigration rates and emigration rates in the period $(t, t+5)$, and $\mathbf{N}_a(t)$ is a vector of net migrations from the rest of the world. The resulting projection equation is analogous to the equation derived by Willekens and Drewe (1984) for multiregional population projections of a single country. The full specification of MULTIPOLES can be found in Kupiszewska and Kupiszewski (2005).

The MULTIPOLES model requires, as an input, data on population stocks in the base year and assumptions on the components of population change in all projection intervals. The preparation of international migration data for MULTIPOLES may be split into two tasks: (i) the estimation of migration between the countries covered in the study; and (ii) the estimation of net migration between each of the countries in the system and the rest of the world. For the estimation of emigration rates for intra-system international migration flows, emigration numbers by origin, destination and age are needed. The age structures of the migrants may be obtained from existing data or estimated by using the Rogers–Castro model (see Chapters 8 and 9), or borrowed from countries for which data are available. The estimation of net international migration between the modelled countries and the rest of the world can be obtained for each country as the difference between the total net migration and the net intra-system migration. Finally, the allocation of international migrants to regions in each country can be done internally in the model based on the shares of regional populations.

MULTIPOLES has been applied in a variety of projection, forecast and simulation studies. These include, for example, population projections for Central and Eastern European countries (Kupiszewski and Kupiszewska 1997), elderly population forecasting (Kupiszewski and Kupiszewska 1999), forecasting labour force supply in Central and Eastern European countries (Kupiszewski 2001), assessment of the international migration impact on the development of regional populations in Central and Eastern Europe (Kupiszewski 2002) and forecasting the regional population in Austria and the surrounding countries (work not published). The most recent applications have modelled the impact of international migration in the context of policy response to population ageing in Europe (Bijak *et al.* 2008) and calculated replacement migration for Europe (Bijak *et al.* 2005, 2007).

14.6 Conclusion

The change in the perception of the importance of international migration for population dynamics was noted by policy-makers and academics in the late 1980s and early 1990s. However, it has taken a long time for population forecasters to address this in their models. In the 1980s, international migration was incorporated into multiregional models at the national level, but the issue of incorporating international migration flows in multinational models was left unresolved. A successful proposal on how to solve the problem was offered by Rees (Rees *et al.* 1992; Rees 1996). Kupiszewski and Kupiszewska followed Rees' concepts and proposed a hierarchical multiregional population projection model called MULTIPOLES. In this model, populations of individual countries are not independent, but form a system and are connected through international migration.

The multiregional methodology is both versatile and useful for the modelling of regional populations of multinational population systems. In 1989, Rees noted that 'The multistate model has proved to be an adaptable beast and is likely to live on into the 1990s' (Rees 1989:36). Today, we can add that it is likely to thrive in the 21st century, in parallel with the new modelling approaches. However, there are still some problems to address. A better method for allocating international migrants to regions is required. Here, spatial choice modelling (Pellegrini and Fotheringham 2002) and hierarchical spatial modelling (Nazra *et al.* 2006) may offer some solutions. More research is needed to devise a methodology that can overcome the inconsistencies in the reported international migration data and to estimate missing patterns. This research could build on and extend the ideas of Poulain (1993), as well as those in Chapters 6, 7 and 10 in this volume. Finally, research is needed to ensure consistency between migration flow data and population stock data. The reliable estimates of international migration flows and population stocks are crucial for a significant improvement of the reliability of population projections and forecasts.

In conclusion, it is clear that inconsistent data on international migration hamper the successful application of contemporary sophisticated population projection models. Developments in population dynamics modelling and in the handling of international migration in the models are well ahead of the developments in international migration statistics.

Acknowledgements

The authors are grateful to the BMU Foundation of Zurich for partially supporting this research. We are also grateful to our employers over the past 10 years, who have supported our work at various stages of development: the International Organization for Migration, Institute of Geography and Spatial Organisation of the Polish Academy of Sciences, and the School of Geography of the University of Leeds. We are very grateful to Professor Philip Rees, who was very helpful and supportive during our work on MULTIPOLES. We are also grateful to the editors of this book for their very thoughtful comments and suggestions.

References

Bijak J, Kupiszewska D, Kupiszewski M and Saczuk K. 2005. *Impact of international migration on population dynamics and labour force resources in Europe.* CEFMR Working Paper 1/2005. Central European Forum for Migration Research, Warsaw.

Bijak J, Kupiszewska D, Kupiszewski M, Saczuk K and Kicinger A. 2007. Population and labour force projections for 27 European countries, 2002–2052: impact of international migration on population ageing. *European Journal of Population* 23(1):1–31.

Bijak J, Kupiszewska D and Kupiszewski M. 2008. Replacement migration revisited: simulations of the effects of selected population and labour market strategies for the ageing Europe, 2002–2052. *Population Research and Policy Review* forthcoming.

Booth H. 2006. Demographic forecasting: 1980 to 2005 in review. *International Journal of Forecasting* 22:547–581.

Bos E, Vu MT, Massiah E and Bulatao R. 1994. *World population projections 1994–95: Estimates and projections with related demographic statistics.* Baltimore, MD: Johns Hopkins University Press, for the World Bank.

de Beer J and de Jong A. 1996. National population scenarios for countries of the European Economic Area. *Maandstatistiek van de bevolking* 44(July):7–19.

de Beer J and van Wissen L, eds. 1999. *Europe: one continent, different worlds. Population scenarios for the 21st century.* Dordrecht: Kluwer Academic.

de Jong A. 1997. *Analysing and forecasting of fertility by generation and birth order.* Paper presented to the Working Party on Demographic Projections, Eurostat, Luxembourg.

de Jong A and Visser H. 1997. *Long-term international migration scenarios for the European Economic Area.* Eurostat Working Paper, E4 1997–6. Eurostat, Luxembourg.

Eding H, Willekens F and Cruijsen H. 1996. *Long-term demographic scenarios for the European Union.* Groningen Demographic Report no. 20. University of Groningen.

Espenshade TJ. 1987. Population dynamics with immigration and low fertility. In *Below-replacement fertility in industrial societies.* Davis K, Bernstam MS and Ricardo-Campbell R, eds, pp. 248–261. Cambridge: Cambridge University Press.

Espenshade TJ, Bouvier L and Arthur WB. 1982. Immigration and the stable population model. *Demography* 19:125–133.

Eurostat. 1991. Two long-term population scenarios for the European Community. Principal assumptions and main results. In *Background papers on fertility, mortality and international migration under two long term population scenarios for the European Community.* Conference on Human resources in Europe at the dawn of the 21st century. Luxemburg: Eurostat.

Eurostat. 1993. *Two long-term population scenarios for the European Free Trade Association.* Luxemburg: Eurostat.

Eurostat. 2006. *EUROPOP 2004 Regional population projections.* Paper presented to the Working Group on Population Projections, Eurostat, Luxembourg.

Haverkate R and van Haselen H. 1990. *Demographic evolution in time in European regions (DEMETER 2015).* Final Report to the European Commission. Netherlands Economic Institute, Rotterdam.

Haverkate R and van Haselen H. 1992. *Demographic evolution in time in European Regions (DEMETER 2015).* Regional Development Studies no. 01. Office for Official Publications of the European Communities, Luxembourg.

Herm A. 2006. Recommendations on international migration statistics and development of the data collection on international level. In *THESIM: Towards Harmonised European Statistics on International Migration.* Poulain M, Perrin N and Singleton A, eds, pp. 77–106. Louvain-la-Neuve: Presses Universitaires de Louvain.

Johnson P. 1999. US Census Bureau international population projections: focus on the ECE area. Presented at *Joint ECE–Eurostat Work Session on Demographic Projections*, Perugia, 3–7 May.

Keilman N. 1988. Methods of national population projections in developed countries. Presented at *Conference on future changes in population age structures*, International Institute of Applied Systems Analysis, Sopron.

Keilman N. 2001. Data quality and accuracy of United Nations population projections, 1950–95. *Population Studies* 55(2):149–164.

Kelly J. 1987. Improving the comparability of international migration statistics: contributions by the Conference of European Statisticians from 1971 to date. *International Migration Review* 21:1017–1037.

Korcelli P. 1994. On interrelations between internal and international migration. *Innovation* 2:151–163.

Kupiszewska D and Kupiszewski M. 2005. *A revision of the traditional multiregional model to better capture international migration: the MULTIPOLES model and its applications.* CEFMR Working Paper 10/2005. Central European Forum for Migration Research, Warsaw.

Kupiszewski M. 2001. *Prognoza dynamiki przemian siły roboczej w Europie Środkowowschodniej i jej konsekwencje dla procesów integracji europejskiej – podejście regionalne.* Raport z grantu Komitetu Badań Naukowych no. 6P04E05217.

Kupiszewski M. 2002. *Modelowanie dynamiki przemian ludności w warunkach wzrostu znaczenia migracji miedzynarodowych.* Prace Geograficzne no. 181. IGiPZ PAN, Warszawa.

Kupiszewski M and Kupiszewska D. 1997. *Projection of Central and East European populations – the model, the data and preliminary results.* Working Paper 97/11. School of Geography, University of Leeds.

Kupiszewski M and Kupiszewska D. 1999. Forecasts of regional structures of the elderly populations in Central and Eastern Europe. Paper presented to the *ECE UN Conference on Status of the older population: Prelude to the 21st century*, Sion.

Kupiszewski M and Kupiszewska D. 2003. *Internal migration component in sub-national population projections in the member states of the European Union.* CEFMR Working Paper 2/2003. Central European Forum for Migration Research, Warsaw

Leslie PH. 1945. On the use of matrices in certain population mathematics. *Biometrika* 33:183–212.

Liaw K-L. 1979. *Analysis and projections of the Canadian interregional population system.* Department of Geography, McMaster University, Hamilton, Ontario.

Lutz W, Sanderson W, Scherbov S and Goujon A. 1996a. World population scenarios in the 21st century. In *The future population of the world: What can we assume today?* Revised edition, Lutz W, ed., pp. 361–396. London: Earthscan.

Lutz W, Sanderson W and Scherbov S. 1996b. Probabilistic population projections based on expert opinion. In *The future population of the world: What can we assume today?* Revised edition, Lutz W, ed., pp. 397–428. London: Earthscan.

Lutz W, Sanderson WC and Scherbov S, eds. 2004. *The end of world population growth in the 21st century: New challenges for human capital formation and sustainable development.* London: Earthscan.

Nazra S, Hewings GJD and Sonis M. 2006. An exploratory analysis of hierarchical spatial interaction: the case of regional income shares in Indonesia. *Journal of Geographical Systems* 8(3):253–268.

Netherlands Economic Institute 1986. *Long-term regional demographic developments up to the beginning of the next century and regional policy.* Rotterdam: Commission of the European Communities and Netherlands Economic Institute.

Netherlands Economic Institute. 1990. *Demographic evolution through time in European regions.* Rotterdam: Netherlands Economic Institute.

Netherlands Economic Institute. 1994. *Regional population and migration scenarios for the European Union. Part I: Two long term population scenarios.* Netherlands Economic Institute, Department of Regional and Urban Development, Department of Public Health, Erasmus University; Netherlands Interdisciplinary Demographic Institute, Rotterdam.

Nowok B and Kupiszewska D. 2005. *Official European statistics on international migration flows: Availability, sources and coverage.* CEFMR Working Paper 5/2005. Central European Forum for Migration Research, Warsaw.

Nowok B, Kupiszewska D and Poulain M. 2006. Statistics on international migration flows. In *THESIM: Towards Harmonised European Statistics on International Migration*. Poulain M, Perrin N and Singleton A, eds, pp. 203–231. Louvain-la-Neuve: Presses Universitaires de Louvain.

O'Neill BC, Balk D, Brickman M and Ezra M. 2001. A guide to global population projections. *Demographic Research* 4(8). Available at www.demographic-research.org

Pellegrini PA and Fotheringham AS. 2002. Modelling spatial choice: a review and synthesis in a migration context. *Progress in Human Geography* 26(4):487–510.

Poulain M. 1993. Confrontation des statistiques de migrations intra-européenes: vers plus d'harmonisation? *European Journal of Population* 9(4):353–381.

Poulain M. 1996. Migration between the countries of the European Union: current trends. In *Population migration in the European Union*. Rees P, Stillwell JSC, Convey A and Kupiszewski M, eds, pp. 51–66. Chichester: John Wiley & Sons, Ltd.

Poulain M, Debuisson M and Eggerickx T. 1991. *Proposals for the harmonization of European Community statistics on international migration*. Institute of Demography, Catholic University of Louvain, Louvain-la-Neuve.

Rees PH. 1984. *Spatial population analysis using movement data and accounting methods: theory, models, the 'MOVE' program and examples*. Working Paper 404. School of Geography, University of Leeds.

Rees PH. 1986. Choices in the construction of regional population projections. In *Population structures and models*. Woods RI and Rees PH, eds, pp. 126–159. London: Allen and Unwin.

Rees PH. 1989. *Old model faces new challenges: a review of the state of the art in multistate population modelling*. Working Paper 531. School of Geography, University of Leeds.

Rees PH. 1996. Projecting the national and regional populations of the European Union using migration information. In *Population migration in the European Union*. Rees PH, Stillwell JSC, Convey A and Kupiszewski M, eds, pp. 331–364. Chichester: John Wiley & Sons, Ltd.

Rees PH and Willekens F. 1981. *Data bases and accounting frameworks for IIASA's comparative migration and settlement study*. Collaborative Paper CP-81-39. International Institute for Applied Systems Analysis, Laxenburg, Austria.

Rees PH and Willekens F. 1986. Data and accounts. In *Migration and settlement. A multiregional comparative study*. Rogers A and Willekens F, eds, pp. 19–58. Dordrecht: Reidel.

Rees PH and Wilson AG. 1977. *Spatial population analysis*. London: Arnold.

Rees PH, Stillwell JSC and Convey A. 1992. *Intra-community migration and its impact on the demographic structure at the regional level*. Working Paper 92/1. School of Geography, University of Leeds.

Rees PH, Kupiszewski M, Eyre H, Wilson T and Durham H. 2001. *The evaluation of regional population projections for the European Union*. Eurostat Working Paper 3/2001/E/n 9. Eurostat, Luxembourg.

Rogers A. 1971. *Matrix methods in urban and regional analysis*. San Francisco: Holden-Day.

Rogers A. 1975. *Introduction to multiregional mathematical demography*. New York: John Wiley & Sons, Inc.

Rogers A. 1990. Requiem for the net migrant. *Geographical Analysis* 22:283–300.

Rogers A and Castro LJ. 1981. *Model migration schedules*. RR-81-30. Laxenburg, Austria: International Institute for Applied Systems Analysis.

Rogers A, Willekens F and Raymer J. 2003. Imposing age and spatial structures on inadequate migration flow datasets. *The Professional Geographer* 55(1):56–69.

Scherbov S and Grechucha V. 1988. *DIAL – A system for modelling multidimensional demographic processes.* WP-88-36. Laxenburg, Austria: International Institute for Applied Systems Analysis.

Scherbov S, Yashin A and Grechucha V. 1986. *Dialog system for modelling multidimensional demographic processes.* WP-86-029. Laxenburg, Austria: International Institute for Applied Systems Analysis.

Stillwell JCH, Eyre H and Rees PH. 1999. *Regional international migration and interregional migration within the European Union: a feasibility study.* Report ERDF 98/00/27/174 prepared for Commission of the European Communities and Eurostat.

Stone R. 1971. *A demographic accounting and model-building.* Paris: OECD.

Tabeau E, Ekamper P, Huisman C and Bosh A. 1997. Forecasting of mortality by gender, age/generation and main cause of death. Paper presented to the *Working Party on Demographic Projections*, Eurostat, Luxembourg.

United Nations. 2005. *World population prospects: 2004 revision.* New York: United Nations.

van der Gaag N and van Wissen L. 1999. The subnational distribution of immigration and migrants within European countries, and the impact of the size of migration stocks upon immigration flows. Presented at the *European Population Conference*, The Hague.

van der Gaag N and van Wissen L. 2001. Modelling regional immigration: using stocks to predict flows. *Tijdschrift voor Economische en Sociale Geografie* 92(1):27–41.

van der Gaag N and van Wissen L. 2002. Modelling regional immigration: using stocks to predict flows. *European Journal of Population* 18:387–409.

van der Gaag N, van Imhoff E and van Wissen L. 1997a. *Regional population projection in the countries of the European Economic Area. Update of the 1992 Questionnaire.* Working Paper 97/1. Netherlands Interdisciplinary Demographic Institute, The Hague.

van der Gaag N, van Imhoff E and van Wissen L. 1997b. *Long-term internal migration scenarios for the European Economic Area.* Eurostat Working Paper, E4 1997–5. Eurostat, Luxembourg.

van Imhoff E and Keilman N. 1991. *LIPRO 2.0: an application of a dynamic demographic projection model to household structure in the Netherlands.* Netherlands Interdisciplinary Demographic Institute (NIDI) and Flemish Centre for Population and Family Studies (CBGS), Publication no. 23. Lisse, Netherlands: Swets and Zeitlinger.

van Imhoff E, van Wissen L and Spiess K. 1994. *Regional population projections in the countries of the European Economic Area.* Netherlands Interdisciplinary Demographic Institute (NIDI) and Flemish Centre for Population and Family Studies (CBGS), Publication no. 31. Amsterdam: Swets and Zeitlinger.

Vu MT and Bos E. 1992. *Latin America and the Caribbean region (and North America) population projections, 1992–93 edition.* Working Paper WPS 1033. Population and Human Resources Department, World Bank.

Willekens F. 1995. *Multiregional demographic forecasting.* Working Paper 95–1. Population Research Centre, Faculty of Spatial Sciences, University of Groningen.

Willekens F. 2005. Bridging the micro–macro gap in population forecasting. Project outline proposed for Framework 6 STREP budget line.

Willekens F and Drewe P. 1984. A multiregional model for regional demographic projection. In *Demographic research and spatial policy. The Dutch experience.* ter Heide H and Willekens F, eds, pp. 309–334. London: Academic Press.

Willekens F and Rogers A. 1978. *Spatial population analysis: Methods and computer programs.* RR-78-18. Laxenburg, Austria: International Institute for Applied Systems Analysis.

Wilson T and Rees PH. 2005. Recent developments in population projection methodology: a review. *Population, Space and Place* 11:337–360.

World Bank. 2005. *World development indicators 2005.* Available at http://www.worldbank.org/data/wdi2005/wditext/Section2.htm

15

What happens when international migrants settle? Projections of ethnic groups in United Kingdom regions

Philip Rees

School of Geography, University of Leeds

15.1 Introduction

This chapter examines the demographic consequences of sustained flows of international migrants into a country and its regions. The main demographic consequence is the growth of the populations of immigrants and their descendants and, if the settled or native population has low rates of growth, the subsequent changes in ethnic composition of the population. This, in turn, leads to changes in national identity and culture. Coleman (2006a,b) has labelled this sequence of events the 'Third Demographic Transition' and reviewed population projections of a set of European countries to establish the pace of the transition.

15.1.1 Migrant and ethnic classifications

It is useful to clarify the relationships between classifications used for migrants and those used for settled population groups. In this chapter we use a classification of

---

International Migration in Europe: Data, Models and Estimates   Edited by J. Raymer and F. Willekens

'origin status', attributes of a person that are acquired through birth. People can be classified as follows: by place of birth, i.e. born in the United Kingdom (UK) or born overseas (native-born or foreign-born); by race, which is based in part on physical characteristics such as skin colour, facial form and stature (note that race is really a social construct, given that there are few genetic differences between races); and by citizenship or nationality, which conveys both rights and obligations on individuals. Other elements which are often employed to arrive at an ascribed (as opposed to achieved) classification of people include religion, language and customs, though these are attributes which can and do change during a person's lifetime.

In the UK, place of origin and racial classifications are combined to yield the official classification of ethnic groups (ONS 2003a). Place of origin refers not only to place of previous residence of international migrants but also to the origins of a person's parents, grandparents or great-grandparents. One challenge in studying ethnicity is that many people are of mixed origins and have multiple identities. National statistical offices handle this issue differently either by allowing multi-ticking of race or ethnic categories on a census or survey form or by inventing a new set of mixed categories. The UK has adopted this second approach.

A person's migration status cross-cuts ethnicity. Persons from any group can migrate into the UK or migrate out of the UK in a time interval. This demographic construction is essential for the estimation of population change, past or future. But this conflicts with the popular perception that 'immigrants' are always 'foreign'. In fact, many immigrants are British by nationality: in the 1970s, 45 % of immigrants had British citizenship; while in the 2000s, 20 % were British citizens (ONS 2006d). In this chapter, the international migrant status of all ethnic groups is estimated.

15.1.2 Ethnic groups

There is a continuous debate about the nature and meaning of 'ethnic group', which is summarised in ONS (2003a). The official definition is that an 'Ethnic group is self-assigned – that is, chosen by the respondent from a list of categories (including an "other" option)' (Large and Ghosh 2006a:22). The list of official categories recognises elements of race (colour – White, Black), world region origin (Asian, Caribbean), country origin (Indian, Pakistani, Chinese), nationality (British, Irish), often in combination. Most official analyses use the 16 summary categories employed in the 2001 Census or aggregations of these (Large and Ghosh 2006b; National Statistics 2006b). In this chapter, five summary groups are used (see Table 15.3). It should be recognised that these groups are very heterogeneous.

15.1.3 The topic and its importance

International migrants move from one country to another. Some are circulatory migrants and return to their countries of origin. Many, however, stay and settle into the population of their destination country. Because these immigrant groups and their descendants may have different demographic behaviours and geographic locations

Table 15.1 Ethnic change in England, 1981–2001.

Ethnic group	Population (1000s)			Change indices (1981 = 100)	
	1981	1991	2001	1981–1991	1981–2001
White	44 682	44 848	44 925	100.4	100.5
Black	707	917	1 286	129.7	181.9
South Asian	1 031	1 487	2 102	144.2	203.9
Chinese and other	414	626	825	151.2	199.3
All ethnic minorities	2 152	3 028	4 213	140.7	195.1
All groups	46 834	47 876	49 139	102.2	104.9

Source: Rees and Butt (2004, Table 3, p. 178); author's computations based on 1981 Census Small Area Statistics, 1991 Census Local Base Statistics, 2001 Census Key Statistics for Local Authorities, estimates from Rees and Phillips (1996), and estimates from the Linking Censuses Through Time project (Dorling *et al.* 2003, and www.lct.mimas.ac.uk). All Census data are Crown Copyright.

from the long-resident population, the structure and composition of the population is changed. So, when we project the population of ethnic groups in a country by region, as is done in this chapter, we are analysing the consequences of past international migration and the likely effects of assumptions about future flows of international migrants classified by ethnicity. We need also to monitor and project fertility and mortality differences between the groups and the different patterns of migration within the country. International immigrants concentrate in the UK's biggest cities. They have added more to the populations of cities in the southern part of the UK in the 1980s and 1990s than to the populations of northern cities. International migrants also experience redistribution through internal migration, with some signs of net shifts to suburban and metropolitan rings in the London region and of shifts from the less vigorous economies of northern cities to southern (Rees and Butt 2004). The ethnic nature of UK society is the subject of a vigorous debate, increasingly informed about the demographic processes occurring (Simpson 2004).

Table 15.1 shows how fast ethnic minority populations in England have been growing between 1981 and 2001. Without this growth the population of England would have stagnated rather than growing by 5 % over the two decades. The White British population (i.e. the major component of the White group but the least demographically dynamic) probably decreased by a small amount over the 20 years. Ethnic minorities grew by 96 % in the same period.

15.1.4 Aims of the chapter

This chapter presents a projection of five ethnic groupings of the United Kingdom population. The aims of the chapter are:

- to describe the methods and model used for population projection;
- to describe the datasets and techniques used to estimate the fertility and migration inputs to the projection;

- to present an overview of the results; and

- to compare the results with other similar projections.

In Section 15.2, the projection model for ethnic groups is described step by step. In Section 15.3, the datasets and estimation methods used to prepare inputs to the projection model are outlined. Projection assumptions are specified in Section 15.4. In Section 15.5, an overview of the main projection results is provided. These results are evaluated in Section 15.6 by comparing them with related work, either for the UK as a whole or for regions. Section 15.7 concludes by making suggestions for improvements in both projection model and inputs.

15.2   Issues and approaches to the projection of ethnic group populations

15.2.1   Trends in international migration

The UK has long been a key player in the international migration system stemming from its role as an imperial power from the 16th to the mid-20th century. In 2000, data from the United Nations (UN) suggested that the UK had 2.3 % of the world's immigrants (defined as persons living in a country outside their country of birth) and 1.8 % of the world's emigrants, ranking the country 11th and 12th respectively in terms of size of these populations (Dorling *et al.* 2006). Large and Ghosh (2006b) estimate that in 2003 some 7 million people out of 50 million in England had 'non-White British' origins (14 % of the total), while Sriskandarajah and Drew (2006) suggested that there are 5.5 million Britons living abroad (11 % of the resident UK population).

For long periods in its past, the UK was a net exporter of people, but since 1990, the balance of inflows and outflows has increasingly resulted in positive net immigration (see 'all countries of birth' graph in Figure 15.1). Since 1999, net immigration has been above 100 thousand per year. Immigration in 2004 (and 2005) was particularly high, resulting in a net immigration total of 223 thousand. In fact, net immigration between 2001 and 2005 contributed to higher rates of population increase than in any equivalent period since the 1960s (ONS 2006a). These recent increases in net immigration have been reflected in an increase of the long-term assumption in the national population projections for net migration to the UK from +130 thousand per year in the 2002-based projections to +145 thousand per year in the 2004-based projections (ONS and GAD 2006:38; Shaw 2006).

The graphs in Figure 15.1 show immigration, emigration and net international migration by country of birth for 11 years bracketing the millennium start. For the UK-born, emigrants exceed immigrants by a large margin, which has grown in the 2000s. There was a generally small net inflow of migrants born in other EU states. However, in 2004, a leap upwards occurred, associated with very large migrant labour flows from the Accession Eight countries of Central and Eastern Europe that joined the EU on 1 May 2004. Note that the inflows recorded in the

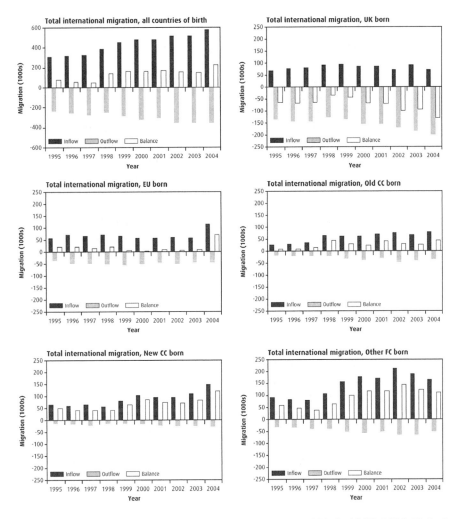

Figure 15.1 Total international migration by country of birth, 1995–2005, United Kingdom.

total international migration statistics are only a fraction (perhaps only a fifth) of the total labour migrant flows (long-term and short-term). The inward balance of migrants from the old Commonwealth (OC) countries (i.e. Australia, Canada, New Zealand and South Africa) was larger than the inflow from the EU until 2004. There is a continuing important net immigration of family members from the new Commonwealth (NC) countries (mainly India, Pakistan, Bangladesh and Nigeria). This balance has risen since 1998. The net inflow from other foreign countries (FC) origins rose between 1998 and 2002, boosted by waves of asylum seekers from failed states, but has receded as tighter controls have been implemented.

15.2.2   From international immigrants to settled UK residents

Only long-term migrants are included for the current mid-year population estimates and projections. Long-term immigrants report an intention to stay at their destination for at least 12 months or are surveyed to have been living outside the country 12 months ago. The UK's Office for National Statistics (ONS) uses a survey of incoming and outgoing travellers, the International Passenger Survey (IPS), to discover how many are immigrants or emigrants by asking a question on intention to stay at the destination for 12 months or more, following the United Nations recommendation. The key difficulty in using the IPS is the small size of the migrant sample (a few hundreds each year). For immigration, other sources are used, such as the decennial Census, which provides nearly complete coverage of migrants, but for only one year in ten, and the Labour Force Survey (LFS), which has a much larger sample of respondents than the IPS and is used to assign IPS national totals to regions. The Census and LFS use a retrospective rather than prospective question, asking where the respondent was living 12 months previously. Note that much international migration is short-term in duration (with migrants spending fewer than 12 months at their destination) but this is poorly measured (National Statistics 2006a). However, until the appropriate population accounting framework recognising the importance of short-term migrants is built, the focus will be on long-term migration.

A proportion of long-term migrants will settle in their destination country, form families or reconstruct a pre-existing but dispersed family at the destination. They may marry natives or other settled migrants and produce children who have UK nationality according to the British Nationality Act 1981 (see Home Office 2007). They will learn or improve their English, apply for and gain British nationality, and over time integrate into UK society. However, they will also retain many elements of their origin country's culture and society, speaking their mother tongue at home, practising their own religion and maintaining strong ties with their origins. They will form communities with people of a similar background. The different immigrant groups thus evolve into ethnic groups that combine features of both origin and destination communities. The speed of this process depends on a number of factors, which include the human and financial capital they bring with them, the distance between their culture and that of the UK, and host community response. The demographic profiles of immigrant communities will be rather different from that of the host community and will influence the rate of population change of settled ethnic groups.

15.2.3   Population projection models: key features and choices

Most national, regional and local population projections use the cohort-component method, which models age–sex–location-specific populations recognising the contribution of births, deaths and migration to population change. Age, sex and location all influence the intensity of fertility, mortality and migration. Age and sex are important also because users want to know about different life course groupings of the population.

There is convergence across national statistical offices and academics in how the cohort-component model is implemented with respect to age. It is agreed that all projection inputs should be decomposed to single year of age detail. There are two reasons for this. The first is that users of projections want flexibility in age aggregation for reporting and further analysis. The second is that, because the age and time intervals in a cohort-component model must be equal, and, if projected, populations are needed for each year over the projection horizon, single years of age must be used. Where projections differ is in the final age used. This has become successively older over the last two decades and needs to be 100 or more to capture anticipated increases in the oldest old populations.

There is agreement that internal (within-country) and external (international) migration must be handled separately. This is the case for the UK national projections (ONS and GAD 2006; Shaw 2006), which involve internal migration flows between the four unit populations of England, Wales, Scotland and Northern Ireland together with net international migration. The model used in this chapter distinguishes four migration streams: outmigration, emigration, inmigration and immigration. The first and third streams involve internal migration; whereas the second and fourth involve international migration.

There are many migration modelling decisions, upon which there is no consensus as to the best approach. These decisions are (15.2.3.1) the extent of region-to-region interaction, (15.2.3.2) the use of gross migration variables or net migration variables, (15.2.3.3) the use of flow measures (migrants or migrations) or flow intensities (rates or probabilities), and (15.2.3.4) the use of constraints. In addition, a decision needs to be made on what system of regions to use. This normally depends on the uses to which the projections will be put. But there is an additional consideration when projecting ethnic group populations: Are the groups large enough in the regions used for reliable estimation of inputs? In the current projections, regions with at least 1.5 million population are used, so this is not an issue. Each of the modelling issues is now briefly explained.

15.2.3.1 Region-to-region interaction

Populations can be handled as separate entities with no explicit interactions. This is called the *single-region* approach, which is recommended for ethnic projections by Simpson (2002). Populations can also be handled as interacting units: one region's outmigrants become another region's inmigrants. This is called the *multiregion* approach. A compromise is the *biregion* approach in which populations are modelled using a single region and the rest of a country, which interact, as in the multiregion. These different approaches have been evaluated by Rogers (1976) and Wilson and Bell (2004). Bijak *et al.* (2005) introduce a more sophisticated model that includes interregional migration within each of 27 European countries modelled, intercountry migration between these countries, and net migration to each country from outside the European set. This builds on a model described in Rees (1996) and discussed in Chapter 14.

15.2.3.2   Gross or net migration

Migration can be handled in the projection as a gross variable (e.g. the number of immigrants to the UK and the number of emigrants from the UK) or a net variable (e.g. the net migration to the UK, which is the difference between immigration and emigration).

15.2.3.3   Flow measures or flow intensities

The migration variable can be entered as an absolute number (e.g. the number of immigrants), as an intensity (e.g. the rate of outmigration from an origin population at risk) or as a ratio of the flow number to a population at destination (e.g. the ratio of immigrants to 'admitting' population). Table 15.2 sets out some choices. Each stream can be modelled using intensities or ratios (multiplied by a population) or as flows with the option of keeping these values constant or giving them certain trajectories.

The first model is the one used in this paper. Migrants leaving a region for either an internal or external destination are projected using migration intensities multiplied by populations at risk. The internal and external inflows are handled as counts as they do not originate from the regional population but will be determined by events in the region and populations elsewhere. The second model also handles internal inmigration using intensities and populations at risk, arguing that opportunities for inmigration are related to the size and development of the regional population. This has some logic for internal migration where there are no restrictions on moving from one region to another. The third model extends intensities to all migration streams, saying that if the region population grows so will the number of immigrants. The fourth model enters counts of migrants for the four streams. This is equivalent, if the flows are held constant, to a net migration model that uses net

Table 15.2   Models for handling migration flow variables in projections.[a]

Migration model	Internal outmigration	Emigration	Internal inmigration	Immigration
Model 1	Intensities	Intensities	Flow counts	Flow counts
Model 2	Intensities	Intensities	Intensities	Flow counts
Model 3	Intensities	Intensities	Intensities	Intensities
Model 4	Flow counts	Flow counts	Flow countss	Flow counts

[a] Intensities are either probabilities (transition case) or occurrence exposure rates (movement case).

migrant counts. However, if the flow trajectories are projected independently, then this will not be true. The flows may be projected using an explanatory migration model.

15.2.3.4 Constraints

Frequently, projections for subnational populations are adjusted to national population projections, either because the latter are felt to be more reliable or because of a need for top-down consistency (e.g. ONS 2004a). More rarely, the subnational components are also adjusted so that the two levels of projection are fully consistent. One consistency check which is occasionally employed is to verify, for a period linked to the jump-off population, that the components of change do produce the observed final population (and are consistent from an accounting point of view).

15.2.4 Critical issues for projecting ethnic group populations

When using a cohort-component model, specified by choosing one set of design features relating to migration set out above, additional issues have to be considered when projecting ethnic groups. A full discussion of these is set out in Haskey and Huxstep (2002). A choice has to be made between modelling the population of each ethnic group separately (as in this chapter) or building in group interactions. These occur principally through marriage or partnership of people of different ethnicities, followed by birth of children with parents of different ethnicities. In censuses or household surveys, parents choose to which group the child should be assigned, which in the case of the UK 2001 Census includes four mixed groups. Rees (2002) also suggests that children be allowed to re-identify at adulthood and change category. This approach has been implemented by Statistics New Zealand. Platt *et al.* (2005) have investigated trends in re-identification between the 1991 and 2001 Censuses using the Longitudinal Study (LS), but Large and Ghosh (2006a) take the view, based on their findings, that the available LS data are not reliable enough to introduce this additional transition. However, Large and Ghosh (2006a) do introduce mixed births to parents of different ethnicities into their 2001–2003 ethnic population estimates. A similar method, though using a different data source, is employed by Coleman and Scherbov (2005) in their UK ethnic projections. A method for generating mixed births is also used by Wilson (2006) in his population projections by indigenous status incorporating exogamy. He also introduces inter-ethnic mobility following Rees (2002).

To sum up, there are a number of important design decisions that need to be made in building a projection model for ethnic group populations by subnational area. However, there is no clear guidance available about which is the optimum combination of choices.

15.3   A projection model for ethnic groups at region scale

15.3.1   Choices of projection models

There are now a suite of models and examples of projections of ethnic or racial or birthplace groups available for scrutiny. These were reviewed in a recent paper (Wilson and Rees 2005:347). Coleman (2006b:Appendix I) reviews the assumptions and results of projections of the foreign-origin populations of Austria, Germany, Denmark, the Netherlands, Norway and Sweden. Naturalisation processes are built in to these projections in different ways, which converts foreign-origin populations into naturalised either explicitly through acquisition of citizenship or implicitly through birth in the country or birth to parents born in the country. The different classifications and rules for change emphasise that ethnic, racial and national groups are social constructs rather than demographic entities. A different approach is taken by Statistics New Zealand (2005), in which ethnic populations are allowed to overlap, because people are allowed to self-identify with more than one group. They are also allowed to change identities. This means that the sum of ethnic group projected populations exceeds the projected national population when the measurement allows multiple identities. Where groups are defined as mutually exclusive, then the sum of group populations should be close to the all-groups projection.

Models proposed for the UK are outlined in three chapters in *Population projections by ethnic group: A feasibility study* (Haskey and Huxstep 2002): Simpson (2002) designs a single-region model with net migration; Murphy (2002) proposes a microsimulation model; and Rees (2002) designs a multiregional model with inter-ethnic births and group change. The National Statistics ethnic population estimates finalised by Large and Ghosh (2006a,b) incorporate elements from each of these proposals, building on the case for such projections made in Haskey (2000).

There are many practical difficulties associated with implementing the Murphy or Rees models for ethnic groups for all regions within the UK. Ethnic minorities are concentrated in some regions – Inner London, Outer London, South East, West Midlands, East Midlands, North West, and Yorkshire and the Humber. Their numbers are very small in the East, South West, North East, Wales and Scotland. If a projection model using single years of age and annual intervals (the only useful framework) is used, then the number of interregional migrants in the vast majority of cells is tiny and unreliable for estimating migration intensities or flows. This small-number problem is further exacerbated if the age range used in projection is extended to 100+, which is necessary to capture the future extent of population ageing of the oldest old.

So a single-region model, the Simpson approach, was adopted for the current task. The model is a component model using period cohorts. Period cohorts are age–time spaces in which demographic components add to or subtract from the starting population. For example, the 20-year-old population at the start of a projection interval becomes the 21-year-old population one year later, having lost members through mortality, internal outmigration and emigration and having gained members through internal inmigration and immigration. For the current model all these variables must be estimated for each ethnic group. Several researchers have suggested that an ethnic group model must have a process for creating a mixed group population. Two processes have been proposed: birth of a mixed ethnicity child to parents of different ethnicities, and transitions during the life course from one self-identified ethnic group to another. Neither process is included in the current projection model given the speed with which it had to be developed and the difficulty of measuring such interactions and transitions at region scale.

15.3.2 The projection model

The cohort-component projection is an implementation of the following model for each region. For ages 0 to 100+, the population of each ethnic group and gender is projected as

$$P_{xt}^{F} = P_{xt}^{S} - s_{xt}P_{xt}^{S} - o_{xt}P_{xt}^{S} - e_{xt}P_{xt}^{S} + M_{xt} + I_{xt}, \tag{15.1}$$

where x = age at start of the time interval t. For flow components t means the interval starting with time t, i.e. t to $t+1$. The combination xt refers to a period cohort space in the classic age–time ('Lexis') diagram. The terms in Equation (15.1) have the following meanings: P_{xt}^{F} is the final (end of interval) population, P_{xt}^{S} is the starting (start of interval) population, s_{xt} is the probability of survival, o_{xt} is the probability of outmigration (within the country), e_{xt} is the probability of emigration (out of the country), M_{xt} is the number of inmigrants from elsewhere in the country, and I_{xt} is the number of immigrants from outside of the country. Note that we have used the person form of the migration variable (i.e. migrants) rather than the events form (i.e. migrations) because most of the migration data are sourced from the 2001 Census where the census question generates counts of migrants not migrations.

Prior to the start of the next time interval, final populations are transferred to become starting populations for ages 0 to 99:

$$P_{x+1,t+1}^{S} = P_{xt}^{F}. \tag{15.2}$$

For the open-ended age 100+, the final populations of the last two period cohorts are added together:

$$P_{100+,t+1}^{S} = P_{100+,t}^{F} + P_{99t}^{F}. \tag{15.3}$$

Births to each ethnic group are projected by multiplying estimated age-specific fertility rates, f_{xt}, by an average population at risk of women in the fertile ages (15 to 46 in this case):

$$B_{t} = \sum_{x=15}^{46} f_{xt} \times 0.5 \times \left(P_{xt}^{S} + P_{xt}^{F} \right). \tag{15.4}$$

Note that the age subscript, x, in this equation refers to period ages rather than period cohorts. Sex proportions (male 0.516, female 0.484) are applied to births to yield male and female births to generate the start population for the first, infant period cohort from birth to age 0, which is labelled with subscript $x = -1$. The survival and migration equation is then used for the infant period cohort:

$$P^F_{-1t} = B_t - s_{-1t}B_t - o_{-1t}B_t - e_{-1t}B_t + M_{-1t} + I_{-1t}. \tag{15.5}$$

The survival probability is derived from a UK life table adjusted to regional mortality conditions by applying a standardised mortality ratio, SMR^r,

$$s^r_{ct} = [1 - (SMR^r/100) \times (1 - s^u_{ct})], \tag{15.6}$$

where c = cohort, r = region and u = UK. This technique shifts the mortality by age schedule up or down uniformly.

15.4   Estimation of projection inputs

To drive a projection of ethnic groups, it is necessary to estimate population, mortality, fertility and migration variables specified for ethnic groups. For the whole population, the necessary demographic components are available as UK National Statistics. The same is not true for those statistics when classified by ethnicity. Estimates must be made from partial or indirect data. This section describes the methods used. We begin with a brief note on the ethnic groups adopted.

15.4.1   Ethnic groups

The projection is carried out for the five major groups identified in the outputs of the 2001 Census for England and Wales. Although there are important differences within these groups in demographic behaviour, it was felt that it would be difficult to carry out a projection for the more detailed 16 ethnic groups in the first instance. The ethnic classification for Northern Ireland as a whole is quite close to that of England and Wales but the Scotland classification is not directly comparable. Table 15.3 sets out how the schemes for Scotland and Northern Ireland were harmonised with that for England and Wales.

15.4.2   Estimation of single-year ethnic group populations

The 2001 Census standard tables provide a 22-age-group breakdown for each ethnic group (mostly five-year ages to 90+ but with more detail at the childhood ages). To achieve a single year of age classification, the proportions in each single year of age in the all-groups regional population were computed and applied to the aggregate age for each ethnicity. The assumption was that the age distribution within each broad age was the same across all ethnic groups in a region.

Table 15.3 Ethnic groups reported in the 2001 Census of population.

Ethnic groups: England and Wales	Ethnic groups: Scotland	Ethnic groups: Northern Ireland

WHITE
  White: British
  White: Irish
  White: Other
	WHITE	WHITE
		  White
		  Irish Traveller

MIXED
  Mixed: White and
    Black Caribbean
 	
  Mixed: White and
    Black African
  Mixed: White and
    Asian
  Mixed: Other
	MIXED Estimated
	  Other × England &
	    Wales proportion
	      Mixed
		MIXED

ASIAN or ASIAN
    BRITISH
  Asian or Asian British:
    Indian
  Asian or Asian British:
    Pakistani
  Asian or Asian British:
    Bangladeshi
  Asian or Asian British:
    Other
	ASIAN
	
	  Indian
	
	  Pakistani and other
		ASIAN
		
		  Indian
		
		  Pakistani
		
		  Bangladeshi
		
		  Other Asian

BLACK or BLACK
    BRITISH
  Black or Black British:
    Caribbean
 	
  Black or Black British:
    African
  Black or Black British:
    Other
	BLACK Estimated
	
	  Other × England &
	    Wales proportion
	      Black
		BLACK
		
		  Black Caribbean
		
		  Black African
		
		  Other Black

CHINESE or OTHER
 	
  Chinese
  Other
	CHINESE or OTHER
	    Estimated
	  Chinese
	  Other × England &
	    Wales proportion
	      Other
		CHINESE or OTHER
		
		  Chinese
		  Other ethnic group

15.4.3 Estimation of single-year survival probabilities

There is no direct source, except in a few local authorities with their own statistical systems, of deaths classified by ethnicity. There are ways of using local deaths information and Census ethnic group populations to impute ethnic mortality differences but the procedure would have been experimental and not verifiable. It was decided to apply the survival probabilities for all groups to each group, as others have done in the UK. The survival probabilities were generated from a life table for 1998 for the UK. This was a full single-year life table estimated from available mortality and population data with ages from 0 to 100+. Published life tables only extend to 90 or 95, which was considered too young a final age. The regional survival probabilities were computed applying published standardised mortality ratios to the national probabilities (ONS 2003c).

15.4.4 Estimation of age-specific fertility rates for ethnic groups

The published birth statistics do not provide a classification by ethnicity of mothers and fathers. A country-of-birth classification is available but these statistics will be biased as estimators of ethnic-specific births because a large proportion of the fertile age ethnic minority population was born in the UK. The age information in the 2001 Census provides one method for estimating regional and ethnic-specific fertility rates by age (f_{xe}) of mother. We computed the child–woman ratio (CWR), the ratio of children aged 0–4 to women aged 15–44, for the population of each of the five ethnic groups and for all groups. The ethnic age-specific fertility rates were computed as follows for each region:

$$f_{xe} = CWR_e / CWR_+ \times f_x, \qquad (15.7)$$

where subscript e refers to ethnic group and subscript + to all groups. A sample calculation for Inner London is shown in Appendix 1.[1]

There is a potential misspecification revealed. The CWR for the mixed group is probably much too large for mixed ethnicity women, because many of the births will be to mothers in other groups. However, the high rates are used to compensate for not having introduced a mixing process in the fertility model.

15.4.5 Migration variables

To forecast regional populations within a country, we need knowledge of two types of migration (internal and international) in two directions (inflow and outflow). The difficulty is to estimate the ethnic composition of these flows. We would

---

[1] The appendices and/or supplementary material for this chapter are available on the book's website at: http://www.wiley.com/go/raymer

have liked to handle internal migration through a multiregional model in which the outflows from one region are the inflows to another and, thus, avoid some of the logical inconsistencies of single-region projection models. However, one glance at the distribution of numbers in groups in the regions with the smallest ethnic minority populations convinced us that a full multiregional approach was infeasible. When projecting the population it is necessary to adopt an annual time interval and therefore a single year of age (period cohort) framework. Therefore, four migration streams in the projection model were used:

- internal inmigration = migration into a region from other regions in the country;

- internal outmigration = migration from a region to other regions in the country;

- immigration = migration into a region from outside the country; and

- emigration = migration from a region to destinations the country.

15.4.6   International migration datasets and trends

The UK's international migration statistics are produced by ONS in collaboration with the Home Office, which is responsible for the control of immigration. Results are published in ONS's *International migration* series, and in this chapter the volumes for 2001, 2002 and 2004 (ONS 2003b, 2004b, 2006b) are used. Results from the last Census in 2001 raised concerns about whether international migration flows had been properly estimated. A major review and re-estimation established a total international migration series from 1991 to the present (ONS 2005). There are likely to be further investments in surveys and embarkation monitoring to improve on emigration estimates in particular (National Statistics 2006a).

Two features of international migration have effects on the composition of the population in terms of ethnicity. The first is that British citizens experience a strong net outflow whereas non-British citizens experience a strong net inflow (ONS 2006b). Some 52 % of emigrants in 2005 were British citizens, while only 16 % of immigrants were returning British citizens (ONS 2006c). The second feature is the importance of student immigration. In 2004, for example, migrants entering for formal study made up 23 % of the total inflow. The continuing strong net inflow of students probably means that large numbers are settling in the UK. They contribute substantial human capital and skills to the UK labour force. In Scotland, maximising the retention of foreign graduates of Scottish universities is now a policy goal (Scottish Executive 2005). In 2004 the Chief Minister, then Jack McConnell, negotiated a derogation of UK immigration rules with the Home Secretary, then David Blunkett, so that all foreign graduates could obtain permission to reside for two years after graduation without the need for an already agreed work permit.

15.4.7 The 2001 Census of population: migration data for ethnic groups

The international migration statistics contain information about both citizenship and country of origin from which some estimates of international migration could be made. However, the 2001 Census did ask questions on migration over a one-year period (2000–2001) and on ethnicity (capturing 16 main ethnicities). This source provides detailed migration data for internal migration (into and out of regions) and on immigration by ethnicity. The standard outputs do not distinguish migrants by ethnicity but a commissioned set of expansions of standard tables have been produced that identify migration by the main ethnic groups, in a way that enables approximate standardisation across English regions, Wales, Scotland and Northern Ireland. A full list of commissioned tables is given in ONS (2007). No detailed age information is available, so migration intensities by age for the UK population as a whole are used. These commissioned tables provide information on three of the regional migration streams, leaving emigration flows by ethnicity, region and age to be estimated. There is also migration by ethnic group available in the special migration statistics from the 2001 Census, but the dataset has holes caused by differences between ethnic classifications in England and Wales, Northern Ireland and Scotland. We now describe how the migration information described above was used to estimate the input variables needed.

15.4.8 Estimation of internal migration and immigration by ethnicity

Commissioned tables M816e (Government Office Regions), M816g (Great Britain), M816f (Wales), M821b (Inner and Outer London), table UV23 (Scotland) and table S387 (Northern Ireland) were used to produce, for each region and ethnic group, tables of all age migration for three migration streams (internal outmigration, internal inmigration and immigration) captured in the 2001 Census. The first subtable in Appendix 2 shows an example for the West Midlands. Emigration estimates were derived from an annual average of the total international migration series for 2000–2003 (the second subtable of Appendix 2) and distributed to ethnic groups in the same proportions as internal outmigration. These totals were then used to control an estimate based on applying rates of internal migration and immigration by single year of age for the UK (supplied by ONS) to the single-year populations of each ethnic group in the region. The calculations are set out in the third and fourth subtables of Appendix 2 for the White population of the West Midlands. In the third subtable, national rates and flows are used to create initial estimates, which are then scaled (adjusted) to ethnic-group all-age person totals estimated from commissioned tables. The estimation equations for internal outmigration, emigration, internal inmigration and immigration are, respectively,

$$O^r_{xgt} = P^r_{xgt} \times m^u_{xg} \times \left[O^r_t / \sum_{xg} P^r_{xgt} \times m^u_{xg} \right], \qquad (15.8)$$

$$E_{xgt}^r = P_{xgt}^r \times m_{xg}^u \times \left[E_t^r / \sum_{xg} P_{xgt}^r \times i_{xg}^u \right], \tag{15.9}$$

$$M_{xgt}^r = P_{xgt}^r \times m_{xg}^u \times \left[M_t^r / \sum_{xg} P_{xgt}^r \times m_{xg}^u \right], \tag{15.10}$$

and

$$I_{xgt}^r = P_{xgt}^r \times m_{xg}^u \times \left[I_t^r / \sum_{xg} P_{xgt}^r \times i_{xg}^u \right], \tag{15.11}$$

where g = gender, m_{xg}^u is the UK internal migration rate, and i_{xg}^u is the UK immigration rate. The age subscript x refers to period cohort.

15.5   Projection assumptions

A very simple set of assumptions were adopted for these projections, given that they would need to be adjusted to the combined Government Actuary's Department (GAD) country and ONS region projections and so are experimental in nature. Fertility rates were assumed to remain constant from 2001 to 2020 at their 2001 levels. This assumption is in line with GAD assumptions for the national projections. The total period fertility rates (TPFRs) for regions and countries and ethnic groups are set out in Table 15.4, based on 2001 regional age-specific fertility and child–woman ratios based on the 2001 Census. The mixed group records apparently high fertility rates (3.82 to 5.58) but these reflect, in large part, births to mothers in other groups who are in mixed marriages or unions. Fertility rates range from 1.27 in Inner London to 1.80 in Northern Ireland for Whites, and from 1.68 in the South West to 2.41 in Inner London for Asians. The TPFRs for the Black population are low (below 1.5 in 10 out of the 13 regions/countries), with highest fertility exhibited in Inner London. The fertility levels estimated for the Chinese and other ethnic group are the lowest, perhaps reflecting both the student status of the younger Chinese population and low fertility rates in origin societies (including the People's Republic of China, with its one-child policy). The bottom two rows of Table 15.4 report unweighted averages for the five groups across regions in England and an estimate based on the 16 ethnicities used by Large and Ghosh (2006a), who used different datasets and techniques. There are considerable differences between the averages for England region TPFRs used here and the ONS estimates. Large and Ghosh have a higher White, Black, and Chinese and other TPFRs, and lower Mixed and Asian TPFRs than those estimated in this chapter. There is clearly considerable scope for further research into ethnic fertility differences, given these uncertainties.

The standardised mortality ratios representing mortality differences between regions in the UK are set out in Table 15.5. These ratios were used to modify UK mortality rates, with no differences recognised for ethnic groups. This assumption is also made by Large and Ghosh (2006a). Age-specific mortality rates were assumed

Table 15.4   Total period fertility rates assumed for the projections.

Region/country	White	Mixed	Asian	Black	Chinese and other
North East	1.55	3.82	2.44	1.39	0.90
North West	1.60	4.35	2.58	1.38	1.13
Yorkshire & Humberside	1.58	4.84	2.64	1.18	0.99
East Midlands	1.58	5.24	1.80	1.17	0.82
West Midlands	1.63	5.51	2.36	1.27	1.08
East	1.64	5.16	1.95	1.24	0.78
South East	1.61	4.62	1.80	1.03	0.70
South West	1.57	4.34	1.68	1.02	0.69
Inner London	1.27	4.27	2.41	2.07	1.02
Outer London	1.52	4.68	1.59	1.77	1.18
Scotland	1.47	4.26	2.10	1.58	1.13
Wales	1.64	3.91	2.14	1.48	1.00
Northern Ireland	1.80	5.58	2.08	1.87	1.32
Unweighted average TPFR England regions[a]	1.56	4.68	2.13	1.35	0.93
Large and Ghosh: England estimates for 2001	1.72	1.72	1.85	1.82	1.50

[a] England TPFRs: computed from Large and Ghosh (2006a), Table 15.2, across Government Office Regions.
Source: Author's computations explained in the text.

to decline at a constant 2 % rate from the 1998 base to 2001, the jump-off year for the projection, and to continue to decline thereafter. This was more optimistic than GAD national projections but probably makes little difference in the 19 years of the current projection.

The two sets of migration intensities and two sets of migration flows, described in the previous section, were assumed to be constant over the projection horizon (2001–2020). The flows by region and country and ethnic group for the year before the 2001 Census are set out in Appendix 3, so that the size and direction of the different streams can be compared. These estimates deserve a thorough analysis, which space here precludes except to point to some general trends. The net external migration column shows that only the White group loses migrants through international migration but there are regions where there are gains as well (e.g. Inner London, Scotland and South East). The internal migration pattern for all groups is one of losses from Northern England regions, and gains to Southern England outside London. The two London zones reveal quite different flow patterns. All groups experience losses from Inner London through internal migration. In Outer London, there are substantial White losses but gains in the Asian, Black, and Chinese and other groups. Note that all groups have positive inflows to the South East, the East, East Midlands and the South West (except for the Chinese and other group).

Table 15.5 Standardised mortality ratios (UK = 100) assumed for the projections.

Region/country	Male	Female
North East	113	111
North West	112	110
Yorkshire & Humberside	104	102
East Midlands	99	102
West Midlands	103	102
East	92	96
South East	91	94
South West	91	91
Inner London	100	97
Outer London	100	97
Scotland	115	115
Wales	104	105
Northern Ireland	103	103

Source: ONS (2003c).

We proceed now to use these assumptions with the model described in Section 15.4 with the understanding that the estimated components of change by ethnic group are rather approximate and deserve much more extensive investigation. So, the results should be regarded, in National Statistics terminology, as experimental projections under development, rather than as definitive forecasts.

15.6 Projection results, 2010 and 2020

The full results for three countries (Scotland, Wales and Northern Ireland) and 10 regions in England (eight England Government Office Regions (GORs) plus one GOR (Greater London) split into two parts, i.e. Inner and Outer London) are provided in the Appendix of Rees and Parsons (2006) for reference. For each area, a three-part table of population counts is given. The first part is an ethnic group by broad age table for the 2001 Census (29 April). The second part is an ethnic group by broad age table for 2010 (29 April). The third part is an ethnic group by broad age table for 2020 (29 April). Here, we describe the main features of the projections.

15.6.1 How much is each group projected to change from 2001 to 2020?

The projections for the different ethnic groups in the UK as a whole are set out in Table 15.6. The results for the 2001–2020 period are generally close to the changes

Table 15.6 Projected ethnic group populations 2010 and 2020, United Kingdom.

Age	Ethnic groups	Populations (1000s)			Change indices (2001 = 100)		Shares (percentages)		
		2001	2010	2020	2010	2020	2001	2010	2020
0–15	White	10459	9635	9444	92	90	88.2	85.6	82.9
	Mixed	336	428	570	127	169	2.8	3.8	5.0
	Asian	669	761	893	114	133	5.6	6.8	7.8
	Black	300	312	329	104	110	2.5	2.8	2.9
	Chinese and other	89	122	157	137	176	0.8	1.1	1.4
	Sum of groups	11855	11258	11393	95	96	100.0	100.0	100.0
16–59	White	32694	33384	33266	102	102	94.1	91.9	90.5
	Mixed	199	322	417	162	210	0.6	0.9	1.1
	Asian	1090	1440	1693	132	155	3.1	4.0	4.6
	Black	517	726	855	140	165	1.5	2.0	2.3
	Chinese and other	245	441	546	180	223	0.7	1.2	1.5
	Sum of groups	34745	36313	36777	105	106	100.0	100.0	100.0
60+	White	11273	12656	14399	112	128	92.1	90.1	88.5
	Mixed	140	218	319	155	227	1.1	1.6	2.0
	Asian	487	671	892	138	183	4.0	4.8	5.5
	Black	239	321	409	134	171	2.0	2.3	2.5
	Chinese and other	98	182	259	185	264	0.8	1.3	1.6
	Sum of groups	12237	14048	16278	115	133	100.0	100.0	100.0
Total	White	54426	55675	57108	102	105	92.5	90.4	88.6
	Mixed	676	968	1306	143	193	1.1	1.6	2.0
	Asian	2246	2872	3479	128	155	3.8	4.7	5.4
	Black	1056	1359	1593	129	151	1.8	2.2	2.5
	Chinese and other	432	745	963	172	223	0.7	1.2	1.5
	Sum of groups	58837	61619	64,449	105	110	100.0	100.0	100.0

Source: Author's computations.

in the 1981–2001 period reported earlier, though the comparisons are not exact because the Rees and Butt (2004) analysis was for England only and did not use a mixed group.

The White population, as a whole, increases by 5 %, more than in the earlier period. Substantial increases occur at old ages and decreases in the childhood ages. The fastest growing population is the Chinese and other group, where the older ages exhibit dramatic increases. The Mixed group grows substantially in all ages. The rates of growth in the Asian and Black groups slow significantly from the earlier period. The Asian group average probably hides greater growth in the Pakistani and Bangladeshi groups and lesser growth in the Indian group. The Black group average hides greater growth in the Black African population and slow growth in the Black Caribbean group. Note that all groups experience the greatest increases in the older ages.

The changing ethnic composition of the population is laid out in the right-hand columns of Table 15.6. Overall, some 11.4 % of the UK population will be ethnic minority by 2020. For children the ethnic minority share will be 17.1 %, in the labour force ages the share will be 9.5 %, while the elderly share will be only 11.5 %, which reflects the movement into retirement ages of migrants who entered the UK in the 1960s and 1970s. The differences in ethnic mix by age show that this process of transition from a largely White population will continue well beyond 2020, as a result of demographic momentum, even if all net immigration ceased.

15.6.2 How do these results vary across regions of the UK?

Appendix 4 provides a summary of the projections across the 13 regions. The regions mirror the UK picture and the well known difference in growth rates between peripheral and core regions. Scotland is projected to lose population, but its ethnic minority population will nevertheless increase, though generally at a lower rate than elsewhere.

Profound changes in the ethnic composition of the populations of UK regions and countries are in prospect. Figure 15.2 presents results for Whites, the largest group, and Asians, the second largest. Cartograms are used in which the area on the map is proportional to the size of the area's population. The key map shows the links between cartogram areas and a more conventional map of UK regions. A population cartogram is used because it weights regions more fairly. Note that an average of the Inner and Outer London ethnic group shares is used to represent the London region in the cartogram.

The maps of the White share of the population show substantial decreases in 2020 compared with 2001. London's White population share will have fallen below two-thirds and the White share in the West Midlands will be 83 %. In the three Northern England regions, White shares will be around 90 %, and in the South East and East GORs between 91 % and 92 %. In the other UK regions and countries, the White share remains above 95 %. A converse process is apparent in the Asian map. The Asian share rises substantially in all regions. In London it reaches 15 %

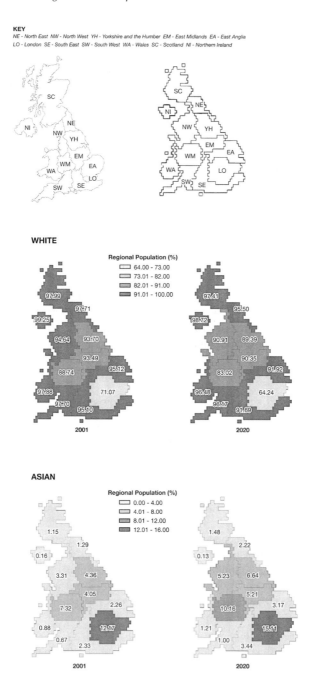

Figure 15.2 Percentages of the population who are White and Asian in 2001 and 2020.

Source: Author's computations and Rees and Parsons (2006).

and in the West Midlands 10 % by 2020. All the percentage increases are smaller in the other regions. In relative terms, they are higher, indicating a more dispersed regional distribution of Asian British by 2020.

One interesting and surprising result is the contrast in projected changes between Inner and Outer London. Inner London is projected to increase by 17 % in population by 2020 and the White population growth at 15 % almost matches this. The Black population of Inner London is forecast to decline. Outer London has lower growth at 12 % and the White population declines by 4 %. Outer London is set to take over from Inner London as the most important region for ethnic minorities. More detailed figures by age for the three regions of South East England are presented in Appendix 5. Note that the greatest gains by ethnic minorities will be in the South East GOR. Ethnic minorities are de-concentrating from the metropolitan core, just as did Whites until recently.

15.7 Comparisons, evaluations and adjustments

Recall that the projection results for ethnic groups were subject to an adjustment that forced them to agree with the 2004-based GAD and ONS projections of the 13 areas. How much impact did the adjustments have? In Table 15.7, some relevant results are assembled for two regions, the North East and Inner London, representing opposite ends of the ethnic diversity continuum. The results show the complexity of demographic processes affecting regional populations of the different ethnic groups. Note that a natural increase only projection was carried out for comparative purposes. (See also Appendix 6 for a comparison with an alternative ethnic projection.)

In the *North East*, the natural increase projection for Whites yields a higher population because Whites experience losses due to both internal and external migration (Appendix 3). For the other groups, the natural increase projection results in smaller numbers because of positive gains from internal (relatively minor) and external migration. When the unconstrained and constrained migration model 1 projections are compared, the constrained results project higher White, Mixed and Asian populations, but marginally lower Black and Chinese and other populations. The increases are due, in part, to higher immigration assumptions.

In *Inner London*, different outcomes are evident. Under the natural increase projections, the White population is lower than in the projections that include migration. This is because there are, on balance, small gains through migration. The net gain from external migration exceeds the net loss from internal migration (Appendix 3). For the Mixed, Asian and Black groups, the migration-based projected populations are lower than the natural increase projections because of net losses, mainly through internal migration not counterbalanced by external migration gains. For the Chinese and other group, the migration-based projection produces much higher numbers, as net immigration is the highest contributor to growth for this group. The results of the constrained migration model are varied between groups: White and Black groups experience increases (probably because of their older age structure), while the projected populations for other groups are reduced (Table 15.7).

Table 15.7 Comparison of natural increase, migration model 1 and projections adjusted to GAD/ONS 2004-based projections, North East and Inner London.[a,b]

Region and Ethnic group	2010			2020		
	Natural increase	Migration model 1		Natural increase	Migration model 1	
		Unconstrained	Constrained		Unconstrained	Constrained
North East						
White	2 423.6	2 392.4	2 449.4	2 391.4	2 339.9	2 419.8
Mixed	15.6	17.3	18.3	20.7	23.0	24.6
Asian	39.2	44.1	44.2	45.2	54.4	56.2
Black	4.2	6.3	6.0	4.3	7.8	7.7
Chinese and other	10.8	20.1	19.0	11.0	26.5	25.5
All groups	2 493.3	2 480.5	2 536.8	2 472.6	2 451.6	2 533.9
Inner London						
White	1 877.3	1 971.8	2 031.2	1 870.8	2 146.4	2 198.9
Mixed	145.5	129.6	125.0	192.0	145.7	143.4
Asian	348.4	341.2	335.3	404.3	387.0	383.9
Black	520.3	458.4	478.5	575.5	453.7	475.5
Chinese and other	100.1	125.3	124.4	103.6	149.0	145.6
All groups	2 991.6	3 026.4	3 094.4	3 146.2	3 281.7	3 347.3

[a] The constrained projections are adjusted to agree with the GAD 2004-based projections and the 2003-based ONS GOR projections adjusted to the GAD 2004-based England projections.
[b] Populations in 1000s.
Source: Author's computations.

Space precludes an exhaustive analysis here of the interplay of growth components, constraints, regions and countries, and ethnic groups, but the discussion of just two of 13 regions and countries has demonstrated the complexity of the demographic dynamics.

15.8   Lessons and further research

This chapter has reported on some recent projections of ethnicity in the regions and countries of the United Kingdom. The results are preliminary but interesting. We now discuss areas where there is agreement, areas where there are disagreements that need resolution and areas where further research needs to be done.

15.8.1   Areas where there is agreement

Ethnic group projections are now feasible. It is now possible to build a population projection model for ethnic groups and to estimate the inputs required. The most sophisticated model is that built by Large and Ghosh (2006a,b) for the local authority populations of England, although the model has only, to date, been applied to estimation forward from the year of the latest Census and not to projection. Coleman and Scherbov (2005) have built an ethnic population projection model for the UK and used probabilistic techniques to establish uncertainty ranges for their projections. In the current chapter, we have built a model that forecasts regional and country ethnic-group populations for the whole UK, using a somewhat simpler model than that of Large and Ghosh (2006a). Ethnic minority populations (non-White groups in the projections of this chapter and Coleman and Scherbov, non-White British in the Large and Ghosh analysis) will continue to grow much faster than the White population and the ethnic mix of all UK regions will change towards *greater diversity*. This extra growth is linked, historically, to immigration in prior decades, which has created age structures for immigrant-origin ethnic groups that are favourable to growth. However, with the passage of time, this *demographic momentum* effect disappears as a result of both demographic ageing and integration (changes in self-identification and mixing). The growth is also linked to fertility rates above the national average for some, but not all, ethnic minority groups.

Chinese and others, which includes a variety of groups including recent asylum seekers, will experience greatest growth, both in the short term (Large and Ghosh 2006b:9 Table 1) and in the medium term (the current projection). This growth is fuelled by the projected continuation of high net international immigration. The *Mixed* group will also grow fast but this will mainly be through mixed marriages and cohabitations leading to mixed offspring. The method used to project the mixed group in the current projections is not optimal but gave results comparable in magnitude to those of Coleman and Scherbov (2005). The *Asian* and *Black* groups will also grow substantially, although there is disagreement between the two sets of projections as to their ranking. This may be due to the exact composition of the broader groups adopted in the projections. Both groupings contain ethnic groups with higher and with lower rates of growth (e.g. the Black Caribbean group experiences low net or negative immigration and fertility; the Black African group experiences high net immigration and higher than average fertility). Finally, there is consensus that the *White* population will experience the lowest growth, but disagreement about whether this will become negative (Coleman and Scherbov 2005) or remain slightly positive (the current projection). The contribution in future of the White non-British population will be crucial. Large and Ghosh (2006b) estimate that between 2001 and 2003 the White British and White Irish populations are decreasing, while the White Other population has grown by 3.3 %.

There is agreement that *international migration* makes a substantial difference to population growth of ethnic minority groups. However, the current projection, which largely relies on estimation from 2001 Census sources, appears to be underestimating net international inflows in 2001. This underestimation is partially corrected through adjustment of the projections to GAD and ONS national and regional all-groups projections. There appears to be no consensus about how international migration flows are distributed by ethnicity. Further careful investigation is indicated.

15.8.2   Areas where there is disagreement

Comparing these three pieces of work (Large and Ghosh 2006a,b; Coleman and Scherbov 2005; and the current projection), it is clear that there is much disagreement about the levels of fertility and international migration associated with each ethnic group. There seems to be little consensus about the levels of *ethnic group fertility* at the start of the projections, which is unsurprising given the different datasets and methods used. The Large and Ghosh and current analyses indicate that differences between ethnic groups in fertility levels have shrunk to quite small levels. The Coleman and Scherbov analysis and the current projections report larger differences (see Appendix 6). Further careful investigation is indicated.

15.8.3   Areas that need further study

There is need for *fuller exposition of the data sources and methods* used in the estimation steps in each projection, so that the best method can be selected. A lesson which we can learn from the projections described in this chapter and from comparisons with other estimates and projections is that the 'devil is in the detail'. Full exposition of the input estimations and model projections is necessary to make progress in improving our understanding of the complex demographic processes, through which international migrants become settled populations, and so change their host societies.

Development of ethnic-specific mortality estimates is needed. There is a need to investigate whether ethnic-group mortality rates can be reliably estimated. None of the estimations or projections attempted this, although we know from health studies that there are considerable differences, linked in part to socio-economic position of each ethnic group. Only when we are confident of reliable estimates of the components of change will it possible to use the projection models sensibly to explore different scenarios and to establish uncertainty ranges.

There is no explicit consideration in either of the projections of the impact of the influx of *migrants from newly joined European Union Member States* (the 2004 Accession Countries, and Romania and Bulgaria in 2007). The influx of migrants from Central and Eastern Europe has been very important in the 2004–2007 period but we do not have definitive statistics about those flows (see National Statistics 2006a; Rees and Boden 2006; Salt and Rees 2006; ONS 2006c).

One of the most neglected variables in national, regional and local projections is *emigration*. There is only one source for this flow, the very small International Passenger Survey migrant sample. Regional estimates are based on very small samples of respondents indeed and local estimates are only 'guestimated'. Two recent estimates of the stock of emigrants from Britain living in the rest of the world were 3.1 million in 2000 (Rees 2006, based on Dorling *et al.* 2006 and UN data) and 5.5 million in 2005 (Sriskandarajah and Drew 2006). The gap is an indication of the level of uncertainty in this crucial flow term.

Acknowledgements and disclaimer

This chapter summarises the results of research carried out for the Joseph Rowntree Foundation in connection with a suite of projects related to Child Poverty by Phil Rees and John Parsons of the University of Leeds (Rees and Parsons 2006). The support of the Joseph Rowntree Foundation is gratefully acknowledged. However, the results and conclusions of the current paper are the responsibility of the author.

References

Bijak J, Kupiszewska D, Kupiszewski M and Saczuk K. 2005. *Impact of international migration on population dynamics and labour force resources in Europe.* CEFMR Working Paper 1/2005. Central European Forum for Migration Research, Warsaw. Available at http://www.cefmr.pan.pl/docs/cefmr_wp_2005-01.pdf (accessed 15 July 2005).

Coleman D. 2006a. The European demographic future: determinants, dimensions and challenges. In *The political economy of global population change, 1950–2050.* Demeny P and McNicoll G, eds. New York: Population Council; *Population and Development Review* 32(Suppl.):52–95.

Coleman D. 2006b. Immigration and ethnic change in low-fertility countries: a third demographic transition. *Population and Development Review* 32(3):401–446.

Coleman D and Scherbov S. 2005. Immigration and ethnic change in low-fertility countries – towards a new demographic transition? Presented at the *Population Association of America Annual Meeting*, Philadelphia. Available at http://www.apsoc.ox.ac.uk/Oxpop/publications%20files/WP29.pdf (accessed 14 June 2005).

Dorling D, Martin D and Mitchell R. 2003. *Linking Censuses through time project website.* Available at http://www.census.ac.uk/cdu/software/lct/ (accessed 5 January 2007).

Dorling D, Newman M, Allsopp G, Barford A, Wheeler B and Pritchard J. 2006. *Worldmapper: the world as you've never seen it before.* Available at http://www.sasi. group.shef.ac.uk/worldmapper/

Haskey J. 2000. Projections of the population by ethnic group: a sufficiently interesting or a definitely necessary exercise to undertake? *Population Trends* 102:34–40.

Haskey J and Huxstep S. eds. 2002. *Population projections by ethnic group: A feasibility study.* ONS Studies in Medical and Population Topics, SMPS no. 67. London: The Stationery Office.

Home Office. 2007. *Chapter 9: Registration of minors at discretion.* Section 3(1) British Nationality Act 1981. Immigration and Nationality Directorate. Available at http://www.ind.homeoffice.gov.uk/documents/nichapter9/

Large P and Ghosh K. 2006a. A methodology for estimating the population by ethnic group for areas within England. *Population Trends* 123:21–31.

Large P and Ghosh K. 2006b. Estimates of the population by ethnic group for areas within England. *Population Trends* 124:8–17.

Murphy M. 2002. Consistency in considerations for population projections by ethnic group, and the possible role of microsimulation models. In *Population projections by ethnic group: A feasibility study.* Haskey J and Huxstep S, eds, pp. 17–26. ONS Studies in Medical and Population Topics, SMPS no. 67. London: The Stationery Office. Available at http://www.statistics.gov.uk/downloads/theme_population/SMPS_67_v2.pdf (accessed 1 January 2007).

National Statistics. 2006a. *Report of the inter-departmental task force on migration statistics.* Available at http://www.statistics.gov.uk/about/data/methodology/specific/population/future/imps/updates/downloads/TaskForceReport151206.pdf (accessed 1 January 2007).

National Statistics. 2006b. *Focus on ethnicity and religion. 2006 Edition.* Dobbs J, Green H and Zealey L, eds. London: Office for National Statistics. Available at http://www.statistics.gov.uk/downloads/theme_compendia/foer2006/FoER_Main.pdf (accessed 1 January 2007).

ONS. 2003a. *Ethnic group statistics: A guide for the collection and classification of ethnicity data.* London: National Statistics. Available at http://www.statistics.gov.uk/about/ethnic_group_statistics/downloads/ethnic_group_statistics.pdf (accessed 31 December 2006).

ONS. 2003b. *International migration: Migrants entering or leaving the United Kingdom and England and Wales, 2001.* Series MN no. 28. London: Office for National Statistics. Available at http://www.statistics.gov.uk/ as MN28.pdf (accessed 29 December 2006).

ONS. 2003c. *Key population and vital statistics, local and health authority areas. Population and vital statistics by area of usual residence in the United Kingdom, 2001.* Series VS no. 28, PP1 no. 24. London: Office for National Statistics. Available at http://www.statistics.gov.uk/downloads/theme_population/KPVS28_2001/KPVS28_PP124.pdf (accessed 10 February 2007).

ONS. 2004a. *2003-based subnational population projections for England: Data and methodology issues.* London: Office for National Statistics. Available at http://www.statistics.gov.uk/STATBASE/Product.asp?vlnk=997 as 2003snpissues.pdf (accessed 29 December 2006).

ONS. 2004b. *International migration: Migrants entering or leaving the United Kingdom and England and Wales, 2002.* Series MN no. 29. London: Office for National Statistics. Available at http://www.statistics.gov.uk/ as MN_no_29_v3.pdf (accessed 29 December 2006).

ONS. 2005. *International migration: Migrants entering or leaving the United Kingdom and England and Wales, 2003.* Series MN no. 30. London: Office for National Statistics. Available at http://www.statistics.gov.uk/ as MN_no_30v3.pdf (accessed 29 December 2006).

ONS. 2006a. Table 1.6, Components of population change. *Population Trends* 125:58.

ONS. 2006b. *International migration: Migrants entering or leaving the United Kingdom and England and Wales, 2004.* Series MN no. 31. London: Office for National Statistics. Available at http://www.statistics.gov.uk/downloads/theme_population/MN31.pdf (accessed 29 December 2006).

ONS. 2006c. *Over 500 a day gained through migration to the UK.* London: Office for National Statistics Available at http://www.statistics.gov.uk/pdfdir/intmigrat1106.pdf (accessed 29 December 2006).

ONS. 2006d. Table 7.3, International migration: Citizenship. *Population Trends* 126:53.

ONS. 2007. *Commissioned output, commissioned output table finder.* London: Office for National Statistics. Available at http://www.statistics.gov.uk/census2001/ commissioned.asp (accessed 10 February 2007).

ONS and GAD. 2006. *National population projections, 2004-based.* Series PP2 no. 25. Basingstoke: Palgrave Macmillan. Available at http://www.statistics.gov.uk/downloads/theme_population/PP2_No25.pdf (accessed 29 December 2006).

Platt L, Simpson L and Akinwale B. 2005. Stability and change in ethnic groups in England and Wales: trends, outcomes and interpretation. *Population Trends* 121:35–36.

Rees P. 1996. Projecting the national and regional populations of the European Union using migration information. In *Population migration in the European Union*. Rees P, Stillwell J, Convey A and Kupiszewski M, eds. pp. 331–364. Chichester: John Wiley & Sons, Ltd.

Rees P. 2002. New models for projecting UK ethnic group populations at national and subnational scales. In *Population projections by ethnic group: A feasibility study*. Haskey J and Huxstep S, eds. pp. 17–26. ONS Studies in Medical and Population Topics, SMPS no. 67. London: The Stationery Office. Available at http://www.statistics.gov.uk/downloads/theme_population/SMPS_67_v2.pdf (accessed 1 January 2007).

Rees P. 2006. What is happening to the UK population in a global mobility system? Presented at the *ESRC/ONS Public Policy Seminar Series, Demographic change and government policy*, Third Seminar on *Globalisation, population mobility and the impact of migration on population*. London: Royal Statistical Society. Available at http://www.esrc.ac.uk/ESRCInfoCentre/Images/Professor%20Phil%20Rees%20-%20Presentation_tcm6-16198.ppt

Rees P and Boden P. 2006. *Estimating London's new migrant population. Stage 1 – review of methodology*. London: Greater London Authority. Available at http://www.london.gov.uk/mayor/refugees/docs/nm-pop.pdf (accessed 5 January 2007).

Rees P and Butt F. 2004. Ethnic change and diversity in England, 1981–2001. *Area* 36(2):174–186.

Rees P and Parsons J. 2006. *Socio-demographic scenarios for children to 2020*. York: Joseph Rowntree Foundation. Available at http://www.jrf.org.uk/bookshop/details.asp?pubID=809 (accessed 29 December 2006).

Rees P and Phillips D. 1996. Geographical spread: the national picture. In *Ethnicity in the 1991 Census:* Volume Three. *Social geography and ethnicity in Britain: Geographical spread, spatial concentration and internal migration*. Ratcliffe P, ed., pp. 23–109. London: HMSO.

Rogers A. 1976. Shrinking large-scale population projection models by aggregation and decomposition. *Environment and Planning A* 8:515–541.

Salt J and Rees P. 2006. Globalisation, population mobility and impacts of migration on population. *ESRC Seminar Series, Mapping the public policy landscape*. Swindon: ESRC. Available at http://www.esrc.ac.uk/ESRCInfoCentre/Images/ESRC_Seminar_Global_tcm6-16062.pdf (accessed 29 December 2006).

Scottish Executive. 2005. *Fresh talent: Working in Scotland*. Available at http://www.scotland.gov.uk/News/Releases/2005/06/16102727 (accessed 2 January 2007).

Shaw C. 2006. 2004-based national population projections for the UK and constituent countries. *Population Trends* 123:9–20.

Simpson L. 2002 Estimating methodologies and assumptions for projections: current practice for population projections by ethnic group. In *Population projections by ethnic group: A feasibility study*. Haskey J and Huxstep S, eds, pp. 53–72. ONS Studies in Medical and Population Topics, SMPS no. 67. London: The Stationery Office. Available at http://www.statistics.gov.uk/downloads/theme_population/SMPS_67_v2.pdf (accessed 1 January 2007).

Simpson S. 2004. Statistics of racial segregation: measures, evidence and policy. *Urban Studies* 41(3):661–681.

Sriskandarajah D and Drew C. 2006 *Brits abroad: Mapping the scale and nature of British emigration*. London: Institute for Public Policy Research. Available at http://www.ippr.org/publicationsandreports/publication.asp?id=509 (accessed 1 January 2007).

Statistics New Zealand. 2005. *National Ethnic Population Projections 2001(base)–2021 update.* Available at http://www.stats.govt.nz/NR/rdonlyres/D9A5C6F4-ABB3-4C6D-BC9C-8A94C58FA98C/0/nationalethnicpopulationprojections01base21updatehotp.pdf (accessed 2 January 2007).

Wilson T. 2006. Population projections by indigenous status incorporating exogamy and inter-ethnic mobility: a case study of the Northern Territory, Australia. Presented at the *European Population Conference*, University of Liverpool.

Wilson T and Bell M. 2004. Comparative empirical evaluations of internal migration models in subnational population projections. *Journal of Population Research* 21(2):127–160.

Wilson T and Rees P. 2005. Recent developments in population projection methodology: a review. *Population, Space and Place* 11:337–360.

16

Conclusion

Frans Willekens[†] and James Raymer[‡]

[†] *Netherlands Interdisciplinary Demographic Institute, The Hague*
[‡] *Division of Social Statistics, School of Social Sciences, University of Southampton*

16.1 Early concerns

In Europe, international migration and asylum-seeking are high on the political agenda. The debates are often heated but the empirical evidence is weak. The European Commission has increasingly realised the need for better information on international migration for purposes of demographic forecasting and migration policy-making.

In 1992, the Maastricht Treaty established the European Union (EU) and, in the same year, the agreement was signed creating the European Economic Area (EEA). The EEA comprises the EU and countries of the European Free Trade Association (EFTA) (except Switzerland). In anticipation of these historic developments, Eurostat organised the international conference *Europe at the Dawn of the 21st Century* in November 1991. For this conference, a team from Eurostat, under the supervision of Harri Cruijsen, cooperated with the Population Research Centre of the University of Groningen in the Netherlands to prepare two long-term demographic scenarios for the European Community (EC), which at that time consisted of 12 countries. It was the first set of demographic scenarios ever prepared by Eurostat. In order to assess the impact of migration, the original intention was to consider, for every country of Europe, migration flows by age, sex and country of destination.

International Migration in Europe: Data, Models and Estimates   Edited by J. Raymer and F. Willekens
© 2008 John Wiley & Sons, Ltd

A multiregional model was to be applied to determine the demographic effects of international migration. However, destination-specific migration rates could not reliably be estimated because of missing and inconsistent data problems. Therefore, it was decided to use net migration. It was a first manifestation of the consequences of the inadequacy of migration flow statistics. The long-term scenarios are included in the background documents of the conference (Eurostat 1991). In 1992, Eurostat decided to extend the study to the countries of the EFTA, several of which were expected to join the European Union (Eurostat 1993; Extercate 1993). At the request of the Directorate-General for Economic and Financial Affairs of the European Commission, the projections were updated in 1995 for pension expenditure projections, using the same methodology (Eding *et al.* 1996; Franco and Munzi 1996). Although migration was an issue, the data were inadequate for accurately quantifying international migration flows in Europe for projection purposes. Today, in 2007, the data situation has not changed much. The most recent long-term population projections prepared by Eurostat (EUROPOP 2004) continue to rely on net migrations (Eurostat 2006). Robert Schuman's vision that 'Europe will not be built at a stroke, nor constructed in accordance with some overall plan; it will be built on concrete achievement' certainly applies to international migration statistics.

The inadequacy of migration statistics is not due to a lack of interest. The measurement of international migration flows in Europe has a long history. For decades the Conference of European Statisticians (CES), a subsidiary body of the United Nations Economic Commission for Europe (UNECE) and the United Nations Statistical Commission had worked towards the improvement and comparability of international migration statistics. As early as 1971, the United Nations *European Seminar on Demographic Statistics*, convened in Ankara and Istanbul, identified serious shortcomings in the statistics of immigration and emigration in that they differed considerably in scope, coverage, definitions, classifications and content. The seminar participants concluded that the improvement and harmonisation of statistics on international migration was an urgent task and recommended that the CES include these activities in its programme of work. It was suggested that the work include tabulations on flows and stocks of migrants.

The CES was also requested to organise an exchange of statistics on international migration among UNECE countries. In 1975, flow data were collected pertaining to the year 1972. The 1972 matrix highlighted the serious lack of comparability in the statistics on international migration flows and demonstrated that in many cases the figures for a particular flow reported by the country of immigration were substantially higher than the figures for the same flow reported by the country of emigration. The total reported number of immigrants was 57 % greater than that of emigrants. The 1972 matrix was published in the *United Nations, Demographic Yearbook 1977* (United Nations 1978). In 1978, the UNECE Secretariat collected statistics for the years 1973–1975 and requested countries to provide data compiled not only on the basis of existing national practices but also adjusted, where necessary, to correspond as far as possible with the definitions of long-term immigrants and emigrants recommended by the United Nations in 1976. In 1979, the UNECE Secretariat started producing migration matrices on a regular basis. The matrices are double entry matrices that show immigration statistics by country of last residence and emigration statistics by country of intended destination. Two double entry matrices were produced, one showing data based on national practices and the other showing data adjusted to correspond as far as possible with the United Nations recommended definition of long-term immigrant and emigrant. Although the data were collected on a regular basis, they were not published regularly. The second international migration matrix published in the *United Nations, Demographic Yearbook* was in 1989. Owing to the lack of resources, the data were not published subsequently (Herm 2006:91). The CES discontinued the collection of the double entry matrix in the mid-1990s.

Eurostat's interest in comparable migration statistics increased with the accession of Spain and Portugal in 1986. The need to coordinate migration statistics in Europe became apparent. In 1989, Eurostat took the initiative to conduct a study on the harmonisation of international migration statistics in the 12 countries of the EC. Poulain and colleagues at the Université Catholique de Louvain (UCL) in Belgium carried out the study and found the double entry matrices useful tools for assessing the comparability of migration statistics (Poulain *et al.* 1990). The study was extended to EFTA countries in 1991 (Poulain and Gisser 1992) and later on to Central European and Baltic countries (Poulain 1997). Additional investigations were carried out to improve international migration data in the EU, using all information included in the double entry matrices with the aim to propose reliable estimation for all cells of the complete intra-EU migration matrix (Poulain 1993, 1999). Moreover, during the first half of the 1990s, Eurostat and UNECE began cooperating on the development of a joint programme of data collection on international migration statistics for EU and EFTA countries. Later, other international organisations joined in a concerted data collection effort (for an overview, see Herm 2006).

16.2 More recent concerns

The production of statistics on international migration entered a new stage with the massive expansion of the scope of EU migration policy.[1] The Treaty of Maastricht, which was signed in 1992 and entered into force in November 1993, made immigration an issue of common interest for the EU. An EU migration policy needs data to effectively plan, evaluate and monitor policies. The European Commission has increasingly realised the need for better information on migration. In 1994, the Commission issued a Communication on Immigration and Asylum Policies stressing that the management of immigration was one of three essential elements of a comprehensive and effective immigration policy. It also stressed the need for accurate information and suggested the establishment of a European Migration Observatory (Kraler *et al.* 2006:46).

---

[1] This section is based on Kraler *et al.* (2006). The authors provide a comprehensive overview of the evolution of the EU migration policy and the implications for data collections.

According to the Amsterdam Treaty, adopted in 1997, the European Community is responsible for laying down uniform procedures for the management of international migration and for the production of community statistics.[2] The latter is specified in the Council Regulation (EC) 322/97 of 17 February 1997 on community statistics.[3] The role of Eurostat is laid down in European Commission Decision 97/281/EC of 21 April 1997. The Tampere European Council of 1999 and the Seville European Council of 2002 reiterated the need to develop a common EU policy on migration and asylum and a common approach to the management of migration flows *at all their stages*. Several measures have been introduced since the Treaty of Amsterdam that aimed to remove the deficiencies in migration data. They include the establishment of a European Migration Network (EMN), the establishment of funding lines within the framework programmes of the Director-General for Research and numerous research projects on migration policies by the Director-General for Justice and Home Affairs. The EMN was set up in 2002 as a pilot project to offer a single reference point for the coordination of the huge amount of information on migration issues being produced throughout the EU. The EMN is a virtual information exchange network on migration and asylum in Europe and provides for the networking of national focal points for exchanging and following up information relating to the political, economic, demographic and social dimensions of migratory phenomena (see http://www.european-migration-network.org/).[4]

In 2003, the European Council of Thessaloniki again ascribed top political priority to migration. The Council concluded that more effective mechanisms were needed for the collection and analysis of EU-wide information on migration and asylum. The Commission issued a Communication detailing an Action Plan for the collection and analysis of Community statistics in the field of migration. The Action Plan is implemented in part by Eurostat, and also included actions leading to the adoption of Community legislation on migration statistics. The Action Plan and the envisaged legislation are geared towards achieving comparable migration data at the European level. Harmonised and comparable Community statistics on migration and asylum are essential for the development and monitoring of Community legislation and policies relating to immigration and asylum, and to the free movement of persons. Member States differ greatly in terms of how migrants are defined, the sources of migration data, and how migration statistics are produced. The European Parliament in its resolution of 6 November 2003 concluded that further progress towards improving migration statistics would require legislation.

---

[2] The Amsterdam Treaty laid down the production of community statistics in paragraph 285 and specified it in the Council Regulation (EC) 322/97 of 17 February 1997 on community statistics. The role of Eurostat as community institution involved in the production of community statistics is laid down in European Commission Decision 97/281/EC of 21 April 1997. Paragraph 285 EC (Article 312a) of the Amsterdam Treaty states that the production of community statistics is made on the basis of impartiality, reliability, objectivity, scientific independence, cost effectiveness and statistical confidentiality: it may not pose an excessive burden on the economic actors. See http://www.europarl.europa.eu/topics/treaty/pdf/amst-en.pdf

[3] For the text of the Council Regulation (EC) 322/97 of 17 February 1997 see http://europa.eu.int/smartapi/cgi/sga_doc?smartapi!celexapi!prod!CELEXnumdoc& lg=EN&numdoc=31997R0322&model=guichett

[4] A public debate on the future role of the EMN was launched by a green paper issued in November 2005 (http://eur-lex.europa.eu/LexUriServ/site/en/com/2005/com2005_0606en01.pdf).

The proposed legislation was prepared jointly by Eurostat and the Directorate-General for Justice, Freedom and Security and involved almost two years of technical consultation (Thorogood 2006). On 14 September 2005, the European Commission adopted the proposal for the 'Regulation of the European Parliament and the Council on Community statistics on migration and international protection'.[5] The objective of the proposed Regulation is to establish a common framework for the collection of Community statistics on international migration and asylum. The systematic production of harmonised Community statistics on international migration and asylum cannot be sufficiently achieved by the Member States acting individually. Therefore, in accordance with the principle of subsidiarity, as set out in Article 5 of the EC Treaty, the Community should take charge. The proposal states that the statistics to be collected under the proposed legislation will, as far as possible, be in accordance with the United Nations recommendations on statistics of international migration. The following statistics will be covered by the Regulation: international migration, usually resident population, acquisition of citizenship, international protection (including asylum), residence permits, returns and prevention of illegal entry and stay (apprehensions and refusals at the border). The proposal obliges Member States to make the best use of available data and to produce statistics that are comparable across Europe. Member States are not required to introduce completely new data sources or change the administrative system for immigration and asylum, as they are considered to be competent in the organisation and operation of their national statistical systems. In accordance with the principle of proportionality, the Regulation confines itself to the minimum required to achieve the objective of harmonised Community statistics on migration and asylum, and does not go beyond what is necessary for that purpose. The proposal relates to the data to be supplied to the Commission (Eurostat). It does not legislate for a specific data source to be used in every Member State. The Member States will, though, have to explain the choice of data source and explain the anticipated effects of data source on the degree of compliance with the harmonised definitions. That requires the supply of detailed metadata that explain the data sources used to allow an assessment of the probable effects of these data sources on the degree to which the statistics comply with the harmonised definitions.

Since the proposal was adopted by the Commission and forwarded to the Parliament and the Council, a number of amendments were added by Committees and Working Parties. On 15 February 2007, the Council approved the final compromise text with a view to a first reading agreement with the European Parliament.[6] This

---

[5] Document COM (2005) 375 final. This paragraph is based on the Explanatory Memorandum accompanying the proposal. The document is available at http://eur-lex.europa.eu/LexUriServ/site/en/com/2005/com2005_0375en01.pdf

[6] For a list of documents in the area of developing a common European immigration policy, including proposed legislation and follow-up, see http://ec.europa.eu/justice_home/doc_centre/immigration/doc_immigration_intro_en.htm The 2007 document is available at the Public Register of Council Documents at http://register.consilium.europa.eu/pdf/en/07/st06/st06351.en07.pdf (Document of the Council of the European Union number 6351/07, date 15 February 2007). See the same document but with text marked: Document number 6351/1/07 date 27 February 2007. Earlier versions of the document may be obtained by searching the Public Register of Council Documents: http://register.consilium.europa.eu/ and search in the register.

proposal included one important element not contained in the 2005 version. Article 9 of the proposed Regulation states that 'As part of the statistics process, scientifically based and well documented statistical estimation methods may be used.' It also states that 'Member States shall report to the Commission (Eurostat) on the data sources used, the reasons for the selection of these sources and the effects of the selected data sources on the quality of the statistics, and on the estimation methods used, and shall keep the Commission (Eurostat) informed of changes thereto.' These statements were absent from the 2005 proposal and have been added later. The new legislation on international migration statistics was adopted by the European Parliament on 14 March 2007, and by the Council of the EU on 12 June 2007. It was published in the *Official Journal* of the European Union on 31 July 2007 as Regulation (EC) no. 862/2007 of the European Parliament and of the Council of 11 July 2007 on Community statistics on migration and international protection and repealing Council Regulation (EEC) no. 311/76 on The compilation of statistics on foreign workers (Legislation L199). For the text of the legislation, the reader is referred to the *Official Journal*.[7]

The call for harmonised and comparable Community statistics on migration and asylum did not go unnoticed in the European research area. The European Commission included international migration statistics as a priority in the 6th Framework Programme of Research (2004–2006). The research involved a description of national systems of data collection, a study of similarities and differences between Member States, and an assessment of activities and measures undertaken by organisations at national and supranational levels to overcome the problems of availability, reliability and comparability of migration data. The THESIM (Towards Harmonised European Statistics on International Migration) project that resulted was the most exhaustive and complete exercise of its kind undertaken to date. The book that was produced and edited by Poulain, Perrin and Singleton (2006) is a unique resource on migration statistics in the European Union.

As documented by the THESIM project, the reasons for the inadequacies of international migration statistics, identified by the CES, continue to exist today. They include the lack of standardisation in (1) national definitions of immigrants and emigrants, (2) the coverage of the statistics, (3) the minimum duration of presence in (or absence from) the country to qualify as an immigrant (or emigrant) and (4) the manner of classifying the migrants by past or future country of residence (Kelly 1987:1034). The CES did not recommend that countries abandon their current practices but encouraged countries to adopt a pragmatic, step-by-step, long-term approach. They should initiate bilateral studies and other types of cooperation between countries to improve the statistics, and they should use supplementary data sources to adjust data compiled for national purposes and to engage in bilateral studies to improve their migration statistics. That approach is also manifest in migration data collection strategies adopted today.

---

[7] At http://eur-lex.europa.eu/LexUriServ/site/en/oj/2007/l_199/l_19920070731en00230029.pdf

During the same month that the European Commission adopted the proposal for a regulation of international migration statistics, the research community organised a workshop on the estimation of international migration in Europe (28–30 September 2005). It was a coincidence but the timing could not have been better. In that workshop, the state of the art was reviewed and new innovative methods for estimating international migration were presented. The meeting was organised as one of the activities of the Research Training Network (RTN) 'Demographic Sustainability and European Integration'. The RTN was coordinated by the Max Planck Institute for Demographic Research in Rostock and the workshop was organised by the Southampton Statistical Sciences Institute (S3RI) of the University of Southampton and the Netherlands Interdisciplinary Demographic Institute (NIDI). The organisers started from the observation that, although the consequences of migration flows are paramount in the public debate, figures on the level and direction of international migration are defective, incomplete or missing entirely. The data that do exist are often not comparable because of differences in definitions and measurement. As a consequence, the debate cannot be fuelled by empirical evidence and trustworthy inferences from facts and evidence-based opinions. Statistical techniques that have been developed for the estimation of migration flows within countries provide adequate starting points for the development of methods for estimating migration flows between countries. This book is an outcome of the S3RI–NIDI workshop.

16.3 This book's contributions

The aim of this book is to present strategies and methods for obtaining estimates of migration flows between countries of Europe that pass the test of scientific rigour. The strategy has three main components. The first is to *use statistical models* of migration to predict migration flows from available (defective) data. The second is to *distinguish data types* and to develop models that can describe data of different types and convert migration data of one type into data of another type. The third component is to combine *data from different sources* and to augment that information by empirically established regularities and expert opinions. The models are therefore generic instruments to harmonise existing data and to estimate missing data by combining data from different sources. Improved current estimates of migration flows provide a foundation for improved migration forecasts. A number of methods that produce forecasts when data are deficient are presented.

The main contribution of this book is the solution to the problem of estimating migration flows from data of different types that are often incomplete and defective. The solution consists of modelling the migration process and using available data to obtain the parameters of the model of migration. The strategy gives priority to the *migration process* rather than to the *migration data*. The migration process that produces the flows is not fully observed. Hence observations provide incomplete information on the process and the information may be of different types. In addition, the data may not be reliable or accurate. Different observation types or data types represent different measurements of the migration process. Issues dealing with data are linked directly to observational deficiencies and they are linked to models of migration. By modelling the migration process rather than the migration data, and by estimating the parameters of the model from the data augmented by expert opinions and other qualitative information, migration flows can be determined that meet particular requirements, e.g. the requirements of the Regulation of the European Parliament and the Council on Community statistics on migration and international protection. The modelling approach presented in this book can be used for the harmonisation of international migration data, i.e. for adjusting data based on national definitions of migration. It can also be used for the estimation of migration flows from defective and incomplete data, i.e. by combining data from different sources, and the forecasting of international migration in the presence of data deficiencies. The approach includes both frequentists and Bayesian models that infer or predict migration flows from available data and other valid but qualitative information. The Bayesian approach is a natural way to combine multiple data sources and prior information on the structure of migration flows and characteristics of migrants (e.g. age structure). This book is the first to systematically include Bayesian estimation techniques to estimate migration flows.

This is also the first volume that draws together modern statistical approaches to incomplete data and migration modelling. The approaches are inspired by models that have been developed over the years for internal migration, where the data problems may be less severe, but the problems of estimation are very similar. Part I of the book covers migration data and issues. The contributors show that the many inconsistencies in migration flow data are caused by differences in concepts, measurements and the processing of migration data.

To tackle the differences and to build a foundation for the harmonisation of migration statistics, one must look beyond the data or observations to intrinsic characteristics of any migration. A migration is an event and the event occurs in time and space. The event is a change in the place of usual residence (address). Many conceptual issues relate to the concepts of address and change. Some people have multiple residences. The UN recommendations and Eurostat are clear. They define the place of usual residence as the place where the person spends most daily rest periods. National statistical offices do not have the common view on what is a place of residence and who is an international migrant. For instance, in some countries of Central Europe, persons may keep a permanent place of residence in these countries while living abroad for many years. The concept of country of usual residence is at the core of the problem of the definition of international migration. A necessary condition for a harmonisation of migration data is a thorough documentation of the conceptualisation of the event of migration and the measurement of migration.

Migration leads to changes in the composition of the population and data on migration should be consistent with data on changes in the composition of the population. Censuses are the main source of the size and characteristics of the population, including the country of birth and the nationality. So, detailed census data on migrant stocks may provide useful benchmarks for the validation of migration data.

Part II of the book discusses models that capture the spatial and age patterns of migration. Spatial patterns are described by spatial interaction models, while age patterns are described by model migration schedules. The models are applied in

Part III to obtain current estimates of international migration flows in Europe and to estimate the impact of various asylum-seeker policies on other potential receiving countries. Finally, Part IV covers the forecasting of immigration and emigration, and Part V, the projection and forecasting of populations.

This book demonstrates that modern statistical techniques have considerable potential for estimating international migration flows and that the research is cutting-edge. It also demonstrates that, in order to be effective, new research should emphasise data types and link country-specific data sources and methods of data collection to the data types. Migration data are manifestations of migration processes and different data types represent different process measures. For instance, the process of international migration in the United Kingdom is not much different from that in France or Germany, but the ways the process is measured are very different. By modelling the process and approaching data as manifestations of the process, a variety of data sources and data types can be dealt with in a consistent fashion. When process models can be improved upon by adding new information that becomes available, a methodology emerges that responds to the challenge of producing a system of timely, reliable and valid migration data for Europe.

The THESIM project resulted in a comprehensive picture of available statistical data sources on international migration in the European Union, and the S3RI–NIDI workshop demonstrated the feasibility of using models for estimating international migration flows and to produce harmonised and comparable migration data. In May 2006 the European Commission (Eurostat) issued an invitation to tender for the development of models of statistical data on migration and migrant populations. The aim of the tender was to improve the completeness and the degree of harmonisation of international migration statistics in Europe. The funding is to be used to develop sound statistical tools for estimating migrant populations and international migration flows, disaggregated by sex and age, and to assist national statistical institutes (NSIs) in Member States in meeting the obligations of the forthcoming European legislation on migration statistics. A consortium of four research centres in Europe are currently working together on the project 'Modelling of statistical data on migration and migrant populations' (MIMOSA). They include the Netherlands Interdisciplinary Demographic Institute, The Hague (NIDI) (coordinator), the Central European Forum for Migration and Population Research, Warsaw (CEFMR), the Southampton Statistical Sciences Research Institute (S3RI), and the Université Catholique de Louvain (UCL). Statistical models will be developed and applied to estimate migration flows between 31 countries of Europe using data on immigration and emigration. In addition, NSIs will be given advice and assistance on applying statistical techniques to estimate migration flow data that are comparable across Europe. The project started in December 2006.

The comprehensive review of statistical data sources on international migration in Europe produced by the THESIM project (Poulain *et al.* 2006), the conceptual and measurement issues discussed in Chapters 2–5, and the statistical models and applications presented in Chapters 6–10 are the point of departure for the modelling of recent migration and migrant populations in the new research tender initiated by the European Commission (Eurostat). The researchers in the THESIM project, in the

MIMOSA project, and those who contributed to this book are working towards a common goal: a European migration information system that comprises harmonised and comparable data on international migration and asylum. The European Parliament, by adopting legislation that accommodates the use of statistical estimation methods in the production of Community statistics on international migration and asylum, established a common framework for the European statistical system that includes scientifically based estimates in addition to direct measurements. Much has been accomplished in the years since 1971, when the Conference of European Statisticians raised the issue of comparability of international migration statistics. However, the issue has yet to be resolved. The legal basis that now exists for Community statistics to allow estimates is expected to stimulate research and development leading to harmonised and comparable international migration statistics in Europe. The migration information system that results will represent a scientific basis for the public debate on migration and for migration policies.

References

Eding JH, Willekens FJ and Cruijsen H. 1996. *Long-term demographic scenarios for the European Union*. Demographic Reports no. 20, Population Research Centre, University of Groningen. Available at http://docserver.repository.knaw.nl/19136.pdf

Eurostat. 1991. Two long-term population scenarios for the European Community. Paper prepared for the *International Conference on Human Resources in Europe at the Dawn of the 21st Century*, Luxembourg, 27–29 November.

Eurostat. 1993. *Two long-term population scenarios for the European Free Trade Association*. Eurostat: Luxembourg.

Eurostat. 2006. Long-term population projections at national level. *Statistics in Focus* 3/2006 (author: Lanzieri G). Available at http://epp.eurostat.ec.europa.eu/cache/ITY_OFFPUB/KS-NK-06-003/EN/KS-NK-06-003-EN.PDF

Extercate M. 1993. *Two long-term status quo population projections for the EC and the EFTA countries*. Population Research Centre, University of Groningen.

Franco D and Munzi T. 1996. Public pension expenditure prospects in the European Union: a survey of national projections. In *Ageing and pension expenditure prospects in the Western world*. European Commission Directorate-General for Economic and Financial Affairs. *European Economy – Reports and Studies* 3:1–126. Available at http://ec.europa.eu/economy_finance/publications/european_economy/1996/eers3_1996en.pdf

Herm A. 2006. Recommendations on international migration statistics and development of data collection at an international level. In *THESIM. Towards Harmonised European Statistics on International Migration*. Poulain M, Perrin N and Singleton A, eds, pp. 77–106. Louvain-la-Neuve: Presses Universitaires de Louvain.

Kelly JJ. 1987. Improving the comparability of international migration statistics: contributions by the Conference of European Statisticians from 1971 to date. *International Migration Review* 21(4):1017–1037.

Kraler A, Jandl M and Hofmann M. 2006. The evolution of EU migration policy and implications for data collection. In *THESIM. Towards Harmonised European Statistics on International Migration*. Poulain M, Perrin N and Singleton A, eds, pp. 35–75. Louvain-la-Neuve: Presses Universitaires de Louvain.

Poulain M. 1993. Confrontation des statistiques de migration intra-européennes: vers une matrice complète? *European Journal of Population* 9(4):353–381.

Poulain M. 1997. *Comparing data sources for measuring international migration in Central and Eastern Europe.* Eurostat Working Paper, no. E4/1997-1. Eurostat, Bruxelles/ Luxembourg.

Poulain M. 1999. International migration within Europe: towards more complete and reliable data? Paper presented at the *Joint ECE–Eurostat Work Session on Demographic Projections*, Perugia, 3–7 May 1999. Louvain-la-Neuve, GéDAP.

Poulain M and Gisser R. 1992. *Migration statistics for the EFTA countries.* (Doc E3/SD/12/92). Luxembourg: Eurostat.

Poulain M, Debuisson M and Eggerickx T. 1990. *Projet d'harmonization des statistiques de migration internationale au sein de la Communauté européenne.* Report to Eurostat. Université Catholique de Louvain, Louvain-la-Neuve, 4 volumes.

Poulain M, Perrin N and Singleton A, eds. 2006. *THESIM. Towards Harmonised European Statistics on International Migration.* Louvain-la-Neuve: Presses Universitaires de Louvain.

Thorogood D. (2006) Improving the quality and availability of migration statistics in Europe. Paper presented at the *United Nations Expert Group on 'Measuring international* migration: concepts and methods', New York, 4–6 December 2006. Available at http://unstats.un.org/unsd/demographic/meetings/egm/migrationegm06/DOC%205%20 Eurostat.pdf

United Nations. 1978. Statistics on international migration. In *United Nations, Demographic Yearbook 1977.* UN Publication, Sales no. E/F.78.XIII.1. New York: United Nations

Index

Page locators in **bold** indicate tables, those in *italics* indicate figures. The suffix 'n' following a page locator indicates a footnote.

International Migration in Europe: Data, Models and Estimates   Edited by J. Raymer and F. Willekens
© 2008 John Wiley & Sons, Ltd